GCSE Physics
Second edition

Cover photo

Abstract representation of an optical fibre
communications system in which messages
are carried over long distances by pulses of
infrared 'light' travelling in glass fibres as
fine as a human hair.
This area of modern technology is one of
the most rapidly developing.

(Photo: Paul Brierley)

GCSE Physics
Second edition

Tom Duncan

John Murray

To F.D.B.

Photo acknowledgements

Thanks are due to the following for permission to reproduce copyright photographs:

Figs. 1a, courtesy Hale Observatories; 1b, Cenco; 2a, *Daily Telegraph* Colour Library; 2b, 70.5, Central Electricity Generating Board; 2c, British Telecom; 2d, IBM (UK) Ltd; 1.2, Central Office of Information; 1.3, 37.2, John Topham Picture Library; 2.5, 49.10, Central Press Photos Ltd; 3.5b, Walden Precision Apparatus Ltd, Cambridge; 6.3b, Ronan Picture Library; 6.7, Rank Precision Industries; 8.7, Science Photo Library; 10.2, 10.4, Kodak Ltd; 10.5, *Watchmaker, Jeweller & Silversmith*; 11.2, Griffin & George; 11.6, Palomar Observatory Photograph; 13.7a,b, 13.9, 13.10, 13.12a, from Llowarch, *Ripple Tank Studies of Wave Motion* (Clarendon Press, Oxford); 13.11, Hydraulics Research Ltd; 14.1, Addison-Wesley Publishing Co. (from *Modern College Physics* by Richards, Sears, Wehr and Zemansky); 14.3, D. G. A. Dyson; 15.4, Barnes Engineering Co.; 15.6, 77.2, The Post Office; 15.7, 35.2b, 73.8, Unilab Ltd; 15.8, Westinghouse; 17.1, Greater London Council; 19.1a,b, British Airways; 19.8, E. Leitz (Instruments) Ltd; 20.3, Mrs Barbara A. Lyle; 21.1, 26.3, 29.4, 55.1, Popperfoto; 22.1, 37.6, (from *PSSC Physics*) reproduced by permission of D. C. Heath and Co., copyright 1965, Education Development Center; 22.5, ICI Mond Division; 26.6a, National Institute of Agricultural Engineering;

26.6b, London Transport Executive; 27.1, British Petroleum Co. Ltd; 28.1, Avon Rubber Co. Ltd; 30.6, National Water Council; 30.9, Ford Motor Co. Ltd; 31.8, Camera Press Ltd; 33.5a, WASAfoto; 33.5b, Sjöhistoriska Museum; 33.6, 49.4a, H.M.S.O., Crown copyright; 33.7, Airship Industries; 33.8b. Oldham Batteries; 35.1, A. G. Spalding & Bros Inc. (first appeared in article 'Dynamics of the Golf Swing' by Dr D. Williams in *Quarterly Journal of Mechanics and Applied Mathematics*, published by Clarendon Press, Oxford); 38.1. British Hovercraft Ltd; 38.5, Parachute Regiment, Aldershot; 39.5a,b, Barnaby's Picture Library; 40.1, 44.3, British Rail; 41.1, Scottish Tourist Board; 43.1, British Steel Corporation; 44.1, British Oxygen Co. Ltd; 45.4, Rutherford Appleton Laboratory; 48.2, Swiss National Tourist Office; 48.4, Leslie Bishop Co. Ltd; 49.4b, Fibreglass Ltd; 50.4, James R. Sheppard; 51.7, Ministry of Defence; 51.8, GEC Turbine Generators Ltd, Rugby; 53.1, US Information Service; 53.5, Hurst Electric Industries Ltd; 56.6a,b,c, Kodansha Ltd; 57.3a,b, 60.2a,b, 64.3, 76.10a,b, 76.14, 76.15a, 76.16a, 76.17a, 81.2, RS Components Ltd; 64.5, General Electric Co. plc; 66.7, Eriez Magnetics; 67.6, Stanley Power Tools Ltd; 74.7a, C. T. R. Wilson; 74.7b, The Royal Society (from *Proceedings*, A. Vol. 104, Plate 16, Fig. 1, 1923); 74.10, Lippke; 76.1a, Casio Electronics Ltd; 76.1b, The London Hospital.

© Tom Duncan 1986, 1987

First published 1986 by
John Murray (Publishers) Ltd
50 Albermarle Street, London W1X 4BD

Second edition 1987
Reprinted 1987, 1988 (twice)

Printed in Hong Kong by
Colorcraft Ltd

British Library Cataloguing in Publication Data

Duncan, Tom
    GCSE physics.——2nd ed.
    1. Physics
    I. Title
    530   QC23

ISBN 0-7195-4380-0

Preface

This book is developed from *Physics for Today and Tomorrow* (2nd edition), modified to bring it into line with the requirements of those taking the first GCSE examinations. There is full coverage of the S.E.C. approved syllabuses of all six GCSE Examining Groups. *Check lists* of specific objectives have been added to help students monitor their progress and to aid revision.

There are over 800 questions, for which a two-level grading system has been adopted. This is designed to meet the needs of a wide range of students and enable questions of an appropriate level to be found readily. The *Questions* at the end of every topic are intended for class use and, along with the *Additional questions* after each group of related topics, for setting as homework. Each group of *Additional questions* is organized into the two-level system: 'Core' level (for all students for whom the material is relevant), while the rest are at the higher 'Further' level (for those seeking higher grades).

For quick but comprehensive revision of basic material before examinations, a set of over 100 mostly objective-type *Core level revision questions*, of a straightforward nature, have been added. The *Further level revision questions* are suitable mainly for students hoping to obtain higher grades.

Second edition. In this second edition, *which is completely compatible with the first*, the opportunity has been taken to make some useful amendments in a number of chapters. Also, material required by only one or two syllabuses has been grouped into three additional *Other topics*.

The author is grateful to Peter Knight for his painstaking analysis of GCSE syllabus requirements.

Acknowledgement is made to the following boards (answers given being the sole responsibility of the author):

Cambridge Local Examination Syndicate (*C.*)
Joint Matriculation Board (*J.M.B.*)
University of London (*L.*)
Northern Ireland G.C.E. Examination Board (*N.I.*)
Oxford Local Examinations (*O.L.E.*)
Oxford and Cambridge Examination Board (*O. and C.*)
Southern Universities Joint Board (*S.*)
Welsh Joint Education Committee (*W.*)
Associated Lancashire Schools Examining Board (*A.L.*)
East Anglian Examinations Board (*E.A.*)
East Midland Regional Examination Board (*E.M.*)
North West Regional Examinations Board (*N.W.*)
Southern Regional Examinations Board (*S.R.*)
South-East Regional Examinations Board (*S.E.*)
West Midlands Examinations Board (*W.M.*)

To the student

During your course one way of using this book is to:

1 study the *topic* concerned (e.g. *Light rays*),
2 use the *check list for that topic* to test that you can satisfy the objectives stated (page no. of check list is given at end of topic),
3 try the *questions* at the end of the topic,
4 try the *additional questions at core level* for the topic, and
5 try the *additional questions at further level* if you are hoping for higher grades.

When revising before an examination you might find it better to:

1 use the *check lists for each group of related topics* (e.g. *Light and sight*) to test yourself and refer back to the text when necessary,
2 try the *revision questions at core level* for the group of topics, and
3 try some of the *revision questions at further level* if you are aiming for higher grades.

Contents

Physics and technology

Physicists explore the Universe. Their investigations range from stars that are millions and millions of kilometres away to particles that are smaller than atoms, Fig. 1*a*, *b*.

As well as having to find the *facts* by observation and experiment, they also must try to discover the *laws* that summarize (often as mathematical equations) these facts. Sense has then to be made of the laws by thinking up and testing *theories* (thought-models) to explain the laws. The reward, apart from a satisfied curiosity, is a better understanding of the physical world. Engineers and technologists use physics to solve *practical problems* for the benefit of mankind, though in solving them social, environmental and other problems may arise.

In this book we will study the behaviour of *matter* (the stuff things are made of) and the different kinds of *energy* (such as light, sound, heat, electricity). We will also consider the applications of physics in the home, transport, medicine, industry, communications and electronics, Fig. 2*a*, *b*, *c*, *d*.

Mathematics is an essential tool of physics and a 'reference section' of some of the basic mathematics is given at the end of the book along with a suggested procedure for solving physics problems.

Fig. 1a. Astronomers have found that the many millions of stars in the Universe, of which the Sun is just one, are in widely separated groups called galaxies. The spiral galaxy M81 near the Plough is shown at the left. The Sun and its planets belong to the galaxy called the Milky Way.

The number of galaxies is huge. Most appear as tiny blurred specks. It is estimated that the farthest are 5000 million light-years from the Sun. One light-year is the distance travelled by light in 1 year—about 10 000 000 000 000 km.

Fig. 1b. The photograph below shows the atoms in the tip of a tungsten needle (the metal used for lamp filaments) magnified about 2 million times. It was taken by an instrument called a field ion microscope.

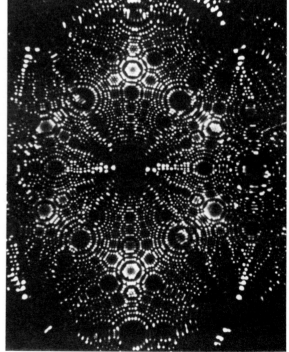

Fig. 2a. The space shuttle *right* is the world's first, and as yet only, re-usable 'Space Transport System' (STS). After being launched like a rocket it glides back to earth like an aeroplane to be used again. Satellites and other materials can be carried in its cargo bay to be placed into orbit when the shuttle is in space. The black tiles on the underside of the vehicle protect it against the fierce heat caused by friction as it re-enters the earth's atmosphere.

Fig. 2b. In the search for alternative energy sources, giant windmills like that *below* are being developed to drive electrical generators where wind power is sufficiently reliable.

Fig. 2c. Communications satellites, like the one shown *above* with its power-generating solar panels, can handle two television channels plus 12 000 telephone circuits. They are launched either by the American *Space Shuttle* (Fig. 2a) or by the European rocket *Ariane* into a geostationary orbit 36 000 km (22 500 miles) above the equator where they circle the earth in 24 hours and appear to be at rest. Microwave signals are sent to it and received from it by earth stations with large dish aerials like those in Fig. 77.2.

Fig. 2d. The production worker *left* is showing a silicon wafer containing computer memory 'chips' that can store over one million bits of information. The process, which can require up to three months, must be done in a controlled, absolutely clean environment.

Light and sight

1 Light rays

You can see an object only if light from it enters your eyes. Some objects such as the sun, electric lamps and candles make their own light. We call these *luminous* sources.

Most things you see do not make their own light but reflect it from a luminous source. They are *non-luminous* objects. This page, you and the moon are examples. Fig. 1.1 shows some others.

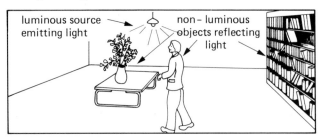

Fig. 1.1

Luminous sources radiate light when their atoms become 'excited' as a result of receiving energy. In a light bulb, for example, the energy comes from electricity. The 'excited' atoms give off their light haphazardly in most luminous sources.

A light source that works differently is the *laser*, invented in 1960. In it the 'excited' atoms act together and emit a narrow, very bright beam of light which can cut a hole through a key 2 mm thick in a thousandth of a second, Fig. 1.2. Other uses are being found for the laser in industry, telecommunications and medicine.

Fig. 1.2

10

Fig. 1.3

Rays and beams

Sunbeams streaming through trees, Fig. 1.3, and light from a cinema projector on its way to the screen both suggest that *light travels in straight lines*. The beams are visible because dust particles in the air reflect light into our eyes.

The direction of the path in which light is travelling is called a *ray* and is represented in diagrams by a straight line with an arrow on it. A *beam* is a stream of light and is shown by a number of rays; it may be parallel, diverging (spreading out) or converging (getting narrower), Fig. 1.4.

Fig. 1.4

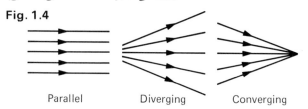

Parallel   Diverging   Converging

Experiment: the pinhole camera

One is shown in Fig. 1.5a. Make a small pinhole in the centre of the black paper. Half-darken the room. Hold the box at arm's length so that the pinhole end is nearest to and about 1 metre from a luminous

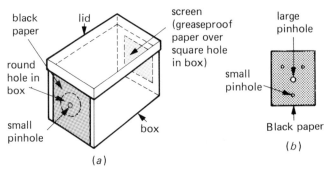

Fig. 1.5

object, e.g. a carbon filament lamp or a candle. Look at the *image* on the screen (an image is a likeness of an object and need not be an exact copy).

Can you see *three* ways in which the image differs from the object? What is the effect of moving the camera closer to the object?

Make the pinhole larger. What happens to the (i) brightness, (ii) sharpness, (iii) size of the image?

Make several small pinholes round the large hole, Fig. 1.5b, and view the image again.

The forming of an image is shown in Fig. 1.6.

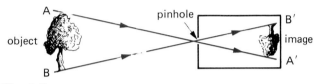

Fig. 1.6

Shadows

Shadows are formed because light travels in straight lines. A very small source of light, called a *point* source, gives a sharp shadow which is equally dark all over. This may be shown as in Fig. 1.7a where the small hole in the card acts as a point source.

If the card is removed the lamp acts as a large or *extended* source, Fig. 1.7b. The shadow is then larger and has a central dark region, the *umbra*, surrounded by a ring of partial shadow, the *penumbra*. You can see by the rays that some light reaches the penumbra but none reaches the umbra.

Eclipses

There is an eclipse of the sun by the moon when the sun, moon and earth are in a straight line. The sun is an extended source (like the bulb in Fig. 1.7b). People

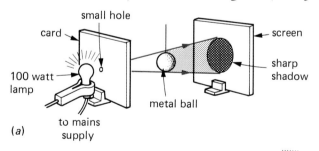

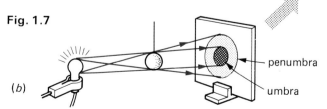

Fig. 1.7

in the umbra of the moon's shadow, at B in Fig. 1.8, see a *total* eclipse of the sun (that is, they can't see the sun at all). Those in the penumbra, at A, see a *partial* eclipse (part of the sun is still visible).

Sometimes the moon is farther from the earth (it does not go round the earth in a perfect circle), and then the tip of the umbra does not reach the earth, Fig. 1.9. In that case people at A still see a partial eclipse, but those at B see an *annular* eclipse (only the central region of the sun is hidden).

A total eclipse seen from one place may last for up to 7 minutes. During this time, although it is day, the sky is dark, stars are visible, the temperature falls and birds stop singing.

A lunar eclipse occurs when the moon passes into the earth's shadow and the light it reflects from the sun is cut off.

Ⓠ Questions

1. How would the size and brightness of the image formed by a pinhole camera change if the camera was made longer?

2. What changes would occur in the image if the single pinhole in a camera was replaced by (a) four pinholes close together, (b) a hole 1 cm wide?

3. A long narrow bench has two small identical lamps mounted one at each end as in Fig. 1.10. A vertical rod is

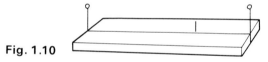

Fig. 1.10

placed on the bench. Copy the diagram and draw the shadows formed, showing the correct size and position. State, giving a reason, which shadow is the darker.

(E.A.)

4. Draw a diagram to show a possible position of the moon for a lunar eclipse to be seen on earth.

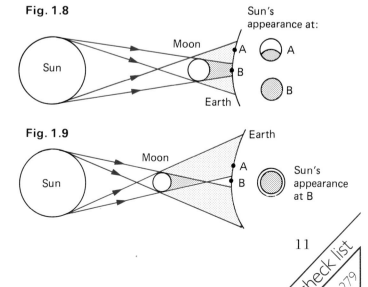

check list p.279

2 Reflection of light

If we know how light behaves when it is reflected we can use a mirror to change the direction in which it is travelling. This happens when a mirror is placed at the entrance of a concealed drive to give warning of approaching traffic.

An ordinary mirror is made by depositing a thin layer of silver on one side of a piece of glass and protecting it with paint. The silver—at the *back* of the glass—acts as the reflecting surface.

Experiment: reflection by a plane mirror

Draw a line AOB on a sheet of paper and using a protractor mark angles on it. Measure them from the perpendicular ON, which is at right angles to AOB. Set up a plane (flat) mirror with its reflecting surface on AOB.

(a) Ray method. Shine a narrow ray of light along say the 30° line, onto the mirror, Fig. 2.1.

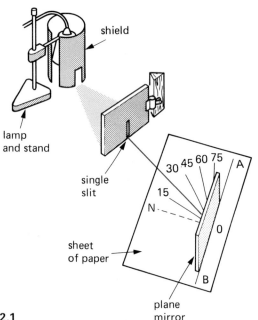

Fig. 2.1

Mark the position of the reflected ray, remove the mirror and measure the angle between the reflected ray and ON. Repeat for rays at other angles. What can you conclude?

(b) Pin method. Insert two pins P_1 and P_2 on the 30° line, Fig. 2.2, to indicate a 'ray' of light falling at this angle on the mirror. Look into the mirror and insert two sighting pins P_3 and P_4 so that they are in line

12

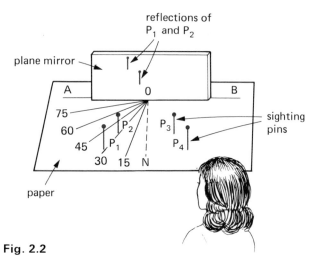

Fig. 2.2

with the reflections (images) of P_1 and P_2. P_3P_4 gives the path of 'ray' P_1P_2 after it is reflected.

Remove P_3 and P_4 and mark their positions with crosses (lettered P_3 and P_4). Remove the mirror and draw a straight line through P_3 and P_4 to meet the mirror; this should be at O. Measure angle P_4ON. Repeat for other angles. What do you conclude?

Laws of reflection

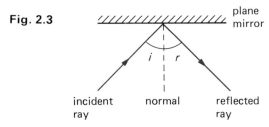

Fig. 2.3

Terms used in connection with reflection are shown in Fig. 2.3. The perpendicular to the mirror at the point where the incident ray strikes it is called the *normal*. Note that the angle of incidence i is the angle between the incident ray and the *normal*; similarly for the angle of reflection r. There are two laws of reflection.

1. *The angle of incidence equals the angle of reflection.*

2. *The incident ray, the reflected ray and the normal all lie in the same plane.* (This means that they can all be drawn on a flat sheet of paper.)

Periscope

A simple periscope consists of a tube containing two plane mirrors, fixed parallel to and facing one another. Each makes an angle of 45° with the line

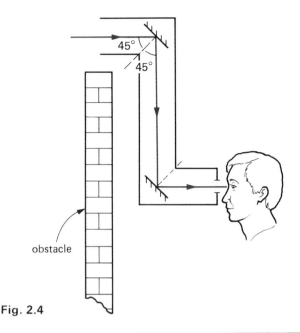

Fig. 2.4

Fig. 2.5

joining them, Fig. 2.4. Light from the object is turned through 90° at each reflection and an observer is able to see over a crowd, for example, Fig. 2.5, or over the top of an obstacle.

In more elaborate periscopes like those used in submarines, prisms replace mirrors (see p. 21).

Regular and diffuse reflection

If a parallel beam of light falls on a plane mirror it is reflected as a parallel beam, Fig. 2.6a, and *regular* reflection is said to occur. Most surfaces however reflect light irregularly and the rays in an incident parallel beam are reflected in many directions, Fig. 2.6b.

Irregular or *diffuse* reflection is due to the surface of the object not being perfectly smooth like a mirror. At each point on the surface the laws of reflection are obeyed but the angle of incidence and so the angle of

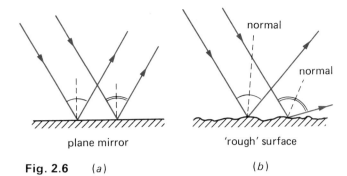

Fig. 2.6　(a)　　　　　(b)

reflection varies from point to point. The reflected rays are scattered haphazardly. Most objects, being rough, are seen by diffuse reflection.

Q Questions

1. Fig. 2.7 shows a ray of light PQ striking a mirror AB. The mirror AB and the mirror CD are at right angles to each other. QN is a normal to the mirror AB.

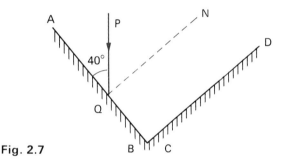

Fig. 2.7

(a) What is the value of the angle of incidence of the ray PQ on the mirror AB?
(b) Copy the diagram and continue the ray PQ to show the path it takes after reflection at both mirrors.
(c) Mark on your diagram the value of the angle of reflection on AB, the angle of incidence on CD and the angle of reflection on CD.
(d) What do you notice about the path of the ray PQ and the final reflected ray? (E.M.)

2. When a ray of light incident on a plane mirror at an angle of incidence of 70° is reflected from the mirror it subsequently strikes a second plane mirror placed so that the angle between the mirrors is 45°. The angle of reflection at the second mirror, in degrees, is

   A 20　　B 25　　C 45　　D 65　　E 70　　(N.I.)

3. A person is looking into a tall mirror in front, Fig. 2.8. What part of the mirror is actually needed to see from eye to toes? (S.R.)

Fig. 2.8

13

check list p.279

3 Plane mirrors

When you look into a plane mirror on the wall of a room you see an image of the room behind the mirror; it is as if there was another room. Restaurants sometimes have a large mirror on one wall just to make them look larger. You may be able to say how much larger after the next experiment.

The position of the image formed by a mirror depends on the position of the object.

Experiment: position of image

Support a piece of thin glass on the bench, as in Fig. 3.1. It must be *vertical* (at 90° to the bench). Place a small paper arrow O about 10 cm from the glass. The glass acts as a poor mirror and an image of O will be seen in it; the darker the bench top the brighter is the image. How do the sizes of O and its image compare?

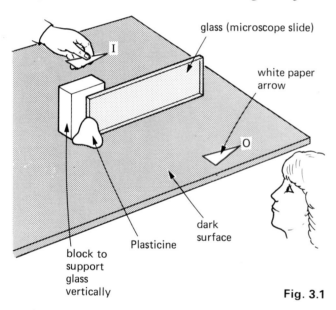

glass (microscope slide)

white paper arrow

I

O

block to support glass vertically

Plasticine

dark surface

Fig. 3.1

Imagine a line joining them. What can you say about it?

Lay another identical arrow I on the bench behind the glass; move it until it coincides with the image of O. Measure the distances of the points of O and I from the glass along the line joining them. How do they compare? Try O at other distances.

Real and virtual images

A *real* image is one which can be produced on a screen (as in a pinhole camera) and is formed by rays that actually pass through it.

A *virtual* image cannot be formed on a screen and is produced by rays which seem to come from it but do not pass through it. The image in a plane mirror is virtual. Rays from a point on an object are reflected at the mirror and appear to come from the point behind the mirror where the eye imagines the rays intersect when produced backwards, Fig. 3.2. IA and

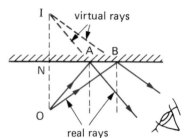

I

virtual rays

A B

N

O

real rays

Fig. 3.2

IB are construction lines and are shown by broken lines.

Lateral inversion

If you close your left eye your image in a plane mirror seems to close the right eye. In a mirror image, left and right are interchanged and the image is said to be *laterally inverted*. The effect occurs whenever an image is formed by one reflection and is very evident if print is viewed in a mirror, Fig. 3.3. What happens if two reflections occur as in a periscope?

Fig. 3.3

Properties of the image

The image in a plane mirror is

(i) as far behind the mirror as the object is in front and the line joining the object and image is perpendicular to the mirror
(ii) the same size as the object
(iii) virtual
(iv) laterally inverted.

Uses of plane mirrors

Apart from their everyday use, plane mirrors can improve the accuracy of measurements in science.

A reading made on a meter which has a pointer moving over a scale is correct only if your eye is directly over the pointer. In any other position there is an error, called the 'parallax' error. (There is parallax between two objects—here the pointer and the scale—if they appear to move in opposite directions when you move your head sideways. It arises when objects do not coincide; if they do coincide they move together.)

If a plane mirror is fitted in the scale the correct position is found by moving your head until the image of the pointer in the mirror is hidden behind the pointer, Fig. 3.4. (A similar error occurs when reading a ruler, due to its thickness, if you do not look at right angles to it.)

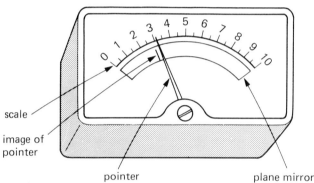

Fig. 3.4

Rotation of a plane mirror

When a mirror is rotated through a certain angle, the reflected ray turns through *twice* that angle. In Fig. 3.5a, when the mirror is in position MM, the ray IO is reflected along OR. Suppose the mirror is rotated through angle θ into position M'M', the direction of IO still being the same; the angle between the reflected ray OR' and OR can be shown to be 2θ.

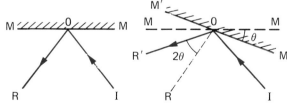

Fig. 3.5a

The idea is used in light-beam galvanometers, Fig. 3.5b, to enable them to measure very small electric currents, i.e. to make them more sensitive.

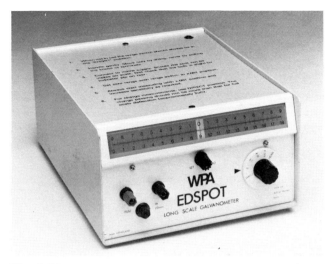

Fig. 3.5b

Questions

1. Fig. 3.6 shows a plan view of a jar of water and a vertical sheet of glass in a box, the inside of which is painted black.

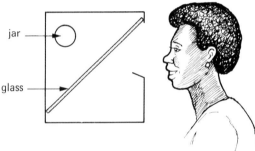

Fig. 3.6

(a) Copy the diagram and show where a candle might be placed so that it appears to the viewer to burn in the jar of water.
(b) Trace the path of two rays of light from the candle to the eye and show with dotted lines how they appear to come from inside the jar.
(c) What does the glass do to the light to get this effect?
(S.E.)

2. The image in a plane mirror of a modern clock (with dots instead of numbers) looks as in Fig. 3.7.
The correct time is

   A 2.25   B 2.35   C 8.35   D 9.25   (W.M.)

Fig. 3.7

3. A girl stands 5 m away from a large plane mirror. How far must she walk to be 2 m away from her image?

15

check list p.279

4 Curved mirrors

Curved mirrors have several uses. There are two main types.

Concave and convex mirrors

A *concave* mirror curves inwards, like a cave; a *convex* one curves outwards. A plane mirror reflects parallel rays of light so that they stay parallel. Curved mirrors reflect each ray in a different direction (but still according to the laws of reflection).

A concave mirror brings parallel rays together to a point, called a real *focus* F, in front of the mirror, as in Fig. 4.1a.

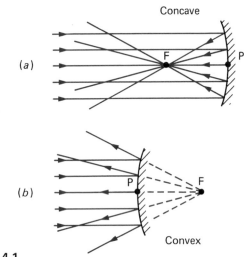

Fig. 4.1

A convex mirror spreads out parallel rays so that they appear to come from a point, called a virtual *focus* F, behind the mirror, as in Fig. 4.1b.

Uses of curved mirrors

(a) Reflectors. Concave mirrors are used as reflectors in, for example, car headlamps and flashlamps, because a small lamp at their focus gives a *parallel* reflected beam. This is only strictly true if the mirror

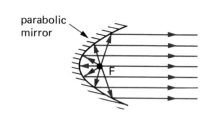

Fig. 4.2

has a parabolic shape (rather than spherical) like that in Fig. 4.2.

(b) Make-up and shaving mirrors. A concave mirror forms a *magnified*, upright image of an object *inside* its focus. This accounts for these two cosmetic uses. The image appears to be behind the mirror and is virtual.

(c) Driving mirrors. A convex mirror gives a wider field of view than a plane mirror of the same size, Fig. 4.3a,b. For this reason and because it always gives an erect (but smaller) image, it is used as a car driving mirror. However it does give the driver a false idea of distance.

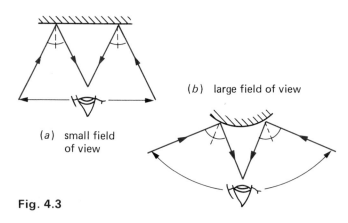

Fig. 4.3

Q Questions

1. A communications satellite in orbit sends a parallel beam of signals down to earth. If they obey the same laws of reflection as light and are to be focused on to a small receiving aerial, what is the best shape for the metal 'dish' used to collect them?

2. Account for the use of a convex mirror on the stairs of a double-decker bus.

5 Refraction of light

If you place a coin in an empty cup and move back until you *just* cannot see it, the result is surprising if someone *gently* pours in water. Try it.

Although light travels in straight lines in one transparent material, e.g. air, if it passes into a different material, e.g. water, it changes direction at the boundary between the two, i.e. it is bent. The *bending of light* when it passes from one material (called a medium) to another is called *refraction*. It causes effects like the coin trick.

Experiment: refraction in glass

Shine a ray of light at an angle on to a glass block (which has its lower face painted white or frosted), Fig. 5.1. Draw the outline ABCD of the block on the sheet of paper under it. Mark the positions of the various rays in air and in glass.

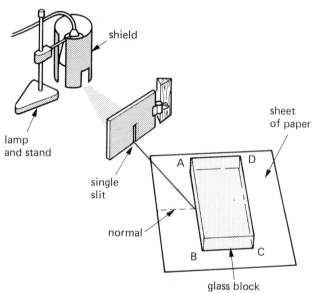

Fig. 5.1

Remove the block and draw the normals on the paper at the points where the ray enters AB (see Fig. 5.1) and where it leaves CD.

What *two* things happen to the light falling on AB? When the ray enters the glass at AB is it bent towards or away from the part of the normal in the block? How is it bent at CD? What can you say about the direction of the ray falling on AB and the direction of the ray leaving CD?

What happens if the ray hits AB at right angles?

Facts about refraction

The previous experiment shows that:

(a) a ray of light is bent *towards* the normal when it enters an optically denser medium at an angle (e.g. from air to glass), i.e. the angle of refraction *r* is less than the angle of incidence *i*, Fig. 5.2*a*,

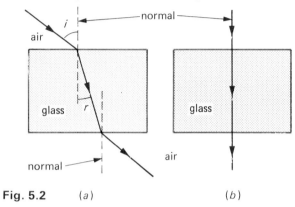

Fig. 5.2   (*a*)   (*b*)

(b) a ray of light is bent *away from* the normal when it enters an optically less dense medium (e.g. from glass to air),

(c) a ray emerging from a parallel-sided block is *parallel* to the ray entering, but is displaced sideways,

(d) a ray travelling along the normal is *not refracted*, Fig. 5.2*b*.

Note. Optically denser means having a greater refraction effect; the actual density may or may not be greater.

Real and apparent depth

Rays of light from a point O on the bottom of a pool are refracted away from the normal at the water surface since they are passing into a less dense medium, i.e. air, Fig. 5.3. On entering the eye they appear to come from a point I *above* O; I is the virtual

Fig. 5.3

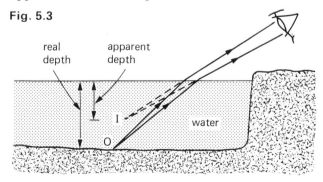

image of O formed by refraction. The apparent depth of the pool is less than its real depth. A straight stick seems bent in water. Why?

Refractive index

Light is refracted because its speed changes when it enters another medium. An analogy helps to explain why.

Suppose three people A, B, C are marching in line, with hands linked, on a good road surface. If they approach marshy ground at an angle, Fig. 5.4a, A is slowed down first, followed by B and then C. This causes the whole line to swing round and change its direction of motion.

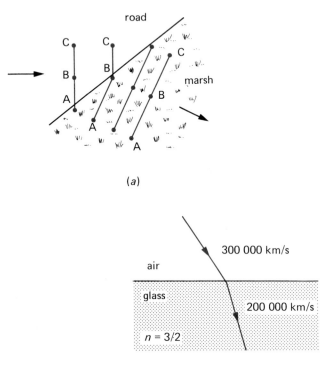

(a)

300 000 km/s

air

glass

200 000 km/s

$n = 3/2$

Fig. 5.4   (b)

In air (and a vacuum) light travels at 300 000 km/s (3×10^8 m/s), in glass its speed falls to 200 000 km/s (2×10^8 m/s), Fig. 5.4b. The *refractive index* of the medium, i.e. glass, is defined by the equation

$$refractive\ index = \frac{speed\ of\ light\ in\ air\ (or\ a\ vacuum)}{speed\ of\ light\ in\ medium}$$

$$= \frac{300\,000\ \text{km/s}}{200\,000\ \text{km/s}} = \frac{3}{2}$$

In symbols

$$n = \frac{c_{\text{air}}}{c_{\text{medium}}}$$

18

The more light is slowed down when it enters a medium from air, the greater is the refractive index of the medium and the more it is bent.

✎ Worked example

What is the speed of light in water if the refractive index of water is 4/3 and the speed of light in air is 300 000 km/s?

We have $n = 4/3$ and $c_{\text{air}} = 300\,000$ km/s

From $n = \dfrac{c_{\text{air}}}{c_{\text{medium}}}$

we get $c_{\text{medium}} = c_{\text{water}} = \dfrac{c_{\text{air}}}{n}$

$$\therefore\ c_{\text{water}} = \frac{300\,000}{(4/3)} = \frac{3}{4} \times 300\,000$$

$$= 225\,000\ \text{km/s}$$

Light travels faster in water than in glass.

⩗ Experiment: *n* by real and apparent depth

Stick a pin O at the end of a glass block, Fig. 5.5, and mount a second pin I with its point on top of the block. (The mounting can be a cork with three small pins as legs.) View O through the block and at the same time move I nearer or farther from O till I and the image of O remain together in the same straight line when you move your head to the right or left. I then occupies the position of the image of O (and there is no parallax between them).

Measure the real and apparent depths of the block. It can be shown for perpendicular viewing (of O in this case) that

$$refractive\ index\ (n) = \frac{real\ depth}{apparent\ depth}$$

Fig. 5.5

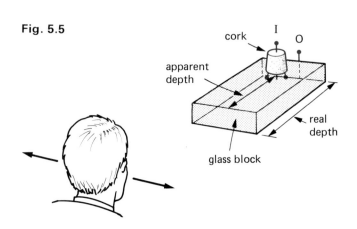

Refraction by a prism

In a triangular glass prism, Fig. 5.6a, the deviation of a ray due to refraction at the first surface is added to the deviation at the second surface, Fig. 5.6b. The

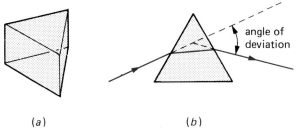

(a)   (b)

Fig. 5.6

deviations do not cancel out as in a parallel-sided block where the emergent ray, although displaced, is parallel to the incident ray.

Q Questions

1. Fig. 5.7 shows a ray of light entering a rectangular block of glass. **(a)** Copy the diagram and draw the normal

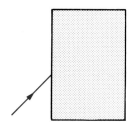

Fig. 5.7

at the point of entry, **(b)** sketch the approximate path of the ray through the block and out of the other side.

2. Draw two rays from a point on a fish in a stream to show where someone on the bank will see the fish. Where must the person aim to spear the fish.

3. If the refractive index from air to glass is 3/2 and from air to water is 4/3, the speed of light decreases as it leaves

    1 water to enter glass
    2 air to enter glass
    3 water to enter air

Which statement(s) is (are) correct?

    A 1, 2, 3 **B** 1, 2 **C** 2, 3 **D** 1 **E** 3

4. What is the speed of light in a medium of refractive index 6/5 if its speed in air is 300 000 km/s?

5. If the refractive index of water is 4/3 how deep will a pond really be if it appears to be 0.6 m when looking vertically downwards? *(A.L.)*

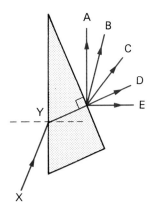

Fig. 5.8

6. Fig. 5.8 shows a ray of light XY striking a glass prism and then passing through it. Which of the rays A to E is the correct representation of the emerging ray?
(J.M.B./A.L./N.W. 16+)

check list
p.279

6 Total internal reflection

When light passes at small angles of incidence from a denser to a less dense medium, e.g. from glass to air, there is a strong refracted ray and a weak ray reflected back into the denser medium, Fig. 6.1a.

Slowly rotate the paper so that the angle of incidence increases until total internal reflection *just* occurs. Mark the incident ray. Measure the angle of incidence; it equals the critical angle.

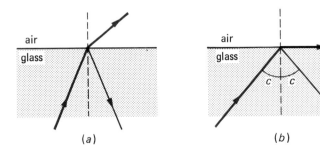
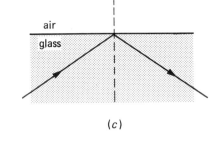

Fig. 6.1

Increasing the angle of incidence increases the angle of refraction.

At a certain angle of incidence, called the *critical angle c*, the angle of refraction is 90°, Fig. 6.1b. For angles of incidence greater than *c*, the refracted ray disappears and *all* the incident light is reflected inside the denser medium, Fig. 6.1c. The light does not cross the boundary and is said to suffer *total internal reflection.*

Experiment: critical angle of glass

Place a semicircular glass block on a sheet of paper, Fig. 6.2, and draw the outline LOMN where O is the centre and ON the normal at O to LOM. Direct a narrow ray (at an angle of about 30°) *along a radius towards O*. The ray is not refracted at the curved surface. Why? Note the refracted ray in the air beyond LOM and also the weak internally reflected ray in the glass.

Multiple images in a mirror

An ordinary mirror silvered at the back forms several images of one object, due to multiple reflection inside the glass, Fig. 6.3a, b. These blur the main image I (which is formed by one reflection at the silvering), especially if the glass is thick. The trouble is absent in front-silvered mirrors but they are easily damaged.

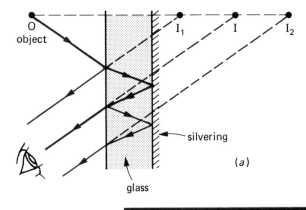

(a)

Fig. 6.2   semicircular glass block

Fig. 6.3   (b)

Totally reflecting prisms

The defects of mirrors are overcome if 45° right-angled glass prisms are used. The critical angle of ordinary glass is about 42° and a ray falling normally on face PQ of such a prism, Fig. 6.4a, hits face PR at 45°. Total internal reflection occurs and the ray is turned through 90°. Totally reflecting prisms replace mirrors in good periscopes.

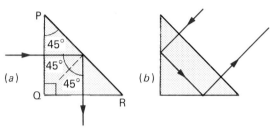

Fig. 6.4

Light can also be reflected through 180° by a prism, Fig. 6.4b; this happens in binoculars.

Mirages

They can often be seen on a hot day as a pool of water on the road some distance ahead. One explanation is that the light from the sky is gradually refracted away from the normal as it passes through layers of warm but less dense air near the hot road. Warm air has a slightly smaller refractive index than cool air and when the light meets a layer at the critical angle, it suffers total internal reflection, Fig. 6.5. To an

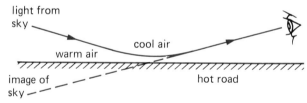

Fig. 6.5

observer the reflection of the sky appears as a puddle in the road.

Light pipes: optical fibres

Light can be trapped by total internal reflection inside a bent glass rod and 'piped' along a curved path, Fig. 6.6. A single, very thin glass fibre behaves

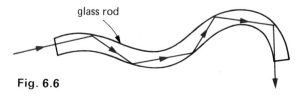

Fig. 6.6

Fig. 6.7

in the same way. If several thousand are taped together a flexible light pipe is obtained that can be used (e.g. by doctors and engineers) to light up some awkward spot for inspection. Fig. 6.7 shows a motorway sign lit by light pipes. The latest telephone 'cables' are optical (very pure glass) fibres carrying pulses of laser light.

Q Questions

1. Fig. 6.8 shows rays of light in a semicircular glass block.

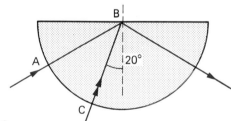

Fig. 6.8

(a) Explain why the ray entering the glass at A is not bent.
(b) Explain why the ray AB is reflected at B and not refracted.
(c) Ray CB does not stop at B. Copy the diagram and draw its approximate path after it leaves B.

2. Copy Fig. 6.9a and b and complete the paths of the rays.

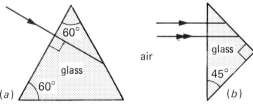

Fig. 6.9

7 Lenses

Lenses are used in many optical instruments (p. 28); they often have spherical surfaces and there are two types. A *convex* lens is thickest in the centre and is also called a *converging* lens because it bends light inwards, Fig. 7.1*a*. You may have used one as a magnifying glass, Fig. 7.2*a*, or as a burning glass. A *concave* or *diverging* lens is thinnest in the centre and spreads light out, Fig. 7.1*b*; it always gives a diminished image, Fig. 7.2*b*.

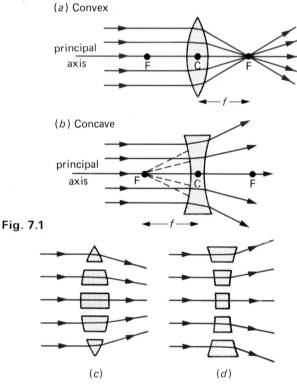

(*a*) Convex

principal axis

(*b*) Concave

principal axis

Fig. 7.1

(*c*)   (*d*)

The centre of a lens is its *optical centre* C; the line through C at right angles to the lens is the *principal axis*.

The action of a lens can be understood by treating it as a number of prisms (most cut-off), each of which bends the ray towards its base, as in Fig. 7.1*c,d*. The centre acts as a parallel-sided block.

Fig. 7.2

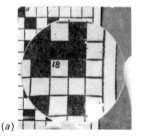

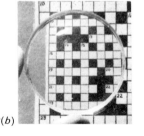

(*a*)   (*b*)

Principal focus

When a beam of light parallel to the principal axis passes through a convex lens it is refracted so as to converge to a point on the axis called the *principal focus* F. It is a real focus. A concave lens has a virtual principal focus behind the lens, from which the refracted beam seems to diverge.

Since light can fall on both faces of a lens it has two principal foci, one on each side, equidistant from C. The distance CF is the *focal length f* of the lens; it is an important property of a lens.

Ray diagrams

Information about the images formed by a lens can be obtained by drawing *two* of the following rays.

1. *A ray parallel to the principal axis which is refracted through the principal focus F.*

2. *A ray through the optical centre C which is undeviated for a thin lens.*

3. *A ray through the principal focus F which is refracted parallel to the principal axis.*

In diagrams a thin lens is represented by a *straight line* at which all the refraction is considered to occur.

Images formed by a convex lens

In the formation of images by lenses two important points on the principal axis are F and 2F; 2F is at a distance of twice the focal length from C.

In each ray diagram in Fig. 7.3 two rays are drawn

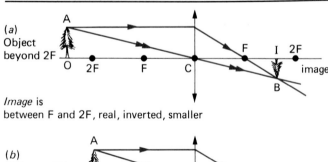

(*a*) Object beyond 2F

Image is between F and 2F, real, inverted, smaller

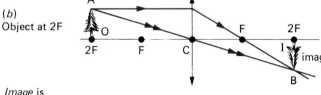

(*b*) Object at 2F

Image is at 2F, real, inverted, same size

Fig. 7.3

from the top A of an object OA and where they intersect after refraction gives the top B of the image IB. The foot I of each image is on the axis since ray OC passes through the lens undeviated. In (d) the dotted rays, and the image, are virtual.

∿ Experiment: *f* of a convex lens

(a) Distant object method. We use the fact that rays from a *point* on a very distant object, i.e. at infinity, are nearly parallel, Fig. 7.4a.

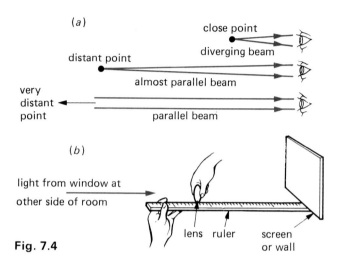

Fig. 7.4

Move the lens, arranged as in Fig. 7.4b, until a *sharp* image of a window at the other side of the room is obtained on the screen. The distance between the lens and the screen is *f* roughly. Why?

(b) Plane mirror method. Using the apparatus in Fig. 7.5a move the lens until a *sharp* image of the object, i.e. the illuminated cross-wire, is formed on the screen beside the object. When this happens, light from the object must travel back along nearly the same path and hit the mirror normally, Fig. 7.5b. The object is then at the lens' principal focus F. Why?

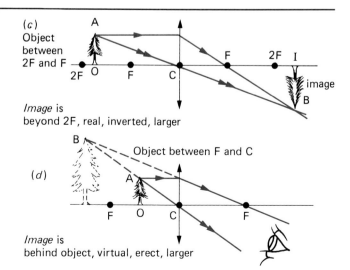

(c)
Object between 2F and F

Image is beyond 2F, real, inverted, larger

(d)
Object between F and C

Image is behind object, virtual, erect, larger

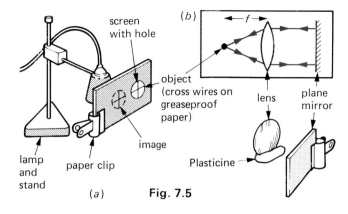

Fig. 7.5

Magnification

The *linear magnification m* is given by

$$m = \frac{\text{height of image}}{\text{height of object}}$$

From Fig. 7.3 it can be shown that in all cases triangles OAC and BIC are similar and so

$$m = \frac{\text{IB}}{\text{OA}} = \frac{\text{IC}}{\text{OC}} = \frac{\text{distance of image from lens}}{\text{distance of object from lens}}$$

Ⓠ Questions

1. A small electric lamp when placed at the focal point of a converging lens will produce a

   A parallel beam of light
   B converging beam of light
   C diffuse beam of light
   D diverging beam of light   *(W.M.)*

Fig. 7.6

2. **(a)** What kind of lens is shown in Fig. 7.6?
   (b) Copy the diagrams and complete them to show the path of the light after passing through the lens.

Fig. 7.7

   (c) Fig. 7.7 shows an object AB 6 cm high placed 18 cm in front of a lens of focal length 6 cm. Draw the diagram to scale and by tracing the paths of rays from A find the position and size of the image formed.
                             (S.E.)

3. Where must the object be placed for the image formed by a convex lens to be
   (a) real, inverted and smaller than the object,
   (b) real, inverted and same size as the object,
   (c) real, inverted and larger than the object,
   (d) virtual, upright and larger than the object?

check list p.280

8 The eye

An image is formed on the *retina* of the eye, Fig. 8.1, by successive refraction at the *cornea*, the *aqueous humour*, the *lens* and the *vitreous humour*. Electrical signals then travel along the *optic nerve* to the brain to be interpreted. In good light, the *yellow spot (fovea)* is most sensitive to detail and the image is automatically formed there.

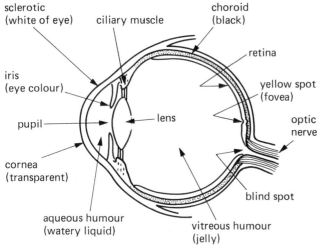

Fig. 8.1

Objects at different distances are focused by the *ciliary muscles* changing the shape and so the focal length of the lens—a process called *accommodation*. The lens fattens to view near objects. The *iris* has a central hole, the *pupil*, whose size it decreases in bright light and increases in dim light.

⋀⋀⋀ Experiment: the eyes

(a) Blind spot. This is the small area of the retina where the optic nerve leaves the eye. It has no light-sensitive nerve endings and in each eye it is closer to the nose than the yellow spot.

To show its existence, close your left eye and look at the cross in Fig. 8.2. You will also see the black dot.

✛   ●

Fig. 8.2

Slowly bring the book towards you. At a certain distance the dot disappears; its image has fallen on the blind spot of your right eye.

(b) Binocular vision. Your eyes see an object from slightly different angles, giving two slightly different images which the brain combines to give a three-dimensional impression. This also helps us to judge distances.

Roll a sheet of paper into a tube and hold it to your right eye with your right hand, Fig. 8.3. Close your left eye and place your left hand halfway along the tube. Open your left eye. Is there a 'hole' in your hand to 'see' through? Explain.

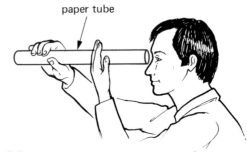

Fig. 8.3

(c) Image inverted on retina, Fig. 8.4*a*. To show this make a pinhole in a piece of paper and hold it about 10 cm away. Close one eye and look at the hole against the sky or something bright. Hold a pin, head up, very close to your eye, Fig. 8.4*b*. Keep looking at the hole and move the pin about slowly until you 'see' it inverted in the hole. Make several pinholes round the first and look again. What do you see?

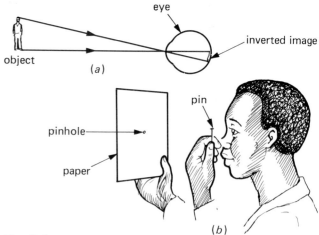

Fig. 8.4

The pin is too close to the eye for a real image (except a large blur) to be formed on the retina, but a sharp *upright shadow* is produced by light from the pinhole and this falls directly (i.e. *still upright*) on the retina. But as you 'see' it upside-down you know that the brain must turn it upside-down. If the brain does that with the shadow it will do the same with any image falling on the retina. So, as you normally see an upright object as upright, the image must be upside-down on the retina.

Defects of vision

The average adult eye can focus objects comfortably from about 25 cm (the *near point*) to infinity (the *far point*). Your near point may be less than 25 cm; it gets farther away with age.

(a) Short sight. A short-sighted person sees near objects clearly but his far point is closer than infinity. The image of a distant object is formed in front of the retina because the eyeball is too long, Fig. 8.5*a*. The defect is corrected by a concave spectacle lens which diverges the light before it enters the eye, to give an image on the retina, Fig. 8.5*b*.

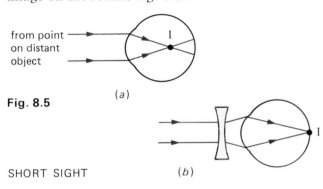

Fig. 8.5

(a)

SHORT SIGHT   (b)

(b) Long sight. A long-sighted person sees distant objects clearly but his near point is beyond 25 cm. The image of a near object is focused behind the retina because the eyeball is too short, Fig. 8.6*a*. A convex spectacle lens corrects the defect, Fig. 8.6*b*.

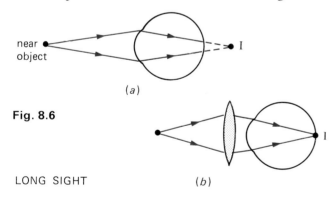

Fig. 8.6

(a)

LONG SIGHT   (b)

Contact lenses

These can be worn instead of spectacles and consist of tiny plastic lenses, Fig. 8.7, held on the cornea by eye fluid.

Fig. 8.7

Persistence of vision

An image lasts on the retina for about one-tenth of a second after the object has disappeared, as can be shown by spinning a card like that in Fig. 8.8. The

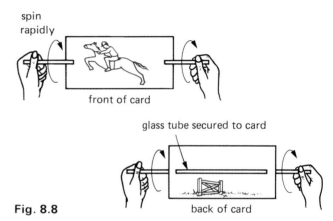

spin rapidly

front of card

glass tube secured to card

Fig. 8.8

back of card

effect makes possible the production of motion pictures. Twenty-four separate pictures, each slightly different from the previous one, are projected on to the screen per second and give the impression of continuity. In a TV receiver twenty-five complete pictures are produced every second.

Q Questions

1. Name the part of the eye **(a)** which controls how much light enters it, **(b)** on which the image is formed, **(c)** which changes the focal length of the lens.

2. Refraction of light in the eye occurs at

  A the lens only   **B** the iris   **C** the cornea only
  D the pupil   **E** both the cornea and the lens

3. A short-sighted person has a near point of 15 cm and a far point of 40 cm.
  (a) Can he see clearly an object at a distance of **(i)** 5 cm, **(ii)** 25 cm, **(iii)** 50 cm?
  (b) To see clearly an object at infinity what kind of spectacle lens does he need?

4. The near point of a long-sighted person is 50 cm from the eye.
  (a) Can he see clearly an object at **(i)** a distance of 20 cm, **(ii)** infinity?
  (b) To read a book held at a distance of 25 cm will he need a convex or a concave spectacle lens?

check list p.280

9 Colour

Colourful clothes, colour TV and the flashing coloured lights in a discotheque all help to make life brighter. It was Newton who, in 1666, set us on the road to understanding how colours may arise. He produced them by allowing sunlight (which is white) to fall on a triangular glass prism, Fig. 9.1. The band

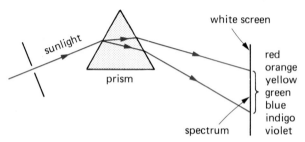

Fig. 9.1

of colours obtained is a *spectrum* and the effect is called *dispersion*. He concluded that (*i*) white light is a mixture of many colours of light which the prism separates out because (*ii*) the refractive index of glass is different for each colour, being greatest for violet.

〰 Experiment: a pure spectrum

A pure spectrum is one in which the colours do not overlap, as they do when a prism alone is used. It needs a lens to focus each colour as in Fig. 9.2a.

Arrange a lens L, Fig. 9.2b, so that it forms an image of the vertical filament of a lamp on a screen at S_1, 1 m away. The filament acts as a narrow source of white light. Insert a 60° prism P and move the screen to S_2, keeping it at the same distance from L, to receive the spectrum; rotate P until the spectrum is pure.

Place different colour filters between P and S_2.

Recombining the spectrum

The colours of the spectrum can be recombined to form white light by

(a) arranging a second prism so that the light is deviated in the opposite direction, Fig. 9.3a, or

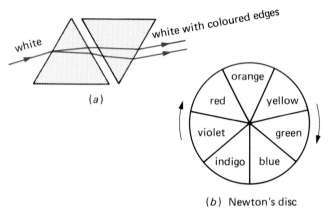

(*b*) Newton's disc

Fig. 9.3

(b) using an electric motor to rotate at high speed a disc with the spectral colours painted on its sectors, Fig. 9.3b. (The whiteness obtained is slightly grey because paints are not pure colours.)

Colour of an object

The colour of an object depends on (i) the colour of the light falling on it and (ii) the colour(s) it transmits or reflects.

(a) Filters. A filter lets through light of certain colours only and is made of glass or celluloid. For example, a red filter transmits mostly red light and absorbs other colours; it therefore produces red light when white light shines through it.

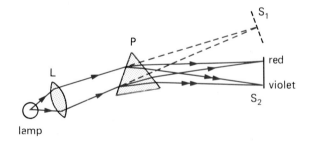

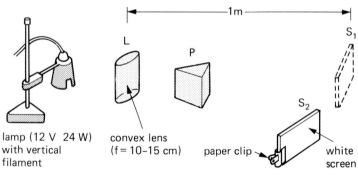

Fig. 9.2   (*a*)   (*b*)

(b) Opaque objects. These do not allow light to pass but are seen by reflected light. A white object reflects all colours and appears white in white light, red in red light, blue in blue light, etc. A blue object appears blue in white light because the red, orange, yellow, green and violet colours in white light are absorbed and only blue reflected. It also looks blue in blue light but in red light it appears black since no light is reflected and blackness indicates the absence of colour.

Mixing coloured lights

In science red, green and blue are *primary* colours (they are not the artist's primary colours) because none of them can be produced from other colours of light. However, they give other colours when suitably mixed.

The primary colours can be mixed by shining beams of red, green and blue light on to a white screen so that they partially overlap, Fig. 9.4a. The results are summarized in the 'colour triangle' of Fig. 9.4b.

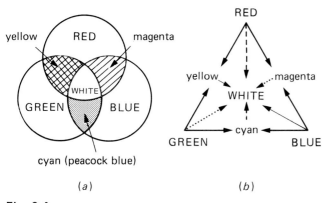

(a)   (b)

Fig. 9.4

The colours formed by adding two primaries are called *secondary* colours; they are yellow, cyan (turquoise or peacock blue) and magenta. The three primary colours give white light, as do the three secondaries. We would also expect a primary colour and the secondary opposite it in the colour triangle to give white. Why? Any *two* colours producing white light are *complementary* colours, e.g. blue and yellow.

Mixing coloured pigments

Pigments are materials which give colour to paints and dyes by reflecting light of certain colours only and absorbing all other colours. Most pigments are

impure, i.e. they reflect more than one colour. When they are mixed the colour reflected is the one common to all. For example, blue and yellow paints give *green* because blue reflects indigo and green (its neighbours in the spectrum) as well as blue, whilst yellow reflects green, yellow and orange, Fig. 9.5. Only *green* is reflected by both.

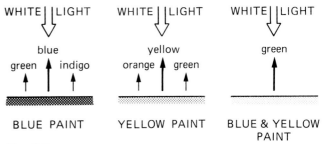

Fig. 9.5

Mixing coloured pigments is a process of *subtraction*; coloured lights are mixed by *addition*.

Ⓠ Questions

1. Copy the diagram, Fig. 9.6, mark and label **(a)** the reflected ray of light at the first surface, **(b)** the path followed by the light through and out of the prism, **(c)** what is seen on the screen AB.   (E.M.)

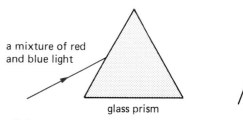

Fig. 9.6

2. A sheet of white paper is viewed through a piece of blue glass and the paper looks blue. This is because

   A the colour of the glass is reflected onto the paper
   B blue light is absorbed by the glass
   C blue light travels faster than red through glass
   D the glass absorbs all colours except blue
   E the glass adds blue light to the light coming from the paper.   (O.L.E./S.R. 16+)

3. **(a)** What colour is formed when the following beams of equally bright light are shone onto a white screen and completely overlap: **(i)** yellow and blue, **(ii)** cyan (turquoise) and red?
   (b) A book which looks red in white light is viewed in magenta light: what colour does it appear?
   (c) White light is viewed through a piece of yellow filter and a piece of red filter held in contact. What colour light is seen?   (N.W.)

10 Simple optical instruments

Camera

A camera is a light-tight box in which a convex lens forms a real image on a film, Fig. 10.1. The film contains chemicals that change on exposure to light; it is 'developed' to give a negative. From the negative a photograph is made by 'printing'.

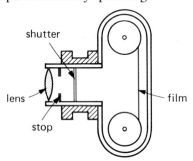

Fig. 10.1

(a) Focusing. In simple cameras the lens is fixed and all distant objects, i.e. beyond about 2 metres, are in reasonable focus. Roughly how far from the film will the lens be if its focal length is 5 cm? In other cameras exact focusing of an object at a certain distance is done by altering the lens position. For near objects it is moved away from the film, the correct setting being shown by a scale on the focusing ring.

(b) Shutter. When a photograph is taken, the shutter is opened for a certain time and exposes the film to light entering the camera. Sometimes exposure times can be varied and are given in fractions of a second, e.g. 1/1000, 1/60, etc. Fast-moving objects require short exposures.

(c) Stop. The brightness of the image on the film depends on the amount of light passing through the lens when the shutter is opened and is controlled by the size of the hole (aperture) in the stop. In some cameras this is fixed but in others, Fig. 10.2, it can be made larger for a dull scene and smaller for a bright one.

The aperture may be marked in *f-numbers*. The diameter of an aperture with f-number 8 is $\frac{1}{8}$ of the focal length of the lens and so the *larger* the f-number the *smaller* the aperture. The numbers are chosen so that on passing from one to the next higher, e.g. from 8 to 11, the area of the aperture is halved.

Projector

A projector forms a real image on a screen of a slide in a slide projector, and a film in a cine-projector. The

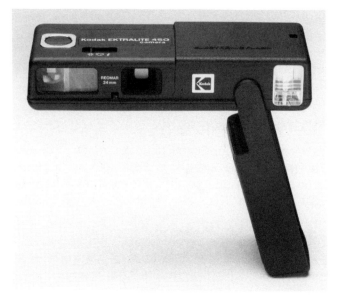

Fig. 10.2

image is usually so highly magnified that very strong but even illumination of the slide or film is needed if the image is also to be bright. This is achieved by directing light from a small but powerful lamp on to the 'object' by means of a concave mirror and a condenser lens system arranged as in Fig. 10.3. The image is produced by the projection lens which can be moved in and out of its mounting to focus the picture.

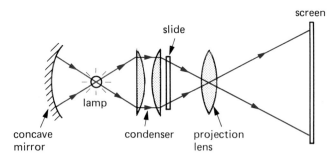

Fig. 10.3

In a projector like that in Fig. 10.4 the slide must be inverted to give an erect image, and must be between 2F and F from the lens (see Fig. 7.3*c*).

Magnifying glass

A watchmaker's magnifying glass is shown in use in Fig. 10.5.

28

Fig. 10.4

The sleepers on a railway track are all the same length but those nearby seem longer. This is because they enclose a larger angle at your eye than more distant ones: as a result their image on the retina is larger so making them appear bigger.

A convex lens gives an enlarged, upright virtual image of an object placed inside its principal focus F, Fig. 10.6a. It acts as a magnifying glass since the

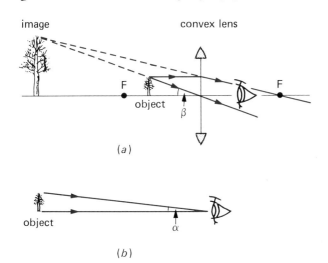

(a)

(b)

Fig. 10.6

angle β made at the eye by the image, formed at the near point, is greater than the angle α made by the object when it is viewed directly at the near point without the magnifying glass, Fig. 10.6b.

The fatter (more curved) a convex lens is, the shorter is its focal length and the more does it magnify. Too much curvature however distorts the image.

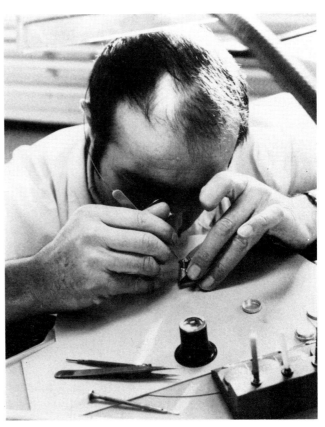

Fig. 10.5

Q Questions

1. Fig. 10.7a shows a camera focused on an object in the middle distance. Is the camera shown in Fig. 10.7b focused on a distant or a close object? Give a reason.

(E.A.)

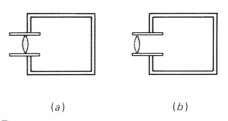

(a)   (b)

Fig. 10.7

2. If a projector is moved farther away from the screen on which it was giving a sharp image of a slide, state (i) *three* changes that occur in the image, (ii) how the projection lens must be adjusted to re-focus the image.

3. (a) Three converging lenses are available having focal lengths of 4 cm, 40 cm and 4 m respectively. Which one would you choose as a magnifying glass?
(b) An object 2 cm high is viewed through a converging lens of focal length 8 cm. The object is 4 cm from the lens. By means of a ray diagram find the position, nature and magnification of the image.

check list p.280

11 Microscopes and telescopes

Compound microscope

A compound microscope gives much greater magnification than a magnifying glass and less distortion. In its simplest form it consists of two *short-focus* convex lenses arranged as in Fig. 11.1. The one

Fig. 11.1

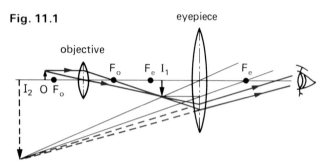

nearer the object, called the *objective*, forms a real enlarged, inverted image I_1 of a small object O placed just outside its principal focus F_o. I_1 is just inside the principal focus F_e of the second lens, the *eyepiece*, which treats I_1 as an object and acts as a magnifying glass to give a further enlarged, virtual image I_2.

The thicker arrowed lines are actual rays from O, by which the eye sees the top of I_2. The thin lines without arrows are construction lines drawn to find the position of I_2.

To reduce distortion in a microscope, Fig. 11.2, the objective and eyepiece each consist of several lenses. The object is seen inverted.

Fig. 11.2

Refracting astronomical telescope

A lens astronomical telescope consists of two convex lenses, a *long-focus* objective and a *short-focus* eyepiece. Rays from a *point* on a distant object, e.g. a star, are nearly parallel on reaching the telescope (see p. 23). The objective forms a real, inverted, diminished image I_1 of the object at its principal focus F_o, Fig. 11.3. The eyepiece acts as a magnifying glass treating I_1 as an object and forming a magnified, virtual image. Normally the eyepiece is adjusted to give this final image at infinity, i.e. a long way off, and so I_1 will be at the principal focus F_e of the eyepiece. That is, F_o and F_e coincide. The object is seen inverted.

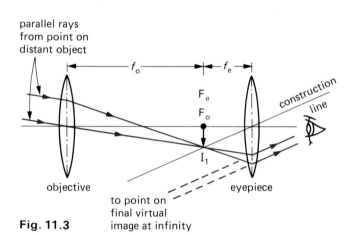

Fig. 11.3

A telescope magnifies because the final image it forms, subtends a much greater angle at the eye than does the distant object viewed without the telescope. Fig. 11.3 shows that the longer the focal length of the objective, the larger is I_1, and so for greatest magnification the objective should have a long focal length and the eyepiece a short one (like any magnifying glass).

Experiment: simple lens telescope

Arrange a carbon filament lamp at the far end of the room. Mount a long-focus convex lens (e.g. 50 cm) at eye-level, Fig. 11.4a. Point it at the lamp and find its image on a piece of greaseproof paper.

Insert a short-focus convex lens (e.g. 5 cm) as the eyepiece and adjust it so that you can see the image on the greaseproof paper clearly, Fig. 11.4b. Remove the paper and view the image of the lamp. Look at a book beside the lamp and outside objects.

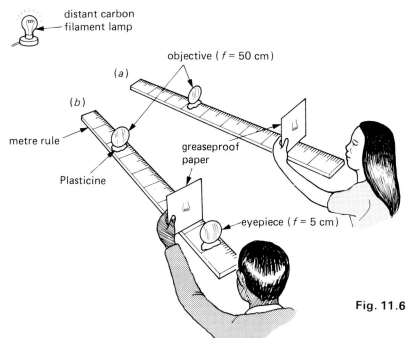

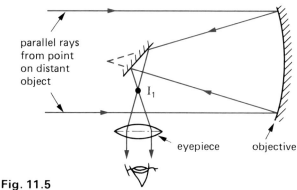

Fig. 11.4

Reflecting astronomical telescope

The objective of a telescope must have a large diameter as well as a large focal length. This (i) gives

Fig. 11.6

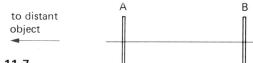

Fig. 11.5

it good light-gathering power so that faint objects can be seen and (ii) enables it to reveal detail.

The largest *lens* telescope has an objective of diameter 1 metre; anything more would sag under its own weight. The biggest astronomical telescopes use long-focus *concave mirrors* as objectives; they can be supported at the back. Fig. 11.5 shows how parallel rays from a distant point object are reflected at the objective and then intercepted by a small plane mirror before they form a real image I_1. This image is magnified by the eyepiece.

The Mount Palomar telescope in California, Fig. 11.6, has a mirror of diameter 5 metres.

Q Questions

1. Select from the list **A** to **E** the pairs of lenses most suitable for **(a)** a compound microscope, **(b)** an astronomical telescope.

> **A** Two concave lenses of focal lengths 100 cm and 5 cm.
> **B** Two convex lenses of focal lengths 100 cm and 5 cm.
> **C** Two concave lenses of focal lengths 5 cm and 3 cm.
> **D** Two convex lenses of focal lengths 5 cm and 3 cm.
> **E** A convex lens of focal length 5 cm and a concave lens of focal length 3 cm. (*J.M.B.*)

2. A and B are two convex lenses correctly set up as a telescope to view a distant object, Fig. 11.7. One has focal length 5 cm and the other 100 cm.

Fig. 11.7

(a) What is A called and what is its focal length?
(b) How far from A is the first image of the distant object?
(c) What is B called?
(d) What acts as the 'object' for B and how far must B be from it if someone looking through the telescope is to see the final image at the same distance as the distant object?
(e) What is the distance between A and B with the telescope set up as in **(d)**?

3. **(a)** What is used as the objective in **(i)** an optical refracting telescope, **(ii)** an optical reflecting telescope?
(b) Give *two* reasons for an optical telescope having an objective of large diameter. (*J.M.B./A.L./N.W.16+*)

31

check list p.280

12 Additional questions

Core level (all students)

Light rays

1. Which one of the following statements about the image produced in a pinhole camera is correct?

    A The image is bigger if the object is further away.
    B The image is smaller if the screen is nearer the pinhole.
    C The image is brighter if the object is further away.
    D The image is sharper if the pinhole is made bigger.
    E The image is bigger if the object is brighter.

                                               (O.L.E./S.R.16+)

2. Eclipses of the sun can be described as total, partial or annular. Explain with the aid of diagrams, how the three types of eclipse are formed.

Show clearly where observers must be to see each of these eclipses.   *(J.M.B.)*

Reflection of light: mirrors

3. The image in a plane mirror is
    A upright, real with a magnification of 2
    B upright, virtual with a magnification of 1
    C inverted, real with a magnification of $\frac{1}{2}$
    D inverted, virtual with a magnification of 1
    E inverted, real with a magnification of 2

4. A ray of light falls on a plane mirror which is then rotated through 20°. The reflected ray will be turned through

    A 10°   B 20°   C 40°   D 60°   E 80°

5. A parallel beam of light is produced by reflection from a mirror when a point source of light is

    A at the focus of a concave mirror
    B at the focus of a convex mirror
    C inside the focus of a concave mirror
    D inside the focus of a convex mirror
    E in front of a plane mirror

Refraction: total internal reflection

6. Light travels up through a pond of water of critical angle 49°. What happens at the surface if the angle of incidence is **(a)** 30°, **(b)** 60°?

32

7. Which diagram in Fig. 12.1 shows the ray of light refracted correctly?

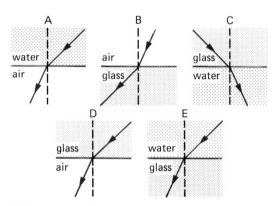

Fig. 12.1

8. In which direction should a spotlight be pointed for a swimming pool to be inspected at point P in Fig. 12.2?

    A 1   B 2   C 3   D 4   E 5

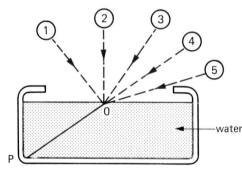

Fig. 12.2

Lenses

9. An illuminated object is placed at a distance of 15 cm from a convex lens of focal length 10 cm. The image obtained on a screen is

    A upright and magnified
    B upright and the same size
    C inverted and magnified
    D inverted and the same size
    E inverted and diminished   *(N.I.)*

10. An object is placed 4 cm in front of a convex lens. A real image is produced 16 cm from the lens.
    (a) What is the magnification produced by the lens?
    (b) By means of a full size diagram determine the focal length of the lens. Mark the principal focus.

                                                 (E.M.)

The eye

11. Fig. 12.3 shows a simplified section of a human eye.
(a) Name the parts labelled A, B and C and explain what happens there.
(b) What is the effect if light goes through the lens on to the nerve ending D?
(c) Why does the eye automatically turn to look straight at an object if possible?   (*S.R.*)

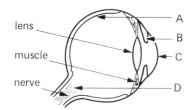

Fig. 12.3

12. What changes occur in the normal eye when, after looking at a dimly lit distant object, one reads a brightly lit book?   (*O. and C.*)

13. A man is obliged to hold a newspaper at arm's length in order to read it.
(a) State a defect of vision which would cause this.
(b) What type of spectacle lenses would he require to correct this defect?

Colour

14. Draw a diagram showing an arrangement of apparatus to project a pure spectrum from white light onto a screen and indicate the paths of rays that produce the blue and red parts of the spectrum.

Account for the separation of the colours.   (*L.*)

15. Why is the result of mixing blue and yellow paints very different from that of mixing blue and yellow lights?

Optical instruments

16. (a) Draw a ray diagram to show the action of the converging (convex) lens in a variable focus camera when it is being used to photograph a very close object. Mark the position of the principal focus of the lens and label it F.
(b) If we now wish to take a photograph of a more distant object we must adjust the focusing. How will this adjustment affect the distance between the lens and the film in the camera?
(*J.M.B./A.L./N.W.16+*)

17. Compare the optical system of the eye with that of a camera. Discuss in particular the focusing of objects at different distances, the regulation of the amount of light entering the system and the nature of the images formed.
(*N.I.*)

18. Draw a ray diagram to show the action of a magnifying glass. Is the image real or virtual? Explain your answer.

19. (a) Draw a simple ray diagram of a reflecting telescope.
(b) What advantages has this type of telescope over a refractor?
(c) Why should the objective of the telescope be as large as possible?
(d) What factors determine the magnification obtained?
(e) What observational advantage is gained by putting an optical telescope in orbit in a satellite?
(*A.L.*)

Further level (higher grade students)

Light rays

20. The diagram (Fig. 12.4) shows an illuminated arrow of length 6 cm placed 40 cm in front of a pinhole camera 16 cm in length having two small holes, H_1 and H_2, 4 cm apart.

Fig. 12.4

(a) Make a scale diagram and draw suitable rays from the luminous arrow to show how two images are formed on the back of the box.
(b) Show why one large blurred image is obtained if the cardboard between the small holes is cut away to form one large circular hole of diameter 4 cm.   (*L.*)

21. A boy finds that by holding an opaque circular disc, 8.0 mm in diameter, at a distance of 90 cm from his eye he can just cover the full moon. Calculate the diameter of the moon if it is at a distance of 380 000 km. Upon what property of light does this test depend?   (*S.*)

Reflection of light: mirrors

22. Show by a ray diagram how a suitably placed eye sees an image of a point object which is placed 10 cm in front of a plane mirror. Show clearly the position of the image and give two reasons why it is described as virtual.   (*L.*)

23. By how far does the distance between a boy and his image decrease if he walks from a position where he is 10 m away from a mirror to one where he is 3 m away?

24. Describe an experiment to determine the position of the image formed by a plane mirror. State the expected result of such an experiment.

Explain the meaning of the term 'parallax errors' as applied to the reading of a pointer on a scale and describe how this error may be reduced by using a plane mirror.

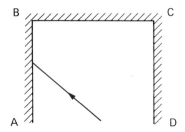

Fig. 12.5

Three plane mirrors are arranged along three sides of a square as shown in Fig. 12.5 and a ray of light is incident on the mirror AB at its mid-point with an angle of incidence of 40° so that the ray is afterwards reflected by BC and CD. Copy the diagram, sketch the approximate path of the ray and calculate the angle through which the ray is turned at each of the three reflections.   (*J.M.B.*)

Refraction: total internal reflection

25. What is the apparent thickness of a glass block of real depth 9 cm if the refractive index of glass is 3/2?

26. Fig. 12.6 shows the end view of a glass prism whose angles are 30°, 60° and 90°. Light passes into the prism in the direction AB and some of it after reflection in the face XY emerges in the direction BA. What is the angle of refraction of the ray AB at B?   (*O. and C. part qn.*)

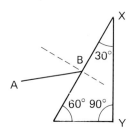

Fig. 12.6

Lenses

27. An object is set up 0.20 m (20 cm) in front of lens *A* and the details of the image are noted and are shown below. The process is repeated for a different lens *B*.

Lens A. Real, inverted, magnified and at a great distance.
Lens B. Real, inverted and same size as the object.

State the type of each lens and give as definite a value as possible for each focal length, explaining how you arrive at the values given.   (*J.M.B.*)

34

28. (a) What is meant by the focal length of a converging (convex) lens?
(b) Define magnification.
(c) How would you determine by an experiment the focal length of a converging lens, first approximately and then more accurately?
(d) An object 1.5 cm high stands upright on the axis of a converging lens of focal length 4.0 cm. The object is placed in turn at distances 8.0 cm, 7.0 cm and 2.0 cm from the lens. For each of these positions determine the position, nature and size of the image.
(*J.M.B./A.L./N.W.16+*)

29. Fig. 12.7 shows a pin AB, a converging lens L and a plane mirror M. The pin and the mirror are held perpendicular to the principal axis of the lens, and the point A, of the pin, is at the principal focus of the lens.

Fig. 12.7

Make two copies of this arrangement. On one copy, trace the paths of *two* rays of light from the point A of the pin, passing through the lens and reflected by the mirror through the lens again. On the second copy, trace the paths of *two* rays from the head B of the pin, passing through the lens and reflected by the mirror through the lens again.   (*C.*)

Colour

30. Three lamphouses A, B and C, each with a filter of one of the primary colours as shown in Fig. 12.8, are set up equal distances apart on a line parallel to a white screen so as to direct beams of equal intensity onto the centre of the screen. An opaque rectangular object painted white is placed halfway between the line of the lamps and the screen as shown. If the width of this object is slightly less than the distance between adjacent lamps, describe and explain the appearance of the object and the screen, using a diagram to indicate the parts of the screen referred to in your description.   (*J.M.B.*)

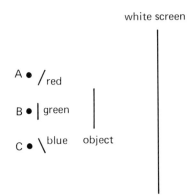

Fig. 12.8

Optical instruments

31. Draw a diagram of a slide projector and use it to explain the function of the condenser lens and the concave mirror placed behind the light source.

What advantage has a matt white surface for the screen over a glossy surface?

A slide projector using a slide 5 cm × 5 cm produces a picture 3 m × 3 m on a screen placed at a distance of 24 m from the projection lens. How far from the lens must the slide be? Make an approximate estimate of the focal length of the projection lens. If this lens gets broken and the only substitute available is one of about half its focal length, what would you do to arrange that the picture is still the same size? *(O. and C.)*

32. Draw a ray diagram to show how two lenses may be used to construct a compound microscope. State the type of each lens and its approximate focal length. Show on the diagram the positions of the principal foci of the lenses in relation to the object and image positions. *(J.M.B.)*

Waves and sound

13 Water waves

Several kinds of wave occur in physics. A *progressive* or *travelling* wave is a disturbance which carries energy from one place to another. There are two types, *transverse* and *longitudinal* (Topic 16).

In a transverse wave, the direction of the disturbance is at *right angles* to the direction of travel of the wave. One can be sent along a rope (or a spring) by fixing one end and moving the other rapidly up and down, Fig. 13.1. The disturbance generated by the hand is

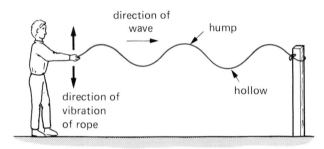

Fig. 13.1

passed on from one part of the rope to the next which performs the same motion but slightly later. The humps and hollows of the wave travel along the rope as each part of the rope vibrates transversely about its undisturbed position.

Describing waves

Terms used to describe waves can be explained with the aid of a *displacement-distance* graph, Fig. 13.2. It shows the distance moved sideways from their undisturbed positions, of the parts vibrating at different distances from the cause of the wave, at a *certain time*.

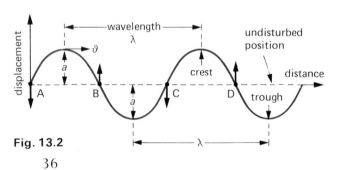

Fig. 13.2

36

(a) *Wavelength*, represented by the Greek letter λ (lambda), is the distance between successive crests.

(b) *Frequency f* is the number of complete waves generated per second. If the end of a rope is jerked up and down twice in a second, two waves are produced in this time. The frequency of the wave is 2 vibrations per second or *2 hertz* (2 Hz; the hertz being the unit of frequency) as is the frequency of jerking of the end of the rope. That is, the frequencies of the wave and its source are equal.

The frequency of a wave is also the number of crests passing a chosen point per second.

(c) *Speed v* of the wave is the distance moved by a crest or any point on the wave in 1 second.

(d) *Amplitude a* is the height of a crest or the depth of a trough measured from the undisturbed position of what is carrying the wave, e.g. a rope.

(e) *Phase.* The arrows at A, B, C, D show the directions of vibration of the parts of the rope at these points. The parts at A and C have the same speed in the same direction and are *in phase*. At B and D the parts are also in phase but they are *out of phase* with those at A and C because their directions of vibration are opposite.

The wave equation

The faster the end of a rope is waggled the shorter is the wavelength of the wave produced. That is, the higher the frequency of a wave the smaller its wavelength. There is a useful connection between f, λ and v which is true for all types of wave.

Fig. 13.3

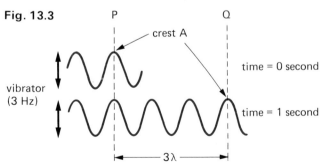

Suppose waves of wavelength $\lambda = 20$ cm travel on a long rope and three crests pass a certain point every second. The frequency $f = 3$ Hz. If Fig. 13.3 represents this wave motion then if crest A is at P at a particular time, 1 second later it will be at Q, a distance from P of three wavelengths, i.e. $3 \times 20 = 60$ cm. The speed of the wave $v = 60$ cm per second (60 cm/s), obtained by multiplying f by λ. Hence

speed of wave = frequency × wavelength

or, $v = f\lambda$

〰 Experiments: the ripple tank

The behaviour of water waves can be studied in a ripple tank. It consists of a transparent tray containing water, having a lamp above and a white screen below to receive the wave images, Fig. 13.4.

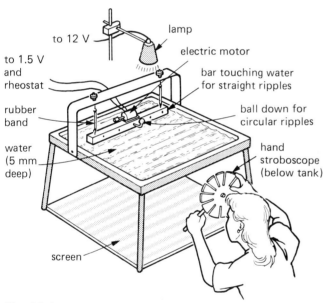

Fig. 13.4

Pulses (i.e. short bursts) of ripples are obtained by dipping a finger in the water for circular ones and a ruler for straight ones. *Continuous* ripples are generated by an electric motor on a bar which gives straight ripples if it just touches the water and circular ripples if the bar is raised and a small ball fitted to it so as to be in the water.

Continuous ripples are studied more easily if they are *apparently* stopped ('frozen') by viewing the screen through a disc, with equally spaced slits, that can be spun by hand, i.e. a stroboscope. If the disc speed is such that the waves have advanced one wavelength each time a slit passes your eye, they appear at rest.

Reflection

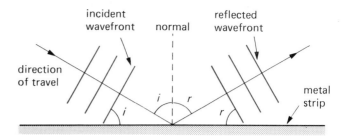

Fig. 13.5

In Fig. 13.5 *straight* water waves are represented falling on a metal strip placed in a ripple tank at an angle of 60°, i.e. the angle i between the direction of travel of the waves and the normal to the strip is 60°, as is the angle between the wavefront and the strip. The angle of reflection r is 60°. Incidence at other angles shows that the angles of reflection and incidence are always equal. (The straight lines representing wavefronts can be thought of as the crests of the waves.)

Reflection at a concave surface is shown in Fig. 13.6. The incident straight wave, which is another way of representing a parallel beam, is converged after reflection to the focus F as a circular wave.

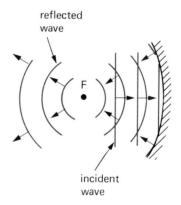

Fig. 13.6

Water waves evidently behave in the same way as light when reflected.

Refraction

If a glass plate is placed in a ripple tank so that the water is about 1 mm deep over it but 5 mm elsewhere, continuous straight waves in the shallow region are found to have a shorter wavelength than those in the deeper parts, Fig. 13.7a. Both sets of waves have the frequency of the vibrating bar and since $v = f\lambda$, then if λ has decreased so has v, since f is *fixed*. Hence *waves travel more slowly in shallow water.*

37

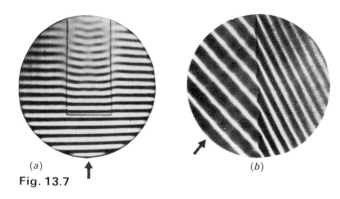

(a)　　　　　　　　　　(b)

Fig. 13.7

When the plate is at an angle to the waves, Fig. 13.7b, their direction of travel in the shallow region is bent towards the normal, Fig. 13.8, i.e. refraction occurs. We saw earlier (Topic 5) that light is

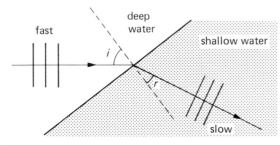

Fig. 13.8

refracted because its speed changes when it enters another medium. The refraction of water waves for the same reason seems to suggest that light itself may be a kind of wave motion.

Diffraction

In Figs 13.9 and 13.10, straight water waves in a ripple tank are falling on gaps formed by obstacles. In Fig. 13.9 the gap width is about the same as the wavelength of the waves (1 cm); those passing

Fig. 13.10

Fig. 13.9

through are circular and spread out in all directions. In Fig. 13.10 the gap is wide (10 cm) compared with the wavelength and the waves continue straight on; some spreading occurs but it is less obvious.

The spreading of waves at the edges of obstacles is called *diffraction*; when designing harbours, engineers use models like that in Fig. 13.11 to study it.

Interference

When two sets of continuous circular waves cross in a ripple tank, a pattern like that in Fig. 13.12a is obtained.

At points where a crest from S_1 arrives at the same time as a crest from S_2, a bigger crest is formed and the waves are said to be *in phase*. At points where a crest and a trough arrive together, they cancel out (if their amplitudes are equal); the waves are exactly *out of phase* (due to travelling different distances from S_1 and S_2) and the water is undisturbed, Fig. 13.12b. The dark 'spokes' radiating from S_1 and S_2 join such points.

Interference or *superposition* is the combination of waves to give a larger or a smaller wave. Fig. 13.12c shows how the pattern in Fig. 13.12a is formed. All points on AB are equidistant from S_1 and S_2 and since these vibrate in phase, crests (or troughs) from S_1 arrive at the same time as crests (or troughs) from S_2. Hence along AB reinforcement occurs by superposition and a wave of double amplitude is obtained. Points on CD are half a wavelength nearer to S_1 than to S_2, i.e. there is a path difference of half a wavelength. Therefore crests (or troughs) from S_1 arrive simultaneously with troughs (or crests) from S_2 and the waves cancel. Along EF the difference of distances from S_1 and S_2 to any point is one wavelength making EF a line of reinforcement.

Fig. 13.11

38

Fig. 13.12

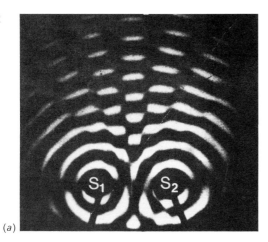

(a)

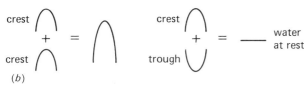

(b)

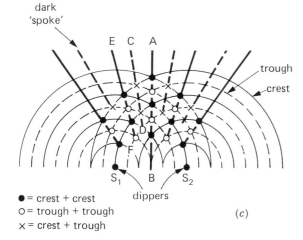

● = crest + crest
○ = trough + trough
× = crest + trough

(c)

Study this effect with the two ball 'dippers' about 3 cm apart on the bar. Also observe the effect of changing (i) the frequency and (ii) the separation of the 'dippers'; use a stroboscope when necessary. You will find that if the frequency is increased, i.e. the wavelength decreased, the 'spokes' are closer together. Increasing the 'dipper' separation has the same effect.

Similar patterns are obtained if straight waves fall on two small gaps: interference occurs between the emerging (circular), diffracted waves.

Q Questions

1. The lines in Fig. 13.13 are crests of straight ripples.
   (a) What is the wavelength of the ripples?
   (b) If ripple A occupied 5 seconds ago the position now occupied by ripple F, what is the frequency of the ripples?
   (c) What is the speed of the ripples?

Fig. 13.13

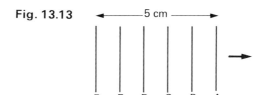

2. A straight ripple ABC is shown in Fig. 13.14 moving towards a wall XY. Draw one diagram to show the position of the ripple when B reaches the wall and another when C reaches it. On each diagram mark the angles which are 30°.

Fig. 13.14

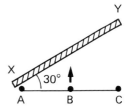

3. One side of a ripple tank ABCD is raised slightly, Fig. 13.15, and a ripple started at P by a finger. After a second the shape of the ripple is as shown.
   (a) Why is it not circular?
   (b) Which side of the tank has been raised?

Fig. 13.15

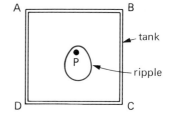

4. Fig. 13.16 gives a full-scale representation of the water in a ripple tank 1 second after the vibrator was started. The dark lines represent crests.

Fig. 13.16

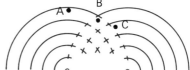

(a) What is represented at A at this instant?
(b) Estimate (i) the wavelength, (ii) the speed of the waves, and (iii) the frequency of the vibrator.
(c) Sketch a suitable attachment which could have been vibrated up and down to produce this wave pattern.
(d) Explain how the waves combine (i) at B and (ii) at C.

5. Copy Fig. 13.17 and show on it what happens to the waves as they pass through the gap when the water is much shallower on the right-hand side than on the left.

Fig. 13.17

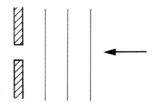

39

14 Light waves

Although we cannot see how light travels it displays the properties of waves.

Diffraction

Fig. 14.1

Diffraction occurs when light passes the edge of an object but it is not easy to detect. This suggests that it has a very small wavelength since we saw that diffraction of water waves is most obvious at a gap when its width is comparable with the wavelength of the waves. Fig. 14.1 is the diffraction pattern of a very narrow vertical slit (1/100 mm or less) and shows how light has spread into regions that would be in shadow if it went exactly in straight lines.

Interference

A steady interference pattern is obtained with water waves in a ripple tank because both sets of waves have the same frequency and wavelength and are exactly in phase when they leave the sources S_1 and S_2. They are said to be *coherent*. It is impossible to obtain a steady interference pattern with two separate lamps since most light sources emit light waves of many wavelengths in short erratic bursts, each out of phase with the next. The two sets of waves are not coherent.

These difficulties were overcome by Young in 1801 by allowing light from *one* lamp in a darkened room to fall on two narrow parallel slits very close together (about 0.5 mm separation), as shown in the modern

Fig. 14.3

arrangement of Fig. 14.2. A pattern of equally spaced bright and dark bands, called *fringes*, is obtained on a screen, Fig. 14.3. The waves leaving the slits are coherent since any phase changes in the bursts of light from the lamp affect both sets of waves at the same time.

The fringes can be explained by assuming that diffraction occurs at each slit and in the region where the two diffracted beams cross, there is interference, Fig. 14.4*a*. At points on the screen where a 'crest' from one slit arrives at the same time as a 'crest' from the other, the waves are in phase and there are bright bands. Dark bands occur where 'crests' and 'troughs' arrive simultaneously and the waves cancel. We then have: light + light = darkness. This makes sense only if we regard light as having a wave nature. Fig. 13.12*c*, which we used to explain the water wave interference pattern, is also a help when thinking about how light produces interference effects.

The bright bands are coloured, except for the centre

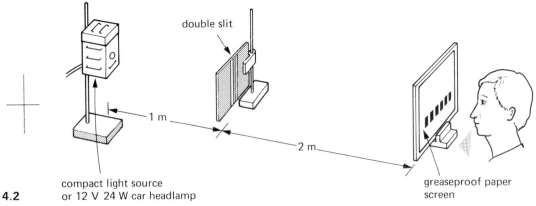

double slit

1 m

2 m

compact light source
or 12 V 24 W car headlamp

greaseproof paper
screen

Fig. 14.2

40

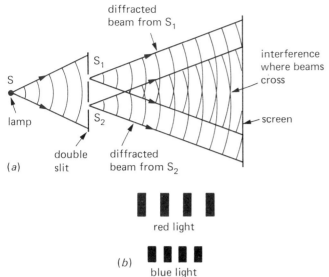

red light

(b)

blue light

Fig. 14.4

one which is white. If a red filter is placed in front of the lamp, the bright bands are red and are spaced farther apart than those given by a blue filter, Fig. 14.4*b*.

The wavelength of the light used can be measured and it can be shown that it is given by

$$\frac{\text{wavelength } (\lambda)}{\text{fringe separation } (y)} = \frac{\text{slit separation } (a)}{\text{distance of slits from screen } (D)}$$

In symbols, $\lambda = \dfrac{a \times y}{D}$

Experiment: wavelength of light by double slit

Arrange the apparatus as in Fig. 14.2, ensuring that the filament of the lamp is vertical, parallel to the double slit and at the same height. Darken the room and shield the screen from stray light. View the fringes from the behind the screen.

Three measurements must be made.

(i) *Bright band or fringe separation (y).* The distance from the centre of one bright band to the centre of the next has to be found. This is done by measuring across as many bands as can be seen (they are equally spaced). To do this make two marks at appropriate places on the screen while looking at the bands. Measure the distance afterwards in daylight with a mm scale, then find *y*.

(ii) *Distance of slits from screen (D).* Measure this with a metre rule (about 2 m).

(iii) *Slit separation (a).* Measure this in daylight by

one of the following methods. Lay the double slit and a $\frac{1}{2}$ mm transparent scale on white paper and compare them through a magnifying lens; or, better, examine them together under a microscope. A third method is to hold both double slit and scale in a slide projector and view their images on a distant wall. Substitute your measurements in $\lambda = ay/D$ and obtain an estimate for the average wavelength of white light in m.

Colour, wavelength and frequency

Red light has the longest wavelength of about 0.000 7 mm (7×10^{-7} m = 0.7 μm), while violet light has the shortest of about 0.0004 mm (4×10^{-7} m = 0.4 μm). Colours between those in the spectrum of white light have intermediate values. Light of one colour and so of one wavelength is called *monochromatic* light.

Remembering that $v = f\lambda$ (for all waves including light), it follows that red light has a lower frequency (f) than violet light since (i) the wavelength (λ) of red light is greater and (ii) all colours travel with the same speed (v) of 3×10^8 m/s in air (strictly a vacuum). It is the *frequency* of light which decides its *colour*, rather than its wavelength which is different in different media, as is its speed (p. 18).

The *amplitude* of a light (or any other) wave determines its *intensity* and depends on the brightness of the source.

Diffraction grating

A diffraction grating consists of a piece of glass or plastic with a large number of parallel lines marked on it. The lines scatter light and in effect are opaque. The thin clear strips between the lines transmit light and act as slits.

When white light from a straight filament lamp is viewed through a grating a number of spectra (plural of spectrum) are seen, each like that given by a prism. The spectra are said to be of various 'orders'.

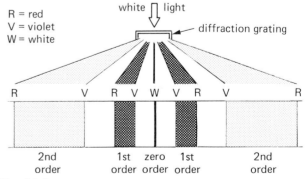

Fig. 14.5

A 'fine' grating with say 300 lines per mm gives the zero, first and second orders, Fig. 14.5. A 'coarse' grating, e.g. 100 lines per mm, gives more orders closer together and the 'coarser' the grating the more does the pattern resemble the two-slits one.

The slits are only a few wavelengths wide and when light falls on them, diffraction occurs causing cylindrical waves to spread out from each one. In certain *directions* the diffracted waves are in phase and if brought to a focus by a lens or the eye, they reinforce. In other directions they more or less cancel, i.e. destructive interference occurs.

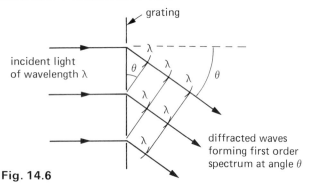

Fig. 14.6

In Fig. 14.6 suppose a first order spectrum is formed in a direction making an angle θ with the incident light of wavelength λ. The diffracted waves from each slit must travel a distance λ more (or less) than those from the next slit, i.e. the path difference between the waves is one wavelength. What will it be for (i) the zero order spectrum, (ii) a second order spectrum?

Red sunset: blue sky

The wavelength of red light is about twice that of blue light. The longer the wavelength the more penetrating is the light, while the shorter the wavelength the more easily is it scattered.

When the sun is setting the light from it has to travel through a greater thickness of the earth's atmosphere and only the longer wavelength red light is able to get through. Sunsets are therefore red.

The shorter wavelengths, like blue, are scattered in all directions by the atmosphere, which is why the sky looks blue.

Lasers and holograms

A laser produces a very intense beam of light which is monochromatic, coherent and parallel. A holo-

gram is a three-dimensional image of an object recorded on a special photographic plate. The image is formed by the interference of two beams of light from a laser, one of which has been reflected from the object. When developed and illuminated, the image looks 'real', appears to float in space and moves when the viewer does.

In cardphones, the card (used instead of money) contains a hologram which keeps a check on the number of units used. Holograms are also used to make safety checks on car tyres and aircraft engines, to design artificial limbs and to display and market goods.

Q Questions

1. In Fig. 14.7, L is a source of light, wavelength λ, S_1S_2 a double slit and O, O_1 and O_2 are points on a screen.

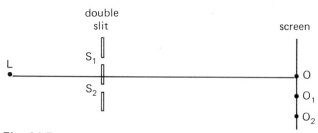

Fig. 14.7

(a) If the central bright band is formed at O, how do the distances S_1O and S_2O compare?
(b) If the first dark band next to the central bright band is formed at O_1, how do S_1O_1 and S_2O_1 compare?
(c) If the first bright band occurs at O_2 next to the first dark band, how do S_1O_2 and S_2O_2 compare?

2. (a) Why is the diffraction of light not easy to detect?
(b) Why is it not possible to produce a steady interference pattern using two light sources?

3. In the double slits experiment using monochromatic light how would the fringe pattern be affected if
(a) the separation of the slits was increased,
(b) the screen was moved farther away,
(c) light of longer wavelength was used, and
(d) white light was used instead of monochromatic light?

4. Give the approximate wavelength in micrometres (μm) of (a) red light, (b) violet light.

5. In a double-slit experiment with white light the slit separation is 0.25 mm, the fringe separation is 4 mm and the distance from the slits to the screen is 2 m. Find the average wavelength of white light in micrometres (μm). (1 μm $= 10^{-6}$ m.)

check list p.281

15 Electromagnetic spectrum

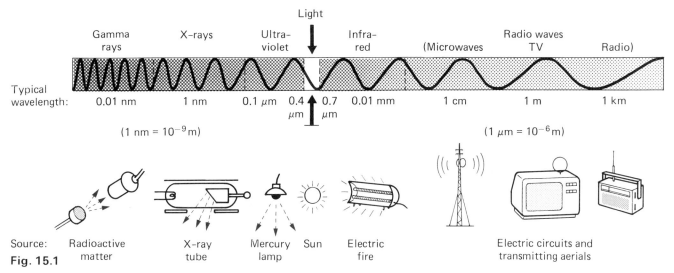

Fig. 15.1

Light is one member of a large family of waves called the electromagnetic spectrum and shown in Fig. 15.1. While each is produced differently, all result from electrons in atoms (Topic 73) undergoing an energy change. For example, radio waves are created when electrons are accelerated in an aerial. Although they differ greatly in their wavelengths and effects, all *electromagnetic waves* have the following basic properties.

(i) They travel through space at 300 000 km/s (3×10^8 m/s), i.e. with the speed of light.

(ii) They exhibit diffraction and interference as well as reflection and refraction.

(iii) They obey the equation $v = f\lambda$ where v is the speed of light, f is the frequency of the wave and λ is the wavelength. Since v is constant (for a given medium), it follows that the higher the frequency of a wave, the smaller is its wavelength.

(iv) Because of their electrical origin and ability to travel in a vacuum (e.g. from the sun to the earth),

they are regarded as *progressive transverse* waves. The wave is a combination of travelling electric and magnetic forces, carrying energy from one place to another. The forces vary in value and are directed at right angles to each other and to the direction of travel of the wave, hence transverse. Fig. 15.2 is a representation of an electromagnetic wave.

Infrared radiation

The presence of infrared (i.r.) in the radiation from the sun or from the filament of an electric lamp can be detected by placing a phototransistor just beyond the red end of the spectrum formed (as explained on p. 26) by a prism, Fig. 15.3. Our bodies detect i.r. (also called 'radiant heat' or 'heat radiation') by its heating effect.

Anything which is hot but not glowing, i.e. below 500 °C, emits i.r. alone. At about 500 °C a body becomes red-hot and emits red light as well as i.r.; the heating element of an electric fire is an example. At

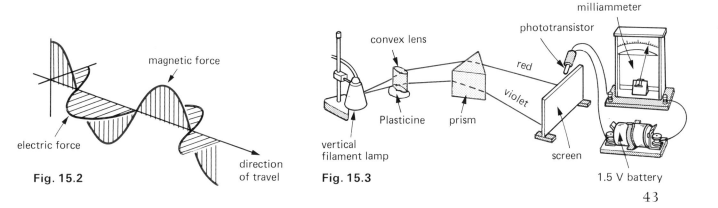

Fig. 15.2

Fig. 15.3

Fig. 15.4

about 1000 °C things such as lamp filaments are white-hot and radiate i.r. and white light, i.e. all the colours of the spectrum.

Infrared is also detected by special photographic films and pictures can be taken in the dark like that in Fig. 15.4 of a car; the white parts are hottest. Infrared lamps are used to dry the paint on cars during manufacture and in the treatment of muscular complaints. Television receiver remote control keypads contain a small infrared transmitter for changing programmes and making other adjustments.

Ultraviolet radiation

Ultraviolet (u.v.) rays have shorter wavelengths than light and can be detected just beyond the violet end of the spectrum (due to the sun or a filament lamp) by fluorescent paper. They cause sun-tan and produce vitamins in the skin but an overdose can be harmful, especially to the eyes.

Ultraviolet causes teeth, finger nails, fluorescent paints and clothes washed in detergents to fluoresce, i.e. they glow by reradiating as light the energy they absorb as u.v. The shells of fresh eggs fluoresce with a reddish colour, those of 'bad' eggs appear violet.

A u.v. lamp used for scientific or medical purposes contains mercury vapour and this emits u.v. when an electric current passes through it. Fluorescent tubes also contain mercury vapour and their inner surfaces are coated with powders which radiate light when struck by u.v.

Radio waves

Radio waves have the longest wavelengths in the electromagnetic spectrum. They are radiated from aerials and used to 'carry' sound, pictures and other information over long distances.

(a) Long, medium and short waves (2 km to 10 m) can bend (diffract) round obstacles so they can be received when hills etc. are in their way, Fig. 15.5a. This allows them to be used for local radio broadcast-

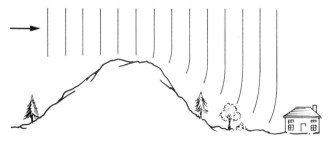

(*a*) Diffraction of radio waves

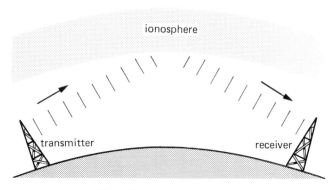

Fig. 15.5   (*b*) Reflection of radio waves

ing. They are also reflected by layers of electrically charged particles in the upper atmosphere (the *ionosphere*), which makes long-distance reception possible, Fig. 15.5b.

(b) VHF (very high frequency) and **UHF (ultra high frequency)** waves have shorter wavelengths and need a clear, straight-line path to the receiver. Local radio and television use them. They pass through the ionosphere.

(c) Microwaves (wavelengths of a few centimetres) are used for radar and also for international (as well as national) telephone and television links. The international links are via geostationary communications satellites (e.g. INTELSAT V, Fig. 2c, page 9) which go round the equator at the same rate as the earth spins and so appear to be at rest. Signals are beamed through the ionosphere by large dish aerials like that in Fig. 77.2, to the satellite where they are amplified and sent back to a dish aerial in another part of the world. Fig. 15.6 shows a tower with dish aerials for microwave links in the U.K.

Microwaves are also used for cooking because, like all electromagnetic waves, they have a heating effect when absorbed.

The wave nature of microwaves can be shown using the apparatus of Fig. 15.7 to demonstrate interference and other effects: the meter reading rises and falls on the receiver as it is moved across behind the metal plates.

Fig. 15.6

Fig. 15.7

X-rays

They are produced when high-speed electrons are stopped by a metal target in an X-ray tube. X-rays have smaller wavelengths than u.v.

X-rays can penetrate solid objects and affect a film; Fig. 15.8 is an X-ray photograph of someone shaving. Very penetrating X-rays are used in hospitals to kill cancer cells. They also damage healthy cells so careful shielding of the X-ray tube with lead is needed. Less penetrating X-rays have longer wavelengths and penetrate flesh but not bone: they are used in dental X-ray photography.

In industry they are used to inspect welded joints and castings for faults.

Gamma rays are more penetrating and dangerous than X-rays and will be studied on pp. 213–14.

Q Questions

1. Name *four* properties common to all electromagnetic waves.

2. Which of the following types of radiation has the largest frequency?

    A u.v.   B radio waves   C light   D X-rays   E i.r.

3. Name one type of electromagnetic radiation which (a) causes sun-tan, (b) is used for satellite communication, (c) passes through a thin sheet of lead, (d) is used to take photographs in haze.

4. A VHF radio station transmits on a frequency of 100 MHz (1 MHz = 10^6 Hz). If the speed of radio waves is 3×10^8 m/s
    (a) what is the wavelength of the waves,
    (b) how long does the transmission take to travel 60 km?

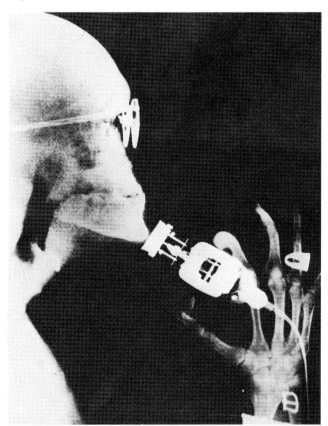

Fig. 15.8

check list
P.281

16 Sound waves

About sound

Sources of sound such as a drum, a guitar and the human voice have some part which *vibrates*. The sound travels through the air to our ears and we hear it. That the air is necessary may be shown by pumping out a glass jar containing a ringing electric bell, Fig. 16.1; the sound disappears though the striker can still be seen hitting the gong. Evidently sound cannot travel in a vacuum like light.

Fig. 16.1

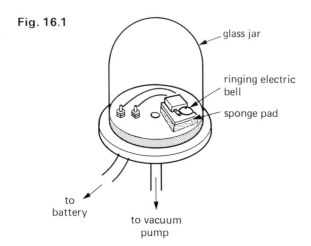

glass jar

ringing electric bell

sponge pad

to battery

to vacuum pump

Other materials, including solids and liquids, transmit sound. 'Cathedral chimes' may be heard if you jingle together some spoons tied to a piece of string with its ends in your ears, Fig. 16.2. Not only does string (a solid) transmit sound but it does so better than air.

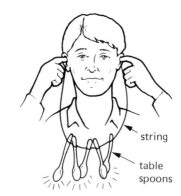

string

table spoons

Fig. 16.2

Sound also gives interference and diffraction effects. Because of this and its other properties, we believe it is a form of energy (as the damage from supersonic booms shows) which travels as a progressive wave, but of a type called *longitudinal*.

Longitudinal waves

(a) Waves on a spring. In a progressive longitudinal wave the particles of the transmitting medium vibrate to and fro along the same line as that in which the wave is travelling and not at right angles to it as in a transverse wave. A longitudinal wave can be sent along a spring, stretched out on the bench and fixed at one end if the free end is repeatedly pushed and pulled sharply. Compressions C (where the coils are closer together) and rarefactions R (where the coils are farther apart), Fig. 16.3, travel along the spring.

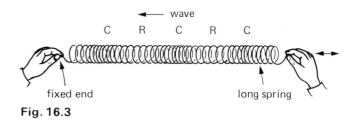

wave

C R C R C

fixed end   long spring

Fig. 16.3

(b) Sound waves. A sound wave produced for example by a loudspeaker consists of a train of compressions and rarefactions in the air, Fig. 16.4.

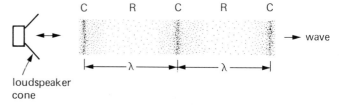

C R C R C

wave

λ λ

loudspeaker cone

Fig. 16.4

The speaker has a cone which is made to vibrate in and out by an electric current. When the cone moves out the air in front is compressed; when it moves in the air is rarefied (goes 'thinner'). The wave progresses through the air but the air as a whole does not move. The air particles vibrate backwards and forwards a little as the wave passes. When the wave enters your ear the compressions and rarefactions cause small, rapid pressure changes on the ear drum and you experience the sensation of sound.

The number of compressions produced per second is the frequency f of the sound wave (and equals the frequency of the vibrating cone); the distance between successive compressions is the wavelength λ. As for transverse waves the speed $v = f\lambda$.

46

Human beings hear only sounds with frequencies from about 20 Hz to 20 000 Hz (20 kHz). These are the *limits of audibility*; the upper limit decreases with age.

Reflection and echoes

Sound waves are reflected well from hard, flat surfaces such as walls or cliffs and obey the same laws of reflection as light. The reflected sound forms an *echo*.

If the reflecting surface is nearer than 15 m, the echo joins up with the original sound which then seems to be prolonged. This is called *reverberation*. Some is desirable in a concert hall to stop it sounding 'dead', but too much causes 'confusion'. In the Royal Festival Hall, London (p. 49), walls and seats are covered with sound-absorbing material.

Speed of sound

The speed of sound depends on the material through which it is passing, being greater in solids than in liquids or gases. Some values are given in the table below.

Material	Air (0°C)	Water	Concrete	Steel
Speed (metres/second)	330	1400	5000	6000

In air the speed *increases with temperature* and at high altitudes, where the temperature is lower, it is less than at sea-level. Changes of atmospheric pressure do not affect it.

An estimate of the speed of sound can be made directly if you stand about 100 metres from a high wall or building and clap your hands. Echoes are produced. When the clapping rate is such that each clap coincides with the echo of the previous one, the sound has travelled to the wall and back in the time between two claps, i.e. an interval. By timing 30 intervals with a stop watch, the time t for one interval can be found. Also, knowing the distance d to the wall, a rough value is obtained from

$$\text{speed of sound in air} = \frac{2d}{t}$$

Diffraction and interference

(a) Diffraction. Audible sounds have wavelengths from about 1.5 centimetres (20 kHz) up to 15 metres

(20 Hz) and so suffer diffraction by objects of similar size, e.g. a doorway 1 metre wide. This explains why we hear sound round corners.

(b) Interference. In Fig. 16.5 sound waves of the same frequency from two loudspeakers (supplied by one signal generator) produce a steady interference pattern. The resulting variations in loudness of the sound, due to the waves reinforcing and cancelling one another, can be heard when you walk past the loudspeakers.

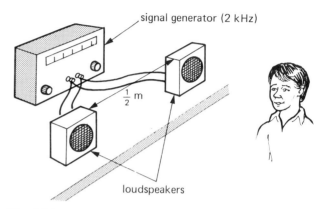

signal generator (2 kHz)

$\frac{1}{2}$ m

loudspeakers

Fig. 16.5

Ultrasonics

Sound waves with frequencies above 20 kHz are called *ultrasonic* waves. They are emitted by bats and enable them to judge the distance of an object from the time taken by the reflected wave to return.

In the echo-sounding system called *sonar*, ships use ultrasonic waves to measure the depth of the sea and to detect shoals of fish, Fig. 16.6. In industry they are

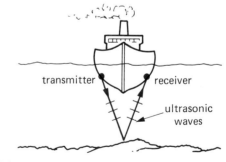

transmitter   receiver

ultrasonic waves

Fig. 16.6

used to reveal flaws in welded joints; also holes of any shape or size are cut in glass and steel by ultrasonic drills. Objects such as street lamp covers can be cleaned by immersion in a tank of water which has an ultrasonic vibrator in the base. Ultrasonic scans are often used as a check on expectant mothers.

47

Ultrasonic vibrators contain a quartz crystal which is made to vibrate electrically at the required frequency. An ultrasonic receiver also consists of a quartz crystal but it works in reverse, i.e. when it is set into vibration by ultrasonic waves it generates an electrical signal which is then amplified.

Q Questions

1. A coastguard sees a distress rocket burst in the sky and 5 seconds later he hears the bang. If the speed of sound is 330 m/s, how far is the exploding rocket from the coastguard?

    A 66 m   B 335 m   C 1100 m
    D 1650 m   E 3300 m.   (S.E.)

2. The time-keeper of a 110 m race stands near the finishing tape and starts his stop-watch on hearing the bang from the starting pistol.

   (a) When should he have started his stop-watch?
   (b) Give a reason for your answer to (a).
   (c) Calculate the error in his timing assuming he makes no further errors. (Speed of sound = 330 m/s.)
                                  (E.M.)

3. (a) What is the relationship connecting the frequency of a sound source with its wavelength and the speed of sound in air?

   (b) An echo sounder in a ship produces a sound pulse and an echo is received from the sea bed after 0.4 s. Assuming the speed of sound in sea water to be 1500 m/s calculate the depth of the sea bed.
   (c) If the echo sounder had produced continuous waves of frequency 6000 Hz (6 kHz), what would have been their wavelength in the sea water? Give your value in cm.   (N.W.)

4. If 5 s elapse between a lightning flash and the clap of thunder how far away is the storm? Speed of sound = 330 m/s.

5. (a) A girl stands 160 m away from a high wall and claps her hands at a steady rate so that each clap coincides with the echo of the one before. If she makes 60 claps in 1 minute, what value does this give for the speed of sound?
   (b) If she moves 40 m closer to the wall she finds the clapping rate has to be 80 per minute. What value do these measurements give for the speed of sound?
   (c) If she moves again and finds the clapping rate becomes 30 per minute, how far is she from the wall if the speed of sound is the value you found in (a)?

6. (a) What properties of sound suggest it is a wave motion?
   (b) How does a progressive transverse wave differ from a longitudinal one? Which type is sound?

check list p. 281

17 Musical notes

Fig. 17.1

Irregular vibrations cause *noise*; regular vibrations such as occur in the instruments of an orchestra, Fig. 17.1, produce *musical notes* which have three properties—pitch, loudness and quality.

Pitch

The pitch of a note depends on the frequency of the sound wave reaching the ear, i.e. on the frequency of the source of sound. A high-pitched note has a high frequency and a short wavelength. The frequency of middle C (in 'scientific pitch') is 256 vibrations per second or 256 Hz and that of upper C is 512 Hz. Notes are an *octave* apart if the frequency of one is twice that of the other. Pitch is like colour in light; both depend on the frequency.

Notes of known frequency can be produced in the laboratory by a signal generator supplying alternating electric current (a.c.) to a loudspeaker. The cone of the speaker vibrates at the frequency of the a.c. which can be varied and read off a scale on the generator. A set of tuning forks with frequencies

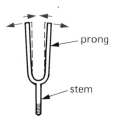

Fig. 17.2

marked on them can also be used. A tuning fork, Fig. 17.2, has two steel prongs which vibrate when struck; the prongs move in and out *together*, generating compressions and rarefactions.

Loudness

A note is louder when more sound energy enters our ears per second than before and is caused by the source vibrating with a larger amplitude. If a violin string is bowed more strongly, its amplitude of vibration increases as does that of the resulting sound wave and the note heard is louder because more energy has been used to produce it.

49

Quality

The same note on different instruments sounds different; we say the notes differ in *quality* or *timbre*. The difference arises because no instruments (except a tuning fork and a signal generator) emit a 'pure' note, i.e. of one frequency. Notes consist of a main or *fundamental* frequency mixed with others, called *overtones*, which are usually weaker and have frequencies that are exact multiples of the fundamental. The number and strength of the overtones decides the quality of a note. A violin has more and stronger higher overtones than a piano. Overtones of 256 Hz (middle C) are 512 Hz, 768 Hz and so on.

The *waveform* of a note played near a microphone connected to a C.R.O. can be displayed on the C.R.O. screen. Those for the *same* note on three instruments are given in Fig. 17.3. Their different shapes show

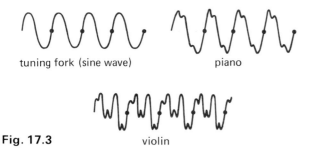

tuning fork (sine wave)   piano

violin

Fig. 17.3

that while they have the same fundamental frequency, their quality differs. The 'pure' note of a tuning fork has a *sine* waveform and is the simplest kind of sound wave. *N.B.* Although the waveform on the C.R.O. screen is transverse it represents a longitudinal sound wave.

Vibrating strings: stationary waves

In a string instrument such as a guitar the 'string' is a tightly-stretched wire or length of gut. When it is plucked transverse waves travel to both ends, which are fixed, and are reflected. Interference occurs between the incident and reflected waves and a *stationary* or *standing* wave pattern is formed. In this, certain points on the string, called *nodes*, are at rest whilst points mid-way between each pair of nodes are in continuous vibration with maximum amplitude; they are called *antinodes*.

A string can vibrate in various ways. If the standing wave has one loop, Fig. 17.4a, the fundamental note is emitted. If there is more than one loop, Figs. 17.4b,c, overtones are produced, but in all cases the *separation of the nodes N and the antinodes A is one-quarter of the wavelength* of the wave on the string

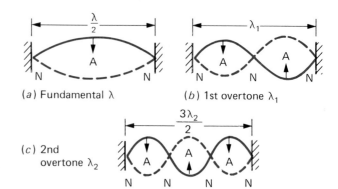

(a) Fundamental λ (b) 1st overtone λ_1

(c) 2nd overtone λ_2

Fig. 17.4

causing the note. In an instrument, a string vibrates in several ways at the same time depending on where it is plucked; this decides the quality of the note.

The standing wave patterns of a vibrating string (or rubber cord) may be viewed as in Fig. 17.5.

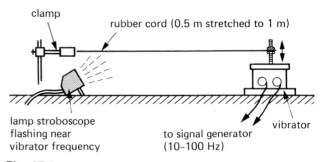

clamp

rubber cord (0.5 m stretched to 1 m)

lamp stroboscope flashing near vibrator frequency

to signal generator (10–100 Hz)

vibrator

Fig. 17.5

The frequency of the fundamental note emitted by a vibrating string depends on its

(i) *length*: short strings emit high notes and halving the length doubles the frequency,
(ii) *tension*: tight wires produce high notes, and
(iii) *mass per unit length*: thin strings give high notes.

Resonance

All objects have a natural frequency of vibration. The vibration can be started and increased by another object vibrating at the same frequency. The

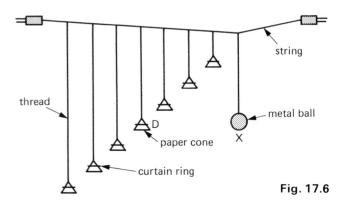

string

thread

D

paper cone

metal ball

X

curtain ring

Fig. 17.6

Fig. 17.7 *(a)*   *(b)*

effect is called *resonance*. For example, when the heavy pendulum X in Fig. 17.6 is set swinging, it forces all the light ones to swing at the same frequency, but D, which has the same length as X, does so with a much larger amplitude, i.e. D resonates with X.

If a large structure such as a suspension bridge starts vibrating at its natural frequency the result can be disastrous. The collapse of the Tacoma Narrows Bridge, USA, in 1940 was due to a cross-wind of just the 'right' speed to cause resonant vibrations, Fig. 17.7a,b. Models of new bridges are now tested in wind tunnels to check that their natural frequencies are not dangerous.

Resonance also occurs in sound when a column of air in a wind instrument is made to vibrate.

Noise 'pollution'

Unpleasant sounds are called noises. High-pitched noises are usually more annoying than low-pitched ones. Noise can damage the ears, cause tiredness and loss of concentration and if it is very loud, result in sickness and temporary deafness. Some of the main noise 'polluters' are aircraft, motor vehicles, greatly amplified music and many types of machinery including domestic appliances.

Ways of reducing unwanted noise may involve designing quieter engines and better exhaust systems. For example, rotating shafts in machinery can be balanced better so that they do not cause vibration. Car engines are often mounted on metal brackets via rubber blocks which absorb vibrations and do not pass them on to the car body.

The use in the home of sound insulating materials such as carpets and curtains, and of double-glazed windows also helps. The farther away the noise originates the weaker it is, so distance is a natural barrier, as are trees between houses and a noisy road. Tractor drivers, factory workers, pneumatic drill operators and others exposed regularly to noise often have to wear ear protectors.

Noise levels are measured in *decibels* (dB) by a noise meter. The sound level the average human ear can just detect, called the *threshold of hearing*, is taken as 0 dB. Normal conversation is about 60 dB, a jet plane overhead is 100 dB and the *threshold of pain* (which explains itself) is 120 dB.

Ｑ Questions

1. **(a)** Draw the waveform of **(i)** a loud, low-pitched note and **(ii)** a soft, high-pitched note.
   (b) If the speed of sound is 340 m/s what is the wavelength of a note of frequency **(i)** 340 Hz, **(ii)** 170 Hz?

2. **(a)** What change does your ear detect when you are listening to a sound if **(i)** the *amplitude* is raised, **(ii)** the *frequency* is raised?
   (b) A certain length of guitar string gives the note middle C. **(i)** What note do you get with half the length? **(ii)** How could you raise the pitch of the note without changing the length?   *(S.R. part qn.)*

check list

p.282

18 Additional questions

Core level (all students)

Water waves

1. A straight vibrator causes water ripples to travel across the surface of a shallow tank. The waves travel a distance of 33 cm in 1.5 s and the distance between successive wave crests is 4.0 cm. Calculate the frequency of the vibrator. (W.)

2. Fig. 18.1 shows a series of waves as might be produced in a laboratory experiment with a ripple tank. The waves are travelling from region A to region B.

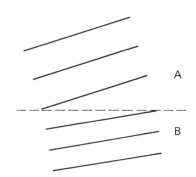

Fig. 18.1

(a) Describe how these waves are produced.
(b) What is happening along the dotted line?
(c) By measurement, determine the wavelength of the waves in the region A and the region B.
(d) What can you say about the frequency of the waves in region A and region B?
(e) What can you say about the speed of waves in region A and region B?
(f) Where is the water deepest—at A or B? (E.M.)

3. Fig. 18.2a and b show ripples approaching metal plates with gaps in them. Gap (a) is narrow compared with the wavelength and gap (b) is wide compared with the wavelength.

Copy the diagrams and draw the shapes of the ripples after passing through the gaps.

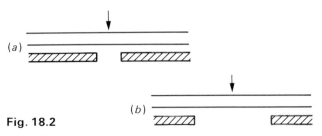

Fig. 18.2

52

Light waves

4. In Fig. 18.3 light waves are incident on an air-glass boundary. Some are reflected and some are refracted in the glass. One of the following is the same for the incident wave and the refracted wave.

  A speed   B wavelength   C direction
  D brightness   E frequency

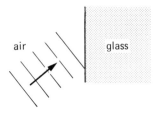

Fig. 18.3

Electromagnetic spectrum

5. The strip in Fig. 18.4 represents part of the electromagnetic spectrum extending on both sides of the region of visible light C. If A is the region of shortest wavelengths and F the region of longest wavelengths, name which section will represent the **(a)** u.v. region, **(b)** X-rays, **(c)** i.r. region.

State one difference between the behaviour of ultraviolet and X-rays. (L.)

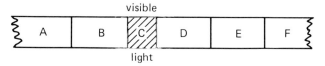

Fig. 18.4

Sound waves

6. Two loudspeakers play the same note from a signal generator. A person moving across the line between them, Fig. 18.5, finds the note is loud in some places and soft in others.
  (a) Why is the note loud in some positions and soft in others?
  (b) What difference would it make to use a higher frequency note? Why? (O.L.E./S.R.16+)

Fig. 18.5

7. The property which distinguishes longitudinal waves from transverse waves is the
   A wavelength
   B velocity
   C ability to be refracted
   D need for a material medium
   E relative directions of oscillation and propagation
   (J.M.B.)

8. An echo-sounder in a trawler receives an echo from a shoal of fish 0.4 s after it was sent. If the speed of sound in water is 1500 m/s, how deep is the shoal?

   A 150 m   B 300 m   C 600 m
   D 7500 m   E 10000 m

Musical notes

9. (a) On what does (i) the loudness, (ii) the pitch of a sound depend?
   (b) Why does middle C on the piano sound different from the same note on the violin?

Further level (higher grade students)

Water waves

10. The continuous line in Fig. 18.6 shows the position of a water wave travelling to the right. X, Y and Z are corks floating on the surface and the dotted line is the undisturbed water surface. As the wave moves forward does (i) X, (ii) Y, (iii) Z, move up or down or stay where it is?

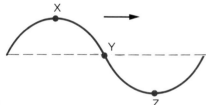

Fig. 18.6

Light waves

11. A double slit is formed on a blackened glass plate by drawing two lines close together. When a blue light is viewed through the double slit, an interference pattern like that in Fig. 18.7a is seen.
   (a) What can be concluded about the nature of light from this pattern?

Fig. 18.7 (a)
(b)

   (b) If the blue light is replaced by one giving a different colour of light, the pattern in Fig. 18.7b is seen. Account for its being more spread out.
   (c) What could be the colour of the light in the second case?

12. If you held a diffraction grating with its lines vertical close to the eye in a darkened room and looked through the grating at a small, bright source of white light some distance away, describe and explain what you might expect to see.

What is meant by the *order of a spectrum?*
(O. and C. part qn.)

Electromagnetic spectrum

13. Light travels at 3×10^8 m/s in air (300 000 000 m/s).
   (a) How fast do radio waves travel in air?
   (b) What is the wavelength of the 80 MHz waves used for broadcasting on VHF? (1 MHz = 1 000 000 Hz.)
   (c) What is the frequency of the 1500 m radio waves used on the Long Wave Band? *(S.R.)*

Sound waves

14. (a) State at least three differences between sound waves and radio waves.
   (b) Explain the following: (i) echo, (ii) reverberation.
   (c) A girl claps her hands at a steady rate of 30 claps a minute and hears the echoes from a high wall. When she is 170 m away from the wall, she hears the echo of each clap mid-way between it and the next clap. What is the speed of sound from these observations?.

Musical notes

15. How is a standing wave produced? How does it differ from a progressive wave?

16. What is meant by resonance? Give an example.

Matter and molecules

19 Measurements

Fig. 19.1a

Fig. 19.1b

Before a measurement can be made, a standard or *unit* must be chosen. The size of the quantity to be measured is then found with an instrument having a scale marked in the unit.

Many types of measuring instruments have been designed. Those on the flight deck of Concorde, Fig. 19.1a,b, provide the pilot with information about the performance of the aircraft.

Three basic quantities we have to measure in physics are *length*, *mass* and *time*. Units for other quantities are based on them. The SI (Système International d'Unités) system is a set of metric units now used in many countries. It is a decimal system in which units are divided or multiplied by 10 to give smaller or larger units.

Powers of ten shorthand

This is a neat way of writing numbers especially if they are large or small. It works like this.

$$
\begin{aligned}
4000 &= 4 \times 10 \times 10 \times 10 &&= 4 \times 10^3 \\
400 &= 4 \times 10 \times 10 &&= 4 \times 10^2 \\
40 &= 4 \times 10 &&= 4 \times 10^1 \\
4 &= 4 \times 1 &&= 4 \times 10^0 \\
0.4 &= 4/10 &= 4/10^1 &= 4 \times 10^{-1} \\
0.04 &= 4/100 &= 4/10^2 &= 4 \times 10^{-2} \\
0.004 &= 4/1000 &= 4/10^3 &= 4 \times 10^{-3}
\end{aligned}
$$

54

The small figures 1, 2, 3 etc., are called *powers of ten* and give the number of times the number has to be multiplied by 10 if it is greater than 1 and divided by 10 if it is less than 1. Note that 1 is also written as 10^0 and for numbers less than 1 the power has a negative sign.

Length

The unit of length is the *metre* (m) and is the distance, believed never to alter, occupied by a certain number of wavelengths of a particular colour of light. Previously it was the distance between two marks on a certain metal bar. Submultiples are:

1 centimetre (cm) $= 10^{-2}$ m
1 millimetre (mm) $= 10^{-3}$ m
1 micrometre (μm) $= 10^{-6}$ m
1 nanometre (nm) $= 10^{-9}$ m

A multiple for large distances is

1 kilometre (km) $= 10^3$ m ($\frac{5}{8}$ mile approx.)

Many length measurements are made with rulers; the correct way to read one is shown in Fig. 19.2. The reading is 76 mm or 7.6 cm. Your eye must be right over the mark on the scale or the thickness of the ruler causes errors due to 'parallax' (p. 15).

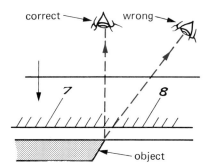

Fig. 19.2

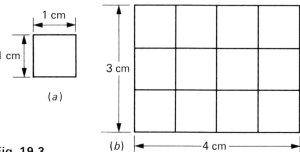

Fig. 19.3

Significant figures

Every measurement of a quantity is an attempt to find its true value and is subject to errors arising from limitations of the apparatus and the experimenter. The number of figures, called *significant* figures, given for a measurement, indicates how accurate we think it is and more figures should not be given than is justified.

For example, a value of 4.5 for a measurement has two significant figures; 0.0385 has three significant figures, 3 being the most significant and 5 the least, i.e. it is the one we are least sure about since it might be 4 or it might be 6. Perhaps it had to be estimated by the experimenter because the reading was between two marks on a scale.

When doing a calculation your answer should have the same number of significant figures as the measurements used in the calculation. For example, if your calculator gave an answer of 3.4185062, this would be written as 3.4 if the measurements had two significant figures. It would be written as 3.42 for three significant figures. Note that in deciding the least significant figure you look at the next figure. If it is less than 5 you leave the least significant figure as it is (hence 3.41 becomes 3.4) but if it equals or is greater than 5 you increase the least significant figure by 1 (hence 3.418 becomes 3.42).

If a number is expressed in powers of ten form (also called *standard form*), the number of significant figures is the number of digits before the power of ten. For example, 2.73×10^3 has three significant figures.

Area

The area of the square in Fig. 19.3*a* with sides 1 cm long is 1 square centimetre (1 cm²). In Fig. 19.3*b* the rectangle measures 4 cm by 3 cm and has an area of $4 \times 3 = 12$ cm² since it has the same area as twelve

squares each of area 1 cm². The area of a square or rectangle is given by

AREA = LENGTH × BREADTH

The SI unit of area is the square metre (m²) which is the area of a square with sides 1 m long. Note that

$$1 \text{ cm}^2 = \frac{1}{100} \text{ m} \times \frac{1}{100} \text{ m} = \frac{1}{10\,000} \text{ m}^2 = 10^{-4} \text{ m}^2$$

An *estimate* of the area of an irregular shape can be made by dividing it up into squares each of area, say 1 cm². Incomplete squares having an area of $\frac{1}{2}$ cm² or more are counted as complete squares and those of area less than $\frac{1}{2}$ cm² are ignored. In Fig. 19.4 a reasonable estimate for the shaded area is 12 cm².

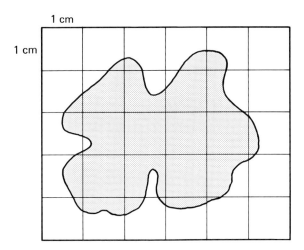

Fig. 19.4

The area of a circle of radius r is πr^2 where $\pi = 22/7$ or 3.14.

Volume

Volume is the amount of space occupied. The unit of volume is the *cubic metre* (m³) but as this is rather large, for most purposes the *cubic centimetre* (cm³) is used. The volume of a cube with 1 cm edges is 1 cm³.

55

Note that

$$1 \text{ cm}^3 = \frac{1}{100} \text{ m} \times \frac{1}{100} \text{ m} \times \frac{1}{100} \text{ m}$$

$$= \frac{1}{1\,000\,000} \text{ m}^3 = 10^{-6} \text{ m}^3$$

For a regularly-shaped object such as a rectangular block, Fig. 19.5 shows that

VOLUME = LENGTH × BREADTH × HEIGHT

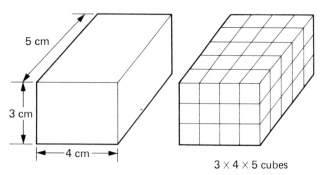

3 × 4 × 5 cubes

Fig. 19.5

The volume of a sphere of radius r is $\frac{4}{3}\pi r^3$ and of a cylinder of radius r and height h is $\pi r^2 h$.

The volume of a liquid may be obtained by pouring it into a measuring cylinder, Fig. 19.6, or a known volume can be run off more accurately from a burette, Fig. 19.7. When making a reading both vessels must be upright and the eye level with the bottom of the curved liquid surface, i.e. the meniscus. The meniscus formed by mercury is curved oppositely to that of other liquids and the top is read.

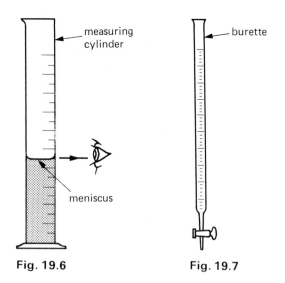

measuring cylinder

burette

meniscus

Fig. 19.6   **Fig. 19.7**

Liquid volumes are also expressed in litres (l); 1 litre = 1000 cm³. One millilitre (1 ml) = 1 cm³.

Mass

The mass of an object is the measure of the amount of matter in it. The unit of mass is the *kilogram* (kg) and is the mass of a piece of platinum alloy at the Office of Weights and Measures in Paris. The gram (g) is one-thousandth of a kilogram.

$$1 \text{ g} = 1/1000 \text{ kg} = 10^{-3} \text{ kg} = 0.001 \text{ kg}$$

The term *weight* is often used when mass is really meant. In science the two ideas are distinct and have different units as we shall see later. The confusion is not helped by the fact that mass is found on a balance by a process we unfortunately call 'weighing'!

There are several kinds of balance. In the *beam balance* the unknown mass in one pan is balanced against known masses in the other pan. In the *lever balance* a system of levers acts against the mass when it is placed in the pan. A direct reading is obtained from the position on a scale of a pointer joined to the lever system. A modern *top pan balance* is shown in Fig. 19.8.

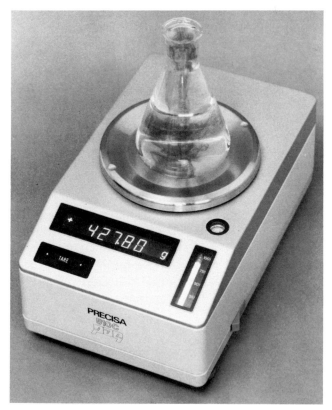

Fig. 19.8

Time

The unit of time is the second which until 1968 was based on the length of a day, this being the time for the earth to revolve once on its axis. However, days are not all of exactly the same duration and the second is now defined as the time interval for a certain number of energy changes to occur in the caesium atom.

Time measuring devices rely on some kind of constantly repeating oscillations. In many clocks and watches a small wheel (the balance wheel) oscillates to and fro; in modern clocks and watches the oscillations are produced by a tiny quartz crystal. A swinging pendulum controls a pendulum clock.

Experiment: oscillations of a mass–spring system

In this investigation you have to make time measurements using a stop watch or clock.

Support a spiral steel spring as in Fig. 19.9a. Hang a mass (e.g. 50 g) from its lower end so that it is stretched several centimetres. Pull the mass *vertically* downwards a few centimetres, Fig. 19.9b, and release it so that it oscillates up and down above and below its rest position.

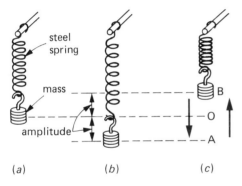

Fig. 19.9

Find the time for the mass to make several complete oscillations; one oscillation is from A to O to B to O to A, Fig. 19.9c. Repeat the timing a few times for the same number of oscillations and work out the average. The time for one oscillation is the *period T*. What is it for your system? The *frequency f* of the oscillations is the number of complete oscillations per second and equals $1/T$. Calculate f.

How does the amplitude of the oscillations, Fig. 19.9b, change with time? Investigate the effect on T of (a) a greater mass, (b) a smaller mass, (c) a different spring.

Mechanical oscillations (vibrations) have their uses (e.g. to produce sound) but if they build up in a system (e.g. in a bridge, p. 51) they can be dangerous.

Q Questions

1. How many millimetres are there in (a) 1 cm, (b) 4 cm, (c) 0.5 cm, (d) 6.7 cm, (e) 1 m?

2. What are these lengths in metres: (a) 300 cm, (b) 550 cm, (c) 870 cm, (d) 43 cm, (e) 100 mm?

3. (a) Write the following in powers of ten form with one figure before the decimal point:

100 000; 3500; 428 000 000; 504; 27 056

(b) Write out the following in full:
10^3; 2×10^6; 6.9×10^4; 1.34×10^2; 10^9

4. (a) Write these fractions in powers of ten form:

1/1000; 7/100 000; 1/10 000 000; 3/60 000

(b) Express the following decimals in powers of ten form with one figure before the decimal point:

0.5; 0.084; 0.000 36; 0.001 04

5. The pages of a book are numbered 1 to 200 and each leaf is 0.10 mm thick. If each cover is 0.20 mm thick, what is the thickness of the book?

6. How many significant figures are there in a length measurement of (a) 2.5 cm, (b) 5.32 cm, (c) 7.180, (d) 0.042?

7. A rectangular block measures 4.1 cm by 2.8 cm by 2.1 cm. Calculate its volume giving your answer to an appropriate number of significant figures.

8. A metal block measures 10 cm × 2 cm × 2 cm. What is its volume? How many blocks each 2 cm × 2 cm × 2 cm have the same total volume?

9. How many blocks of ice cream each 10 cm × 10 cm × 4 cm can be stored in the compartment of a deep freeze measuring 40 cm × 40 cm × 20 cm?

10. A Perspex box has a 6 cm square base and contains water to a height of 7 cm, Fig. 19.10.

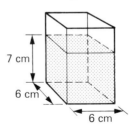

Fig. 19.10

(a) What is the volume of the water?
(b) A stone is lowered into the water so as to be completely covered and the water rises to a height of 9 cm. What is the volume of the stone?

57

check list p.282

20 Density

In everyday language lead is said to be 'heavier' than wood. By this it is meant that a certain volume of lead is heavier than the same volume of wood. In science such comparisons are made by using the term *density*. This is the *mass per unit volume* of a substance and is calculated from

$$\text{DENSITY} = \frac{\text{MASS}}{\text{VOLUME}}$$

The density of lead is 11 grams per cubic centimetre (11 g/cm³) and this means that a piece of lead of volume 1 cm³ has mass 11 g. A volume of 5 cm³ of lead would have mass 55 g. Knowing the density of a substance the mass of *any* volume can be calculated. This enables engineers to work out the weight of a structure if they know from the plans the volumes of the materials to be used and their densities. Strong enough foundations can then be made.

The SI unit of density is the *kilogram per cubic metre*. To convert a density from g/cm³, normally the most suitable unit for the size of sample we use, to kg/m³, we multiply by 10³. For example the density of water is 1.0 g/cm³ or 1.0×10^3 kg/m³.

The approximate densities of some common substances are given above right.

Calculations

Using the symbols *d* for density, *m* for mass and *V* for volume, the expression for density is

$$d = \frac{m}{V}$$

Rearranging the expression gives

$$m = V \times d \qquad \text{and} \qquad V = \frac{m}{d}$$

These are useful if *d* is known and *m* or *V* have to be calculated. If you do not see how they are obtained refer to the *Mathematics for Physics* section on page 270. The triangle in Fig. 20.1 is an aid to remembering them. If you cover the quantity you want to know with a finger, e.g. *m*, it equals what you can still see, i.e. $d \times V$. To find *V*, cover *V* and you get $V = m/d$.

Fig. 20.1

58

Densities of common substances

Solids	Density g/cm³	Liquids	Density g/cm³
aluminium	2.7	meths	0.80
copper	8.9	paraffin	0.80
iron	7.9	petrol	0.80
gold	19.3	pure water	1.0
glass	2.5	mercury	13.6
wood (teak)	0.80	*Gases*	kg/m³
ice	0.90	air	1.3
polythene	0.90	hydrogen	0.09

✎ Worked example

Taking the density of copper as 9 g/cm³, find (a) the mass of 5 cm³ and (b) the volume of 63 g.

(a) $d = 9$ g/cm³, $V = 5$ cm³ and *m* is to be found

$\therefore\ m = V \times d = 5$ cm³ $\times 9$ g/cm³ $= 45$ g

(b) $d = 9$ g/cm³, $m = 63$ g and *V* is to be found

$$\therefore\ V = \frac{m}{d} = \frac{63 \text{ g}}{9 \text{ g/cm}^3} = 7 \text{ cm}^3$$

Simple density measurements

If the mass *m* and volume *V* of a substance are known its density can be found from $d = m/V$.

(a) Regularly-shaped solid. The mass is found on a balance and the volume by measuring its dimensions with a ruler.

(b) Irregularly-shaped solid, e.g. a pebble or a glass stopper. The solid is weighed and its volume measured by one of the methods shown in Fig. 20.2a,b. In (a) the volume is the difference between the first and second readings. In (b) it is the volume of water collected in the measuring cylinder.

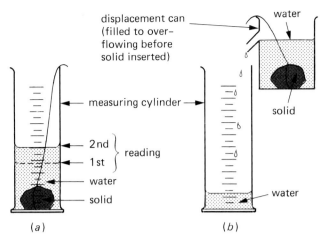

measuring cylinder

displacement can (filled to over-flowing before solid inserted)

water

2nd } reading
1st }

water

solid

solid

water

(a)

(b)

Fig. 20.2

Fig. 20.3

(c) Liquid. A known volume is transferred from a burette or a measuring cylinder into a weighed beaker which is reweighed to give the mass of liquid.

(d) Air. A 500 cm³ round-bottomed flask is weighed full of air and then after removing the air with a vacuum pump; the difference gives the mass of air in the flask. The volume of air is found by filling the flask with water and pouring it into a measuring cylinder.

Floating and sinking

An object sinks in a liquid of smaller density than its own; otherwise it floats, partly or wholly submerged. For example, a piece of glass of density 2.5 g/cm³ sinks in water (density 1.0 g/cm³) but floats in mercury (density 13.6 g/cm³). An iron nail sinks in water but an iron ship floats because its average density is less than that of water. Why is it easy to float in the Dead Sea, Fig 20.3?

Q Questions

1. **(a)** If the density of wood is 0.5 g/cm³ what is the mass of **(i)** 1 cm³, **(ii)** 2 cm³, **(iii)** 10 cm³?
(b) What is the density of a substance of **(i)** mass 100 g and volume 10 cm³, **(ii)** volume 3 m³ and a mass 9 kg?
(c) The density of gold is 19 g/cm³. Find the volume of **(i)** 38 g, **(ii)** 95 g of gold.

2. A piece of steel has a volume of 12 cm³ and a mass of 96 g. What is its density in **(a)** g/cm³, **(b)** kg/m³?

3. What is the mass of 5 m³ of cement of density 3000 kg/m³?

4. What is the mass of air in a room measuring 10 m × 5.0 m × 2.0 m if the density of air is 1.3 kg/m³?

5. A perspex box has a 10 cm square base and contains water to a height of 10 cm. A piece of rock of mass 600 g is lowered into the water and the level rises to 12 cm.
(a) What is the volume of water displaced by the rock?
(b) What is the volume of the rock?
(c) Calculate the density of the rock.   (E.M.)

59

p.282

21 Weight and springs

A *force* is a push or a pull. It can cause a body at rest to move or if the body is already moving it can change its speed or direction of motion. It can also change its shape or size. A weightlifter in action exerts both a pull and a push, Fig. 21.1.

Weight

We all constantly experience the force of gravity, i.e. the pull of the earth. It causes an unsupported body to fall from rest to the ground.

The weight of a body is the force of gravity on it.

The nearer a body is to the centre of the earth the more does the earth attract it. Since the earth is not a perfect sphere but is flatter at the poles, the weight of a body varies over the earth's surface. It is greater at the poles than at the equator.

Gravity is a force which can act through space, i.e. there does not need to be contact between the earth and the object on which it acts as there does when we push or pull something. Other action-at-a-distance forces which like gravity decrease with distance, are:

(i) *magnetic* forces between magnets, and
(ii) *electric* forces between electric charges.

The newton

The SI unit of force is the *newton* (N). It will be defined later (p. 109) but the definition is based on the change of speed a force can produce on a body. Weight is a force and therefore should be measured

Fig. 21.1

in newtons. The weight of an average-sized apple is about 1 newton.

The weight of a body can be measured by hanging it on a *spring balance* marked in newtons, Fig. 21.2, and letting the pull of gravity stretch the spring in the balance. The greater the pull the more does the spring stretch. On most of the earth's surface

the weight of a body of mass 1 kg is 9.8 N.

Often this is taken as 10 N. A mass of 2 kg has a weight of 20 N and so on. The mass of a body is the same wherever it is and does not depend on the presence of the earth as does weight.

Mass is measured by a lever, beam or top-pan balance; weight is measured by a spring balance.

〰 Experiment: stretching a spring

Arrange a steel spring as in Fig. 21.3. Read the scale opposite the bottom of the hanger. Add 100 g loads one at a time (thereby increasing the stretching force by steps of 1 N) and take the readings after each one. Enter the readings in a table for loads up to 500 g.

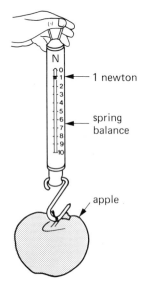

Fig. 21.2

Fig. 21.3

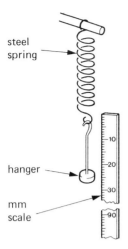

steel spring

hanger

mm scale

Stretching force (N)	Scale reading (mm)	Total extension (mm)

Do the results suggest any rule about how the spring behaves when it is stretched?

Sometimes it is easier to discover laws by displaying the results on a graph. Do this on graph paper by plotting *stretching force* readings along the *y*-axis and *total extension* readings along the *x*-axis. Every pair of readings will give a point; mark them by small crosses and draw a smooth line through them. What is its shape?

Hooke's law

Springs were investigated by Hooke about 300 years ago. He found that the extension was proportional to the stretching force so long as the spring was not permanently stretched. This means that doubling the force doubles the extension, trebling the force trebles the extension and so on. Using the sign for proportionality we can write *Hooke's law*

extension ∝ *stretching force*

It is true only if the *elastic limit* of the spring is not exceeded, i.e. if the spring returns to its original length when the force is removed.

The graph of Fig. 21.4 is for a spring stretched beyond its elastic limit E. OE is a straight line passing through the origin O and is graphical proof that Hooke's law holds over this range. If the force for point A on the graph is applied to the spring, the elastic limit is passed, and on removing the force, some of the extension (OS) remains. Over which part of Fig. 21.4 does a spring balance work?

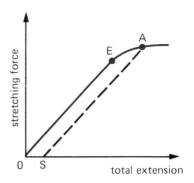

Fig. 21.4

[Q] Questions

1. If a body of mass 1 kg has weight 10 N at a certain place what will be the weight of **(a)** 100 g, **(b)** 5 kg, **(c)** 50 g?

2. The force of gravity on the moon is said to be one-sixth of that on the earth. What would a mass of 12 kg weigh **(a)** on the earth and **(b)** on the moon?

3. A spring is fitted with a scale pan as shown in Fig. 21.5 and the pointer points to the 30 cm mark on the scale.

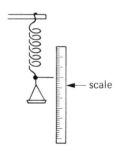

scale

Fig. 21.5

When some sand is placed in the pan the pointer points to the 45 cm mark. When a 20 g mass is placed on top of the sand the pointer points to the 55 cm mark.
   (a) What extension is produced by the sand?
   (b) What extension is produced by the 20 g mass?
   (c) What is the mass of the sand?   (*A.L.*)

4. Fig. 21.6 shows four diagrams, not to scale, of the same spring which obeys Hooke's law. **(a)** What is the length *x*? **(b)** What is the mass *M*?   (*E.A.*)

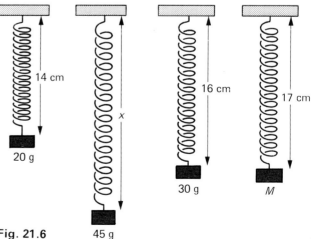

14 cm

20 g

x

45 g

16 cm

30 g

17 cm

M

Fig. 21.6

61

check list p.282

22 Molecules

Fig. 22.1

Matter is made up of tiny particles or *molecules* which are too small for us to see directly. But they can be 'seen' by scientific 'eyes'. One of these is the electron microscope: Fig. 22.1 is a photograph taken with such an instrument showing molecules of a protein. Molecules consist of even smaller particles called *atoms* and are in continuous motion.

⩕ Experiment: Brownian motion

The apparatus is shown in Fig. 22.2a. First fill the glass cell with smoke using a burning drinking straw (made of waxed paper), Fig. 22.2b. Replace the lid on the apparatus and set it on the microscope platform. Connect the lamp to a 12 V supply; the glass rod acts as a lens and focuses light on the smoke.

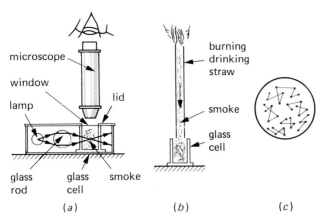

Fig. 22.2

Carefully adjust the microscope until you see bright specks dancing around haphazardly, Fig. 22.2c. The specks are smoke particles seen by reflected light; their random motion is due to collisions with fast-moving air molecules in the cell. The effect is called *Brownian motion*.

62

Kinetic theory of matter

As well as being in continuous motion, molecules also exert strong electrical forces on one another when they are close together. The forces are both attractive and repulsive. (The former hold molecules together and the latter cause matter to resist compression.) The *kinetic theory* can explain the existence of the solid, liquid and gaseous states.

(a) *In solids* the molecules are close together and the attractive and repulsive forces between neighbouring molecules balance. Each molecule vibrates to and fro about one position. Solids therefore have a regular, repeating molecular pattern, i.e. are crystalline, and their shape is definite.

(b) *In liquids* the molecules are usually slightly farther apart than in solids and, as well as vibrating, they can at the same time move rapidly over short distances. However they are never near another molecule long enough to get trapped in a regular pattern and a liquid can flow.

(c) *In gases* the molecules are much farther apart than in solids or liquids (about ten times) and so they are much less dense. They dash around at very high speed (500 m/s for air molecules) in all the space available. It is only during the brief spells when they collide with other molecules or with the walls of the container that the molecular forces act.

A model illustrating the states of matter is shown in Fig. 22.3a. The tray is rotated to and fro on the bench and the motion of the marbles (molecules) observed. Fig. 22.3b is a model of a gas; if a polystyrene ball (1 cm diameter) is dropped into the tube its irregular motion represents Brownian motion.

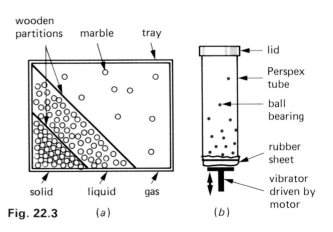

Fig. 22.3

Experiment: size of a molecule

(a) Method. Pour water into a waxed tray till it is overbrimming. Lightly dust the surface with lycopodium.

Obtain a drop of olive oil on a V-shaped loop of thin wire by dipping the end into the oil and *quickly* removing it. Fix the loop in a holder with a $\frac{1}{2}$ mm scale, Fig. 22.4a, and view the drop and the scale through a hand lens. If the diameter of the drop is not about $\frac{1}{2}$ mm, discard it and take up another one.

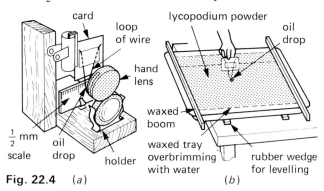

Fig. 22.4 (a) (b)

Remove the loop from the holder and dip the end into the centre of the water, Fig. 22.4b. The oil spreads, pushing the powder before it and leaving a clear circular film. Measure its diameter.

Clean the water surface by drawing two waxed booms across it from the middle of the tray. Repeat the experiment with other $\frac{1}{2}$ mm diameter drops.

(b) Calculation. If d is the diameter of the drop then, assuming it is a sphere

$$\text{volume of drop} = \tfrac{4}{3}\pi(d/2)^3 = \pi d^3/6$$

If D is the diameter of the film, assumed to be a flat cylinder of height h, then

$$\text{volume of film} = \pi(D/2)^2 h = \pi D^2 h/4 \qquad \text{(p. 56)}$$

Also, volume of film = volume of drop

$$\therefore \ \pi D^2 h/4 = \pi d^3/6$$

$$\therefore \ h = \frac{2d^3}{3D^2}$$

But $d = 0.5$ mm $= 0.5 \times 10^{-3}$ m $= 5 \times 10^{-4}$ m
$$\therefore \quad h = 2(5 \times 10^{-4})^3/(3D^2) \text{ m}$$
$$= 250 \times 10^{-12}/(3D^2) \text{ m}$$

Substitute the average value of D and so find h.

Assuming that the film is one molecule thick when the spreading stops, the thickness h of the film equals one dimension of a molecule. Only a rough result is obtained but other methods give values in the region of 2.0×10^{-9} m $= 2$ nm.

Crystals

Crystals have hard, flat sides and straight edges. Whatever their size, crystals of the same substance have the same shape. This can be seen by observing through a microscope, very small cubic salt crystals growing as water evaporates from salt solution on a glass slide, Fig. 22.5.

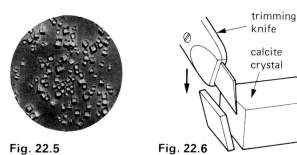

Fig. 22.5 **Fig. 22.6**

A calcite crystal will split cleanly if a trimming knife, held exactly parallel to one side of the crystal, is struck by a hammer, Fig. 22.6.

These facts suggest crystals are made of small particles (e.g. atoms) arranged in an orderly way in planes. Metals consist of tiny crystals.

Q Questions

1. When viewing Brownian motion in a smoke cell the observer sees moving specks of light which are

    A molecules moving in random motion
    B molecules vibrating regularly
    C molecules colliding with each other
    D smoke particles vibrating regularly
    E smoke particles in random motion   *(J.M.B.)*

2. Why are gases more easily squeezed than liquids or solids?

3. **(a)** A tray is 50 cm long and 40 cm wide and has 2.0×10^3 cm^3 of water poured into it. What is the depth of water?
   (b) A drop of oil of volume 1.0×10^{-3} cm^3 spreads out on a clean water surface to a film of area 1.0 m^2 $(1.0 \times 10^4$ cm^2). Calculate the thickness of the film.

4. 100 identical drops of oil, of density 800 kg/m^3, are found to have a total mass of 2×10^{-4} kg. One of these drops is placed on a large clean water surface and it spreads to form a uniform film of area 0.2 m^2.
   (a) What is the mass of one drop?
   (b) What is the volume of one drop?
   (c) What is the thickness of the film?

                       (O.L.E. part qn.)

5. Graphite consists of *layers* of carbon atoms that are $2\frac{1}{2}$ times farther apart than the atoms in the layers. Why is it soft and flaky and used as a lubricant and in pencils?

check list p.283

23 Properties of matter

Diffusion

Smells, pleasant or otherwise, travel quickly and are caused by rapidly moving molecules. The spreading of a substance of its own accord is called *diffusion* and is due to molecular motion.

Diffusion of gases can be shown if some brown nitrogen dioxide gas is made by pouring a mixture of equal volumes of concentrated nitric acid and water on copper turnings in a gas jar. When the action has stopped, a gas jar of air is inverted over the bottom jar, Fig. 23.1. The brown colour spreads into the upper jar showing that nitrogen dioxide molecules diffuse upwards against gravity. Air molecules also diffuse into the lower jar.

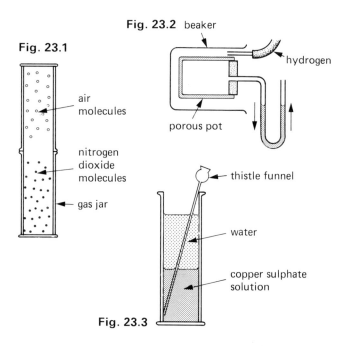

Fig. 23.2 beaker

Fig. 23.1

air molecules

nitrogen dioxide molecules

gas jar

hydrogen

porous pot

thistle funnel

water

copper sulphate solution

Fig. 23.3

The speed of diffusion of a gas depends on the speed of is molecules and is greater for light molecules. The apparatus of Fig. 23.2 shows this. When hydrogen surrounds the porous pot, the liquid in the U-tube moves in the direction of the arrows. This is due to the lighter, faster molecules of hydrogen diffusing into the pot faster than the heavier, slower molecules of air diffuse out. The opposite happens when carbon dioxide surrounds the pot. Why?

Diffusion in liquids can be seen as in Fig. 23.3. After 24 hours the blue copper sulphate solution has diffused upwards into the water.

Surface tension

A needle, though made of steel which is denser than water, will float on a *clean* water surface. A film formed by dipping an inverted funnel in a detergent solution, rises *up* the funnel, Fig. 23.4a. When the film inside the cotton loop in Fig. 23.4b, is broken, the loop forms a *circle*, Fig. 23.4c.

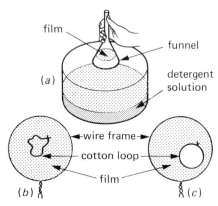

film

funnel

detergent solution

(a)

wire frame

cotton loop

film

(b)

(c)

Fig. 23.4

These facts suggest that the surface of a liquid (in a vessel or on a film) behaves as if covered with an elastic skin that is trying to shrink. The effect is called *surface tension*. It can be reduced if the liquid is 'contaminated' by adding, for example, detergent: a needle floating in water then sinks.

Surface tension is due to the molecules in a liquid surface being slightly farther apart than normal, like those in a stretched wire.

Adhesion and cohesion

The force of attraction (p. 62) between molecules of the same substance is known as *cohesion*, that between molecules of different substances is *adhesion*. The adhesion of water to glass is greater than the cohesion of water and water spilt on clean glass wets it by spreading to a thin film. By contrast, mercury on glass forms small spherical drops or large flattened ones because cohesion of mercury is stronger than its adhesion to glass.

The cleaning action of detergents depends on their ability to weaken the cohesion of water. Instead of forming drops on greasy clothes, the water penetrates the fabric and releases dirt.

Capillarity

If a glass tube of small bore (a capillary tube) is dipped into water, the water rises up the tube a few centimetres, Fig. 23.5a. The narrower the tube the greater the rise. Adhesion between water and glass exceeds cohesion between water molecules, the meniscus curves up and the water rises.

Fig. 23.5

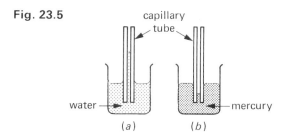

(a)   (b)

The action of blotting paper is due to capillary rise in the narrow spaces between the fibres it contains. The rise of oil up a lamp wick occurs in the same way. The damp-course in a house, i.e. the layer of non-porous material in the walls just above ground level but below the floor, prevents water rising up the pores in the bricks by capillary action from the ground and causing dampness. Mercury is depressed in a capillary tube, Fig. 23.5b.

Mechanical properties of materials

When selecting a material for a particular job we need to know how it behaves when forces act on it, i.e. what its mechanical properties are.

(a) Strength. A strong material requires a large force to break it. The strength of some materials depends on how the force is applied. For example, concrete is strong when compressed but weak when stretched, i.e. in tension.

(b) Stiffness. A stiff material resists forces which try to change its shape or size. It is not flexible. All materials 'give' to some extent although the change may be very small. Steel is strong and stiff, putty is neither. Rope is not stiff but can be strong in tension.

(c) Elasticity. An elastic material such as rubber recovers its original shape and size after the force deforming it has been removed. A material which does not recover but is deformed permanently, like Plasticine, is plastic (Note that man-made plastics are not always 'plastic'; they are so-called because during manufacture they behave in that way.)

(d) Ductility. Materials which can be rolled into sheets, drawn into wires or worked into other useful shapes without breaking, are ductile. Metals owe much of their usefulness to this property.

(e) Brittleness. A brittle material is fragile and breaks suddenly. Bricks, cast iron and glass are brittle.

Stretching a wire[1]

Useful information about mechanical properties is obtained from stretching a long wire of material by gradually loading it as in Fig. 23.6.

The graph obtained by plotting stretching force against extension is similar to that for stretching a spring (Fig. 21.4, p. 61). Between O and the *elastic*

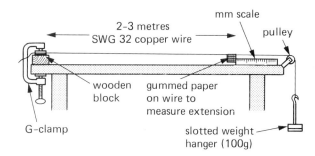

Fig. 23.6

limit E, it is a straight line showing that *extension ∝ stretching force*, i.e. Hooke's law holds. At A, the *yield point*, the wire 'runs' and greater extensions than before are produced by a given increase of load but larger loads can be supported.

Up to E the molecules are pulled slightly farther apart; the deformation is elastic and the wire recovers its original length when the load is removed. Beyond A, layers of atoms slip over each other; the deformation is plastic and if unloaded the wire does not return to its previous length.

Metals used in engineering structures should carry loads which only deform them elastically.

Ⓠ Questions

1. Explain why **(a)** diffusion occurs more quickly in a gas than in a liquid, **(b)** diffusion is still quite slow even in a gas, **(c)** an inflated balloon gradually goes down even when tied.

2. Explain the following observations as fully as you can.
   (a) A small needle may be floated on the surface of water, but if a drop of detergent is added to the water the needle sinks.
   (b) Damp-courses are used in modern houses.
   (c) Gases can be easily compressed but liquids cannot.
   (E.A.)

[1] See Topic 79 for experiments on stretching a wire, a rubber band and a polythene strip.

check list
p.283

24 Additional questions

Core level (all students)

Measurements: density

1. If 200 cm³ of water (density 1.0 g/cm³) is mixed with 300 cm³ of methylated spirit (density 0.80 g/cm³), what is the density of the mixture?

2. **(a)** Why does a piece of wood float and a piece of lead sink in water?
(b) A wooden block, whose volume is 16 cm³, has a hole with a volume of 1.0 cm³ drilled in it. The hole is filled with lead. Will the block sink or float in water? (Give reasons for your answer and show any calculations you make.) Density of lead = 11 g/cm³: density of wood = 0·50 g/cm³: density of water = 1.0 g/cm³.
(c) Which is denser, milk or cream? *(S.E.)*

Weight and springs

3. The springs in Fig. 24.1 are identical. If the extension produced in (*a*) is 4 cm, what are the extensions in (*b*) and (*c*)?

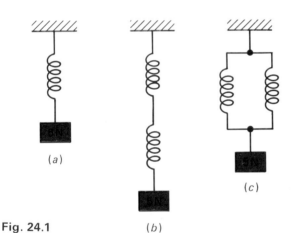

Fig. 24.1   (*b*)

4. A vertical spring of unstretched length 30 cm is rigidly clamped at its upper end. When an object of mass 100 g is placed in a pan attached to the lower end of the spring, its length becomes 36 cm. For an object of mass 200 g in the pan, the length becomes 40 cm. Calculate the mass of the pan. Name and state clearly the law you have assumed.
(S.)

Molecules

5. Describe the differences between solids, liquids and gases in terms of **(a)** the arrangement of the molecules throughout the bulk of the material, **(b)** the separation of the molecules, and **(c)** the motion of the molecules.
(J.M.B.)

6. Smoke particles in an air cell viewed under a microscope exhibit Brownian motion.
(a) Copy Fig. 24.2 and draw a suitable arrangement to show how the particles in the air can be illuminated.
(b) How is the motion of the smoke particles best described?
(c) What accounts for the motion of the smoke particles?
(d) The motion is viewed using bigger smoke particles. What difference in the motion would this lead to? Give the reason for the difference. *(O.L.E./S.R.16+)*

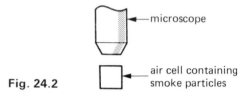

Fig. 24.2

Properties of matter

7. A porous pot holding air is fitted with a tube dipping into water, Fig. 24.3.
(a) When the pot is surrounded by hydrogen, air bubbles rapidly out of the water, but after a short time the bubbles slow down and stop. Explain these stages.
(b) If now the jar of hydrogen is removed, water rises rapidly up the tube, slows down and stops, and very slowly returns. Explain these stages.
(c) What difference does it make when the experiment is repeated using carbon dioxide instead of hydrogen. Why? *(S.R.)*

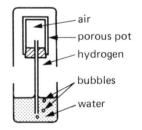

Fig. 24.3

8. **(a)** A piece of thread is carefully placed in a soap film which has been formed on a metal ring, Fig. 24.4a. Copy Fig. 24.4b and show clearly what happens if the soap film inside the thread is pierced by a needle.

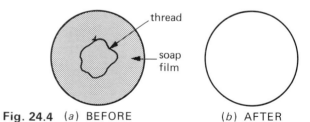

Fig. 24.4 (*a*) BEFORE   (*b*) AFTER

(b) What name is given to the force acting on the thread? *(E.M.)*

9. (a) Name the effect shown in tube A of Fig. 24.5.
(b) Copy the diagram and mark the water levels in B and C and the mercury levels in D and E.

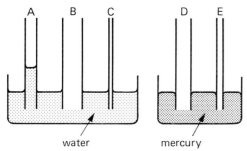

Fig. 24.5

10. The graph shown in Fig. 24.6 represents the behaviour of a wire in an experiment to investigate Hooke's law. Copy the graph. Label each axis and add appropriate labels to the guidelines. *(E.A.)*

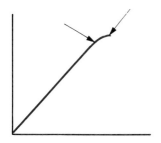

Fig. 24.6

Further level (higher grade students)

Measurements: density

11. Fig. 24.7 shows a measuring cylinder containing some sugar (mass 32.0 g) in paraffin oil (mass 56.4 g, density 0.80 g/cm^3) which does not dissolve the sugar. The total volume of the substances in the cylinder is 90.5 cm^3. Calculate **(a)** the volume of the paraffin oil, **(b)** the volume of the sugar, **(c)** the density of the sugar. *(C.)*

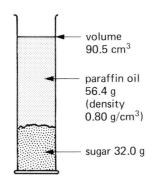

volume
90.5 cm^3

paraffin oil
56.4 g
(density
0.80 g/cm^3)

sugar 32.0 g

Fig. 24.7

12. (a) Name the basic units of: length, mass, time.
(b) What is the difference between two measurements with values 3.4 and 3.42?
(c) Write expressions for **(i)** the area of a circle, **(ii)** the volume of a sphere, **(iii)** the volume of a cylinder.

Weight and springs

13. A light helical spring hangs vertically with its upper end fixed. A light pointer, attached to its lower end, can be used to take readings on a vertical scale when masses of various sizes are attached to the lower end of the spring. The following table gives the scale readings of the pointer for different attached masses.

Mass attached in kg	0	0.2	0.4	0.6	0.8
Scale reading in mm	120	126	132	138	144

(a) Make a table showing corresponding values of the force on the spring in newtons and its resulting extension in metres.
(b) Plot a graph of force (*y*-axis) against extension (*x*-axis).
(c) State Hooke's law. Are the readings for this spring consistent with it? Explain.
(d) Use your graph to find the force that produces an extension of 15 mm.
(Weight of 1 kg is 10 N.) *(O.L.E.)*

Molecules

14. A drop of oil of volume 0.010 cm^3 is allowed to fall on some clean water in a dish and it spreads to form a circle of radius 14 cm. Estimate the upper limit of the diameter of an oil molecule.

In such an experiment describe **(a)** what is done in order that the circle may be clearly visible, **(b)** what apparatus is used and any precautions that are taken. *(L.)*

Properties of matter

15. Describe experiments, one in each case, to show **(a)** diffusion in liquids, **(b)** diffusion in gases.

Use the kinetic theory to explain the result of your experiment demonstrating the diffusion of gases.

(J.M.B.)

Forces and pressure

25 Moments and levers

The handle on a door is at the outside edge so that it opens and closes easily. A much larger force would be needed if the handle were near the hinge. Similarly it is easier to loosen a nut with a long spanner than with a short one.

The *turning effect* or *moment of a force* depends on both the size of the force and how far it is applied from the pivot or *fulcrum*. It is measured by multiplying the force by the *perpendicular* distance of the line of action of the force from the fulcrum. The unit is the *newton metre* (N m).

MOMENT OF A FORCE = FORCE × PERPENDICULAR
DISTANCE FROM FULCRUM

In Fig. 25.1 a force F acts on a gate in (a) at the edge, in (b) at the centre.

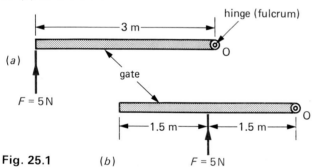

Fig. 25.1   (b)

In (a)

moment of F about $0 = 5 \text{ N} \times 3 \text{ m} = 15 \text{ N m}$

In (b)

moment of F about $0 = 5 \text{ N} \times 1.5 \text{ m} = 7.5 \text{ N m}$

The turning effect of F is greater in (a); this agrees with the fact that a gate opens most easily when it is pushed or pulled at the edge.

Experiment: law of moments

Balance a half-metre rule at its centre, adding Plasticine to one side or the other until it is horizontal.

Hang unequal loads m_1 and m_2 from either side of the ruler and alter their distances d_1 and d_2 from the

68

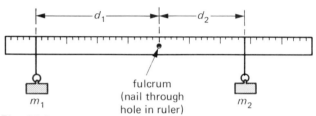

Fig. 25.2

centre until the ruler is again balanced, Fig. 25.2. Forces F_1 and F_2 are exerted by gravity on m_1 and m_2 and so on the ruler; on 100 g the force is 1 N. Record the results in a table and repeat for other loads and distances.

m_1 (g)	F_1 (N)	d_1 (cm)	$F_1 \times d_1$ (N cm)	m_2 (g)	F_2 (N)	d_2 (cm)	$F_2 \times d_2$ (N cm)

F_1 is trying to turn the ruler anticlockwise and $F_1 \times d_1$ is its moment. F_2 is trying to cause clockwise turning and its moment is $F_2 \times d_2$. When the ruler is balanced or as we say in *equilibrium*, the results should show that the anticlockwise moment $F_1 \times d_1$ equals the clockwise moment $F_2 \times d_2$.

The *law of moments* (also called the *law of the lever*) is stated as follows.

When a body is in equilibrium the sum of the clockwise moments about any point equals the sum of the anticlockwise moments about the same point.

Worked example

The see-saw in Fig. 25.3 balances when Sue of weight 320 N is at A, Tom of weight 540 N is at B and Harry of weight W is at C. Find W.

Taking moments about the fulcrum O:

anticlockwise moment
$$= 320 \times 3 + 540 \times 1 = 1500 \text{ N m}$$

clockwise moment $= (W \times 3) \text{ N m}$

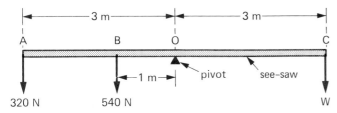

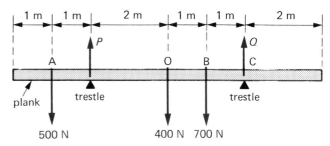

Fig. 25.3

By the law of moments,

clockwise moments = anticlockwise moments

$\therefore 3W = 1500 \qquad \therefore W = 500$ N

Levers

A lever is any device which can turn about a pivot. In a lever a force called the *effort* is used to overcome a resisting force called the *load*.

When we use a crowbar to move a heavy boulder, Fig. 25.4a, our hands apply the effort at one end of the bar and the load is the force exerted by the boulder on the other end. If distances from the fulcrum O are as shown and the load is 1000 N (i.e. the part of the weight of the boulder supported by the crowbar), the effort can be calculated from the law of moments. As the boulder *just begins* to move we can say, taking moments about O, that

clockwise moment = anticlockwise moment
effort × 200 = 1000 × 10
effort = 50 N

The crowbar in effect magnifies the effort 20 times but the effort must move farther than the load.

Other examples of levers are shown in Fig. 25.4b,c. In (b) the load is between the effort and the fulcrum; in this case as in (a) the effort is less than the load. In (c) the effort (applied by the biceps muscle) is between the load and the fulcrum and is greater than the load, which moves farther than the effort.

Conditions for equilibrium

Sometimes a number of parallel forces act on a body so that it is in equilibrium. We can then say:

Fig. 25.5

(i) *the sum of the forces in one direction equals the sum of the forces in the opposite direction*

(ii) *the law of moments must apply.*

As an example consider two decorators of weights 500 N and 700 N standing at A and B on a plank resting on two trestles, Fig. 25.5. In the next topic we will see (p. 71) that the whole weight of the plank (400 N) may be taken to act vertically downwards at its centre, O. If P and Q are the upwards forces exerted by the trestles on the plank (called *reactions*) then we have from (i),

$$P + Q = 500 + 400 + 700 = 1600 \text{ N} \qquad (1)$$

Moments can be taken about any point but taking them about C eliminates the moments due to Q.

Clockwise moment $= P \times 4$ N m

Anticlockwise moments

$$= (700 \times 1 + 400 \times 2 + 500 \times 5) \text{ N m}$$
$$= 700 + 800 + 2500 = 4000 \text{ N m}$$

Since the plank is in equilibrium we have from (ii),

$$4P = 4000 \qquad \therefore P = 1000 \text{ N}$$

From (1), $\qquad Q = 600$ N.

Couples

When you steer a bicycle round a bend with both hands on the handle-bars, two equal and opposite forces are applied, Fig. 25.6. The forces form a *couple* and cause rotation.

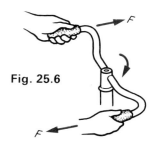

Fig. 25.6

Fig. 25.4

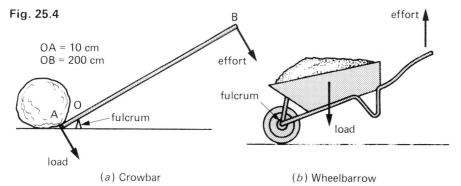

OA = 10 cm
OB = 200 cm

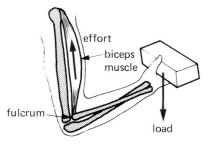

(a) Crowbar   (b) Wheelbarrow   (c) Forearm   69

Beams and structures[1]

(a) Beams. Stretching a rod or bar puts it in *tension*, Fig. 25.7a; squeezing it puts it in *compression*, Fig. 25.7b; twisting it subjects it to *shearing*, Fig. 25.7c. The greater the cross-section area of a rod, the greater are the balancing internal resisting forces it sets up to tensile, compressive and shearing forces. Resistance to such forces also depends on shape, so different beams are made for different purposes.

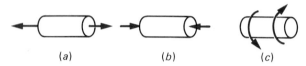

(a)   (b)   (c)

Fig. 25.7

When a beam bends, one side is compressed, the other is stretched and the centre is unstressed (a neutral plane), Fig. 25.8a. I-shaped steel girders, Fig. 25.8b, are used in large structures. They are, in

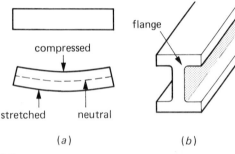

(a)   (b)

Fig. 25.8

effect, beams that have had material removed from the neutral plane and so weigh less. The top and bottom flanges withstand the compression and tension forces due to loading. In a hollow tube the removal of unstressed material gives similar advantages.

(b) Bridges. In an arched stone bridge, the stone is in compression, since it is weak in tension, Fig. 25.9a. A girder bridge has no material in the neutral plane, Fig. 25.9b; it is strengthened by diagonal bars.

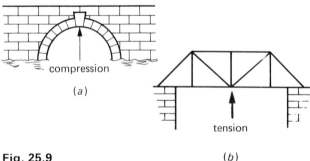

Fig. 25.9   (b)

[1] See Topic 79 for investigations on beams, pillars and bridges.

70

check list
p.283

⟦Q⟧ Questions

1. Fig. 25.10 shows half-metre rules which are marked off at intervals of 5 cm. Identical metal discs are placed on the rules as shown. In each case state whether the rules will turn clockwise, anticlockwise or remain in the horizontal position. Show your working.   *(E.A.)*

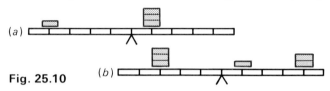

Fig. 25.10

2. The metre rule in Fig. 25.11 is supported at its centre. If the rule is balanced, the values of x and y respectively are

A 3 cm (x) and 5 cm (y) **B** 5 cm (x) and 3 cm (y)
C 6 cm (x) and 10 cm (y) **D** 12 cm (x) and 20 cm (y)
                                                              (W.M.)

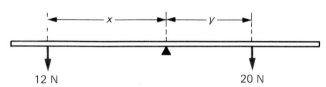

Fig. 25.11   Diagram NOT to scale

3. In Fig. 25.12, the distance AC = CB. Calculate in each case the force X which will keep the system stationary.
                                                              (E.M.)

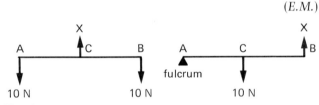

Fig. 25.12

4. Fig. 25.13 shows three positions of the pedal on a bicycle which has a crank 0.20 m long. If the cyclist exerts the same vertically downwards push of 25 N with his foot, in which case is the turning effect **(i)** $25 \times 0.2 = 5$ N m, **(ii)** 0, **(iii)** between 0 and 5 N m? Explain your answers.

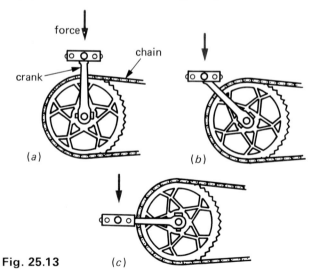

Fig. 25.13   (c)

26 Centres of gravity

A body behaves as if its whole weight were concentrated at one point, called its *centre of gravity* (c.g.) or *centre of mass*, even though the earth attracts every part of it. The c.g. of a ruler is at its centre and when supported there it balances, Fig. 26.1*a*. If it is supported at any other point it topples because the moment of its weight *W* about the point of support is not zero, Fig. 26.1*b*.

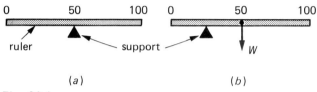

(*a*)   (*b*)

Fig. 26.1

Your c.g. is near the centre of your body and the vertical line from it to the floor must be within the area enclosed by your feet or you will fall over. You can test this by standing with one arm and the side of one foot pressed against a wall, Fig. 26.2. Now try to raise the other leg sideways. Can you do it without falling over?

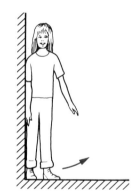

Fig. 26.2

A tight-rope walker has to keep his c.g. exactly above the rope. Some carry a long pole to help them to balance, Fig. 26.3. The combined weight of the walker and pole is then spread out more and if the walker begins to topple to one side he moves the pole to the other side.

The c.g. of a regularly shaped body of the same density all over is at its centre. In other cases it can be found by experiment.

⚡ Experiment: c.g. using a plumb line

Suppose we have to find the c.g. of an irregularly shaped lamina (a thin sheet) of cardboard.

Fig. 26.3

Make a hole A in the lamina and hang it so that it can *swing freely* on a nail clamped in a stand. It will come to rest with its c.g. vertically below A. To locate the vertical line through A tie a plumb line (a thread and a weight) to the nail, Fig. 26.4, and mark its position AB on the lamina. The c.g. lies on AB.

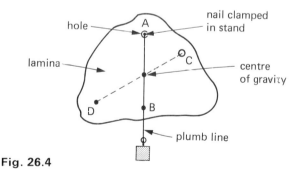

Fig. 26.4

Hang the lamina from another position C and mark the plumb line position CD. The c.g. lies on CD and must be at the point of intersection of AB and CD. Check this by hanging the lamina from a third hole. Also try balancing it at its c.g. on the tip of your forefinger.

Devise a method using a plumb line of finding the c.g. of a tripod.

71

Toppling

The position of the c.g. of a body affects whether or not it topples over easily. This is important in the design of such things as tall vehicles (which tend to overturn when rounding a corner), racing cars, reading lamps and even tea-cups.

A body topples when the vertical line through its c.g. falls outside its base, Fig. 26.5a. Otherwise it remains stable, Fig. 25.6b.

Fig. 26.5

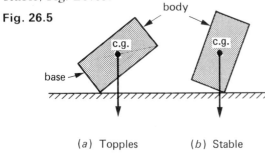

(a) Topples   (b) Stable

Toppling can be investigated by placing an empty can on a plank (with a rough surface to prevent slipping) which is slowly tilted. The angle of tilt is noted when the can falls over. This is repeated with 1 kg in the can. How does this affect the position of the c.g.? The same procedure is followed with a second can of the same height as the first but of greater width. It will be found that the second can with the weight in it can be tilted through the greater angle.

The stability of a body is therefore increased by

(i) lowering its c.g.
(ii) increasing the area of its base.

In Fig. 26.6a the c.g. of a tractor is being found. It is necessary to do this when testing a new design since tractors are often driven over sloping surfaces and any tendency to overturn must be discovered.

The stability of double-decker buses is being tested in Fig. 26.6b. When the top deck only is fully laden with passengers (represented by sand bags in the test), it must not topple if tilted through an angle of 28°. Racing cars have a low c.g. and a wide wheel base.

Fig. 26.6

(a)

Stability

Three terms are used in connection with stability.

(a) A body is in *stable equilibrium* if when slightly displaced and then released it returns to its previous position. The ball at the bottom of the dish in Fig. 26.7a is an example. Its c.g. rises when it is displaced. It rolls back because its weight has a moment about the point of contact.

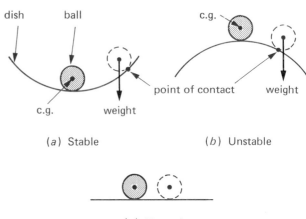

(a) Stable   (b) Unstable

Fig. 26.7   (c) Neutral

(b) A body is in *unstable equilibrium* if it moves farther away from its previous position when slightly displaced. The ball in Fig. 26.7b behaves in this way. Its c.g. falls when it is displaced slightly because there is a moment which increases the displacement.

(c) A body is in *neutral equilibrium* if it stays in its new position when displaced, Fig. 26.7c. Its c.g. does not rise or fall because there is no moment to increase or decrease the displacement.

(b)

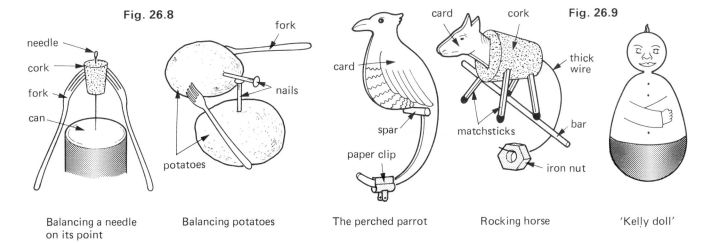

Fig. 26.8

needle
cork
fork
can

fork
nails
potatoes

Balancing a needle on its point   Balancing potatoes

card   cork   Fig. 26.9

card
thick wire

spar
matchsticks   bar

paper clip   iron nut

The perched parrot   Rocking horse   'Kelly doll'

Balancing tricks and toys

Some tricks that you can try or toys to make are shown in Fig. 26.8. In each case the c.g. is vertically below the point of support and equilibrium is stable.

A self-righting toy (a 'Kelly doll') has a heavy base, Fig. 26.9, and when tilted, the weight acting through the c.g. has a moment about the point of contact. This restores it to the upright position.

Ⓠ Questions

1. Fig. 26.10 shows an irregular shape of plywood suspended by a thin thread at two different points, A and B. Copy diagram (b). Mark and label the centre of mass of the plywood. Draw and label across diagram (b) a line which would be vertical if the plywood were suspended at point C.   (E.A.)

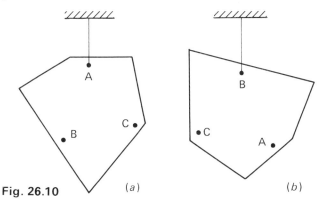

Fig. 26.10   (a)   (b)

2. Fig. 26.11 shows a Bunsen burner in three different positions. State in which position it is in (a) stable equilibrium, (b) unstable equilibrium, (c) neutral equilibrium.

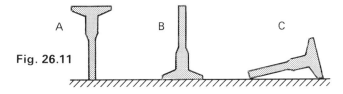

Fig. 26.11

3. The weight of the uniform bar in Fig. 26.12 is 10 N. Does it balance, tip to the right or tip to the left?

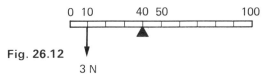

Fig. 26.12

4. (a) A uniform metre rule is balanced at the 30 cm mark when a load of 0.80 N is hung at the zero mark, Fig. 26.13a.
   (i) At what point on the rule is the centre of gravity of the rule?
   (ii) Show with an arrow drawn on a copy of the diagram the weight of the rule acting through the centre of gravity.
   (iii) Calculate the weight of the rule.
   (b) Fig. 26.13b shows the jib of a building site crane. What is the purpose of the concrete block at one end of the jib?

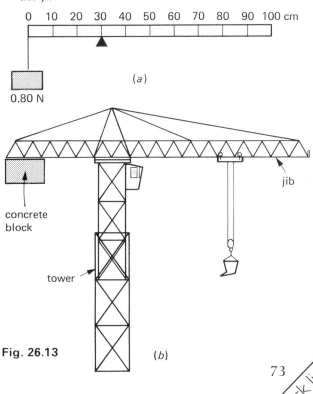

Fig. 26.13

check list p.283

73

27 Adding forces

In structures like a giant oil platform, Fig. 27.1, two or more forces may act at the same point. It is then often useful for the design engineer to know the value of the single force, i.e. the *resultant*, which has exactly the same effect as these forces. If the forces act in the same straight line the resultant is found by simple addition or subtraction, but if they do not they are added by the parallelogram law.

Experiment: parallelogram law

Arrange the apparatus as in Fig. 27.2a with a sheet of paper behind it on a vertical board. We have to find the resultant of forces P and Q.

Read the values of P and Q from the spring balances. Mark on the paper the directions of P, Q and W as shown by the strings. Remove the paper and using a scale of 1 cm to represent 1 N, draw OA, OB and OD to represent the three forces P, Q and W which act at O, Fig. 27.2b. (W = weight of the 1 kg mass = 9.8 N, therefore OD = 9.8 cm.)

P and Q together are balanced by W and so their resultant must be a force equal and opposite to W.

Complete the parallelogram OACB. Measure the diagonal OC; if it is equal in size (i.e. 9.8 cm) and opposite in direction to W then it represents the resultant of P and Q.

Fig. 27.1

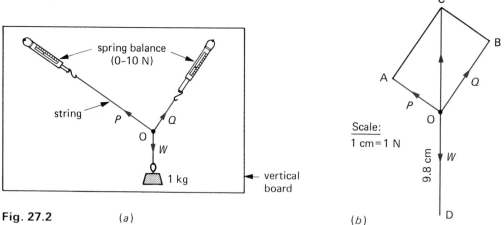

Fig. 27.2   (a)

(b)

The parallelogram law for adding two forces is:

If two forces acting at a point are represented in size and direction by the sides of a parallelogram drawn from the point, their resultant is represented in size and direction by the diagonal of the parallelogram drawn from the point.

Examples

1. *Two people carrying a heavy bucket.* The weight of the bucket is balanced by the resultant F of F_1 and F_2, Fig. 27.3a.

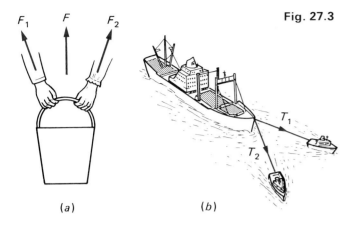

Fig. 27.3

(a)   (b)

2. *Two tugs pulling a ship.* The resultant of T_1 and T_2 is forwards, Fig. 27.3*b*, and so the ship moves forwards.

Vectors and scalars

A *vector* quantity is one such as force which is described completely only if both its size (magnitude) and direction are stated. It is not enough to say, for example, a force of 10 N, but rather a force of 10 N acting vertically downwards.

A vector can be represented by a straight line whose length represents the magnitude of the quantity and whose direction gives its line of action. An arrow on the line shows which way along the line it acts.

A *scalar* quantity has magnitude only. Mass is a scalar and is completely described when its value is known. Scalars are added by ordinary arithmetic; vectors are added geometrically by the parallelogram law which ensures that their directions as well as their magnitudes are considered.

Resolving a force

The parallelogram law enables us to replace two forces by one. Sometimes we have to do the reverse and split or *resolve* one force into two forces called its *components* which together have the same effect as the single force. Usually it is most useful to take the components at right angles to each other.

Suppose a force F is represented by OC in Fig. 27.4 and that we wish to find its components along OX

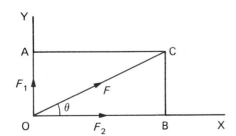

Fig. 27.4

and OY ($\angle$ XOY $= 90°$). If rectangle OACB is drawn, its sides OA and OB represent the required components F_1 and F_2 on the same scale as OC represents F. For example if $F = 200$ N and $\angle$ COB $= 30°$ then $F_1 = 100$ N and $F_2 = 173$ N.

If you have done trigonometry you can show that $F_2 = F \cos \theta$ and $F_1 = F \sin \theta$.

Resolving a force helps to explain why a yacht can sail into the wind. It can be shown (p. 93) that the wind causes a force F at right angles to the sail as it blows over it. This force can be resolved into two components F_1 and F_2, Fig. 27.5. F_1 drives the yacht

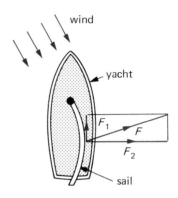

Fig. 27.5

forwards, F_2 would cause it to drift sideways if it did not have a heavy keel. In a wind from the north the yacht could sail alternately NW and NE, i.e. 'tack', and so progress northwards.

Q Questions

1. Using a scale of 1 cm to represent 10 N find the size and direction of the resultant of forces of 30 N and 40 N acting at **(a)** right angles to each other, **(b)** 60° to each other.

2. In Fig. 27.6 the lines represent two forces acting at a point O. Which of the single forces could be the resultant of the two? *(E.A.)*

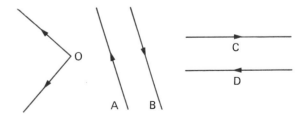

Fig. 27.6

3. Tom, Dick and Harry are pulling a metal ring. Tom pulls with a force of 100 N and Dick with a force of 140 N at an angle of 70° to Tom. If the ring does not move what force is Harry exerting?

4. Explain why a slack clothes line is less likely to break than a taut one.

75

check list
p.283

28 Work, energy, power

Fig. 28.1

In science the word *work* has a different meaning from its everyday one. Work is done when a force moves. A boat pulling a water skier does work, Fig. 28.1, as does a crane when it lifts a load. However, no work is done in the scientific sense by someone holding a heavy pile of books: an upwards force is exerted, but no motion results.

Measuring work

If a building worker carries ten bricks up to the first floor of a building he does more work than if he carries only one brick because he has to exert a larger force. Even more work is required if he carries the ten bricks to the second floor. The amount of work done depends on the *size* of the force applied and the *distance* it moves. We therefore measure work by

WORK = FORCE × DISTANCE MOVED IN DIRECTION
      OF FORCE   (1)

The unit of work is the *joule* (J) and is the *work done when a force of 1 newton* (N) *moves through 1 metre* (m). For example, if you have to pull with a force of 50 N to move a box steadily 3 m in the direction of the force, Fig. 28.2*a*, the work done is 50 N × 3 m = 150 N m = 150 J. That is,

JOULES = NEWTONS × METRES

If you lift a mass of 3 kg vertically through 2 m, Fig. 28.2*b*, you have to exert a vertically upwards force equal to the weight of the body, i.e. 30 N (approximately) and the work done is 30 N × 2 m = 60 N m = 60 J.

Note that we must always take the distance in the direction in which the force acts.

Forms of energy

Energy enables work to be done and people and machines need a supply of it. Experience shows that it exists in many different forms.

(a) Chemical energy. Food and fuels like coal, oil and gas are stores of chemical energy. The energy of food is released by chemical reactions in our bodies. Fuels release their energy when they are burnt (a chemical reaction) in an engine or other device. A car engine uses chemical energy.

(b) Potential energy (p.e.). This is energy which a body has because of its position or condition. A body above the earth's surface has p.e. and can do work. The weight in a wound-up grandfather clock or at the top of a pile driver has p.e. which it loses as it falls. The water in a mountain reservoir is another example of something with p.e.

Wound-up watch springs and stretched elastic bands have p.e. because of their 'strained' condition.

(c) Kinetic energy (k.e.). Any moving body has k.e. and the faster it moves the more k.e. it has. For example as a hammer strikes a nail or an earth-moving machine levels the ground it exerts a force and does work because of the k.e. it possesses as a result of its motion.

Kinetic and potential energy are types of mechanical energy.

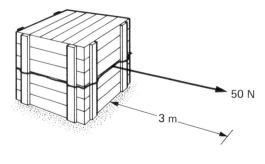

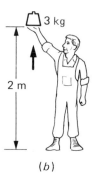

Fig. 28.2   (*a*)   (*b*)

(d) Other forms. These include: electrical energy, heat energy, sound energy, light energy, magnetic energy and nuclear energy.

Energy changes and conservation

A useful thing about energy is that it can be changed from one form to another by suitable devices.

(a) Demonstration. The apparatus in Fig. 28.3 may be used to show a battery changing chemical energy to electrical energy which in the electric motor becomes k.e. The motor raises a weight giving it p.e. If the changeover switch is joined to the lamp and the weight is allowed to fall, what energy changes occur?

(b) Conservation of energy. Some other examples of energy changes are shown in Fig. 28.4. In all cases it is found that *energy cannot be destroyed, it only changes into a different form*. This is known as the *principle of conservation of energy*. If it seems in a change that some energy has disappeared, the 'lost' energy is often converted into heat. For example, when a brick falls its p.e. becomes k.e. As it hits the ground, its temperature rises and heat and sound are produced.

(c) Measuring energy changes. In an energy change work is done. We take the *work done as a measure of the amount of energy changed*. For example, if you have to exert an upwards force of 10 N to raise a stone steadily through a vertical distance of 1.5 m, the work done is 15 J. This is also the amount of

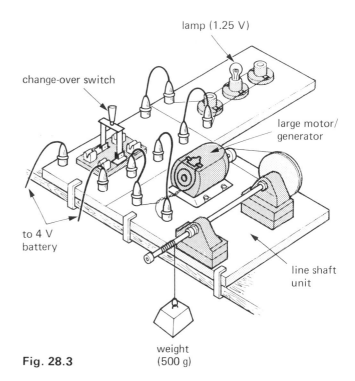

Fig. 28.3

weight (500 g)

chemical energy transferred *from* your muscles *to* p.e. of the stone. All forms of energy, as well as work, are measured in joules.

Power

The more powerful a car is the faster it can climb a hill, i.e the faster it does work. The *power* of a device is the work it does per second, i.e. the *rate at which it*

Fig. 28.4

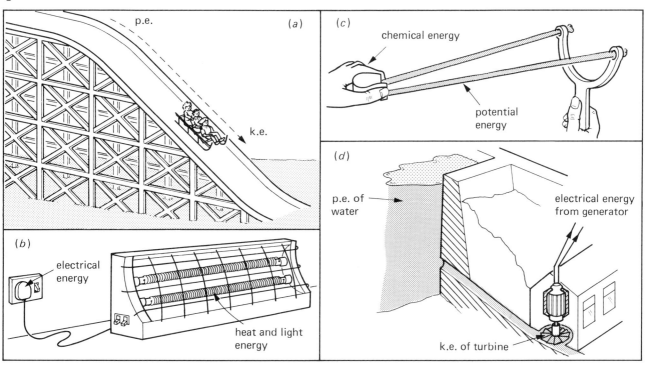

does work. This is the same as the rate at which it changes energy from one form to another.

$$\text{POWER} = \frac{\text{WORK DONE}}{\text{TIME TAKEN}} = \frac{\text{ENERGY CHANGE}}{\text{TIME TAKEN}} \quad (2)$$

The unit of power is the *watt* (W) and is *a rate of working of 1 joule per second*, i.e. 1 W = 1 J/s. Larger units are the kilowatt (kW) and the megawatt (MW).

1 kW = 1000 W = 10^3 W;
1 MW = 1 000 000 W = 10^6 W

If a machine does 500 J of work in 10 s its power is 500 J/10s = 50 J/s = 50 W. A Mini car develops a maximum power of about 25 kW.

Experiment: your own power

Get someone with a stop watch to time you running up a flight of stairs, the longer the better. Find your weight (in newtons). Calculate the total *vertical* height (in metres) you have climbed by measuring the height of one stair and counting the number of stairs.

The work you do (in joules) in lifting your weight to the top of the stairs is (your weight) × (vertical height of stairs). Calculate your power (in watts) from equation (2). About 0.5 kW is good. (1 horse power = 0.75 kW.)

Friction

Friction is the force that opposes one surface moving, or trying to move, over another. It can be a help or a hindrance. We could not walk if it did not exist between the soles of our shoes and the ground. Our feet would slip backwards, as they tend to if we walk on ice. On the other hand engineers try to reduce friction to a minimum in the moving parts of machinery by using lubricating oils and ball bearings.

When a gradually increasing force P is applied through a spring balance to a block on a table, Fig 28.5, it does not move at first. This is because an equally increasing but opposing frictional force F acts where the block and table touch. At any instant P and F are equal and opposite.

If P is increased further the block eventually moves; as it does so F has its maximum value, called *starting* or *static* friction. When it is moving at a steady speed the balance reading is slightly less than that for starting friction. *Sliding* or *dynamic* friction is therefore less than static friction.

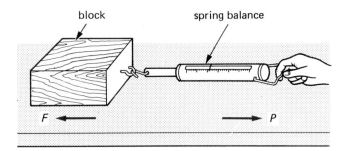

Fig. 28.5

A weight on the block increases the force pressing the surfaces together and increases friction.

When work is done against friction, the temperatures of the bodies in contact rise (as you can test by rubbing your hands together) and some other form of energy is usually changed into heat energy.

Questions

1. How much work is done when a mass of 3 kg (weighing 30 N) is lifted vertically through 6 m?

2. A hiker climbs a hill 300 m high. If he weighs 50 kg calculate the work he does in lifting his body to the top of the hill.

3. Name one device which changes
   (a) electrical energy to light,
   (b) sound to electrical energy,
   (c) chemical energy to electrical energy,
   (d) p.e. to k.e.,
   (e) electrical energy to k.e.

4. A cyclist free-wheels from the top of a hill, gathers speed going down the hill, applies his brakes and eventually comes to rest at the bottom of the hill. Which one of the following energy changes takes place?

   A Potential to kinetic to heat energy.
   B Kinetic to potential to heat energy.
   C Chemical to heat to potential energy.
   D Kinetic to heat to chemical energy. (W.M.)

5. In loading a lorry a man lifts boxes each of weight 100 N through a height of 1.5 m.
   (a) How much work does he do in lifting one box?
   (b) If he lifts 4 boxes per minute at what power is he working?

6. A boy whose weight is 600 N runs up a flight of stairs 10 m high in a time of 12 s. The average power he develops, in W, is
   A 72   B 500   C 720   D 5000   E 7200

7. A 500 kg load is lifted through a vertical height of 10 m in 25 s by a crane. Calculate the power of the crane.

8. How long will it take an electric motor of power output 25 kW to lift a mass of 1000 kg through 20 m?

check list
p.283

29 Machines

A machine is any device which enables a force (the *effort*) acting at one point to overcome another force (the *load*) acting at some other point. A lever (p. 69) is a simple machine as are pulleys, gears, screws, etc. They are used to build more complicated machines like the crane in Fig. 29.4.

Force and distance multipliers

In Fig. 29.1 a lever lifts a *load* of 100 N through 0.50 m when an *effort* is applied at the other end. The effort can be found from the principle of moments by taking moments about the pivot O as the effort *just begins* to raise the load.

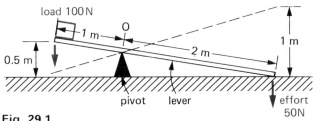

Fig. 29.1

clockwise moment = anticlockwise moment

$$\text{effort} \times 2\text{ m} = \text{load} \times 1\text{ m} = 100\text{ N} \times 1\text{ m}$$
$$\therefore \text{ effort} = 100/2 = 50\text{ N}$$

The lever has enabled an effort (E) to raise a load (L) twice as large, i.e. it is a *force-multiplier*, but E has had to move twice as far as L. The lever has a *mechanical advantage* (M.A.) of 2 and a *velocity ratio* (V.R.) of 2 where

$$\text{M.A.} = \frac{L}{E} \quad \text{and} \quad \text{V.R.} = \frac{\text{distance moved by } E}{\text{distance moved by } L}$$

The forearm (Fig. 25.4c) is a lever which is a *distance-multiplier* because the load (which is smaller than the effort) moves farther than the effort applied by the biceps, i.e. its V.R. (and its M.A.) is less than 1.

Efficiency of a machine

Machines make work easier. No machine is perfect and in practice more work is done by the effort on a machine than by the machine on the load. Work measures energy transfer (or change) and so we can also say that the energy input to a machine is greater than its energy output. Some energy is always wasted to overcome friction and move parts of the machine itself. In Fig. 29.1 there is friction at the fulcrum; the effort will therefore be more than 50 N. We define

$$\text{EFFICIENCY} = \frac{\text{ENERGY OUTPUT}}{\text{ENERGY INPUT}}$$
$$= \frac{\text{WORK DONE ON LOAD}}{\text{WORK DONE BY EFFORT}}$$

This is expressed as a percentage and is always less than 100%.

There is a useful relation between M.A., V.R. and efficiency. From the above equation we can show

$$\text{PERCENTAGE EFFICIENCY} = \frac{\text{M.A.}}{\text{V.R.}} \times 100\%$$

Pulleys

(a) Single fixed pulley, Fig. 29.2. This enables us to *lift* a load L more conveniently by applying a *downwards* effort E. E need be only slightly greater than L and if friction in the pulley bearings is negligible then $E = L$ and M.A. = 1. What is the V.R.?

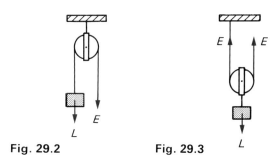

Fig. 29.2   **Fig. 29.3**

(b) Single moving pulley, Fig. 29.3. If the effort applied to the free end of the rope is E the total upward force on the pulley is $2E$ since two parts of the rope support it. A load $L = 2E$ can therefore be raised if the pulley and rope are frictionless and weightless. That is, M.A. = 2 (but less in practice).

To raise the load by 1 m requires each side of the rope to shorten by 1 m. The free end has therefore to take up 2 m of slack and so V.R. = 2.

(c) Block and tackle. This type of pulley system is used in cranes (Fig. 29.4) and lifts. It consists of two blocks each with one or more pulleys. In the

Fig. 29.4

arrangement of Fig. 29.5a the pulleys in the blocks are shown one above the other for clarity; in practice they are side by side on the same axle, Fig. 29.5b. The rope passes round each pulley in turn.

The total upward force on the lower block is 4E since it is supported by four parts of the rope and a load

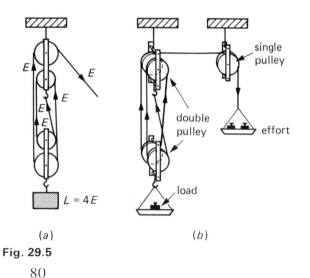

(a)   (b)

Fig. 29.5

80

$L = 4E$ can be raised. Hence M.A. $= 4$ if the pulleys are frictionless and weightless. Using the same reasoning as in (b) we see that that V.R. $= 4$, i.e. the number of times the rope passes from one block to the other.

⩗ Experiment: efficiency of a pulley system

Set up the system of Fig. 29.5b or a similar one. Starting with 50 g in the load pan, add weights to the effort pan until the load just rises steadily. Record the load and effort in a table (100 g has a weight of 1 N) and repeat for greater loads.

The V.R. can be obtained as explained before. Work out the M.A. and the efficiency for each pair of readings of load and effort.

Load N	Effort N	M.A.	Efficiency $= \dfrac{M.A.}{V.R.} \times 100\%$

Notes: 1. The lower pulley block and the load pan are also raised by the effort but are not included as part of the load. They become less important as the load increases and the M.A. and the efficiency both increase for this reason. The V.R. is constant for a particular system.

2. The efficiency is less than 100% because the system is not frictionless and the moving parts are not weightless.

Other simple machines

(a) **Wheel and axle.** A screwdriver and the steering wheel of a car, Fig. 29.6, use the wheel and axle principle. This is shown in Fig. 29.7; the effort is applied to a rope wound round the wheel and the load is raised by another rope wound oppositely on the axle.

Fig. 29.6

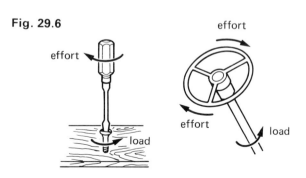

Fig. 29.7

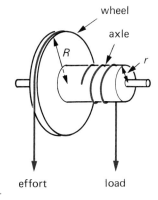

(b) Inclined plane. It is easier to push a barrel up a plank on to a lorry than to lift it vertically. In Fig 29.8, to raise the load L through a *vertical* height h, the smaller effort E moves a greater distance d equal to the length of the incline.

Fig. 29.8

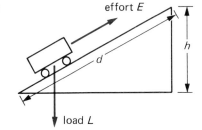

(c) Screw jack. In a car jack a screw passes through a nut carrying an arm that fits into the car chassis, Fig. 29.9*a*. When the effort applied to the lever at the

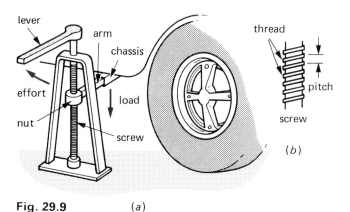

Fig. 29.9   (*a*)

top of the screw makes one complete turn, the screw (and the load) rises a distance equal to its pitch, i.e. the distance between successive threads, Fig. 29.9*b*.

(d) Gears. The V.R. and M.A. of a machine can be changed by gears. In Fig. 29.10 the gear with 10 teeth makes two revolutions for each complete

Fig. 29.10

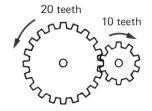

revolution of the one with 20 teeth. The V.R. is therefore 2 if the effort is applied to the small gear which drives the large gear.

Q Questions

1. A load of 500 N is raised 0.20 m by a machine in which an effort of 150 N moves 1.0 m. What is **(a)** the work done on the load, **(b)** the work done by the effort and **(c)** the efficiency?

2. An effort of 250 N raises a load of 1000 N through 5 m in a pulley system. If the effort moves 30 m, what is **(a)** the work done in raising the load, **(b)** the work done by the effort, **(c)** the efficiency?

3. For each pulley system shown in Fig. 29.11*a,b,c*, what is **(i)** the M.A., **(ii)** the V.R. and **(iii)** the efficiency?

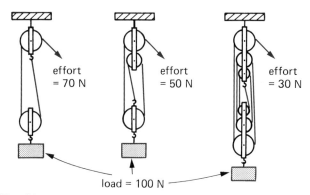

Fig. 29.11

4. A machine has an efficiency of 100% if the

    A work done by the effort is much greater than the work done on the load
    B effort equals the load
    C load is much smaller than the effort
    D work done on the load equals work done by the effort
    E distance moved by the load equals the distance moved by the effort.

5. Is a bicycle a 'distance-multiplier' or a 'force-multiplier'?

6. If a gear A has 30 teeth and drives gear B with 75 teeth, how many times does A rotate for each rotation of B?

check list
p.283

30 Pressure in liquids

To make sense of some effects in which a force acts on a body we have to consider not only the force but also the area on which it acts. For example, wearing skis prevents you sinking into soft snow because your weight is spread over a greater area. We say the *pressure* is less.

Pressure is the force (or thrust) *acting on unit area* (e.g. 1 m²) and is calculated from

$$\text{PRESSURE} = \frac{\text{FORCE}}{\text{AREA}}$$

The pressure exerted on the floor by the same box (*a*) standing on end, (*b*) lying flat, is shown in Fig. 30.1. The unit of pressure is the *pascal* (Pa); it equals 1 newton per square metre (N/m²) and is quite a small pressure. An apple in your hand exerts about 1000 Pa.

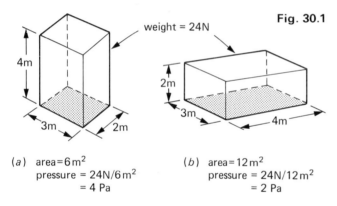

Fig. 30.1

weight = 24N

4m
3m
2m

2m
3m
4m

(*a*) area = 6 m²
pressure = 24N/6 m²
= 4 Pa

(*b*) area = 12 m²
pressure = 24N/12 m²
= 2 Pa

The greater the area over which a force acts the less is the pressure. This is why a tractor with wide wheels can move over soft ground. The pressure is large when the area is small and accounts for a nail being given a sharp point. Walnuts can be broken in the hand by squeezing two but not one. Why?

Liquid pressure

1. *Pressure in a liquid increases with depth* because the farther down you go the greater the weight of liquid above. In Fig. 30.2 water spurts out fastest and furthest from the lowest hole.

2. *Pressure at one depth acts equally in all directions.* The can of water in Fig. 30.3 has similar holes all round it at the same level. Water comes out as fast and as far from each hole. Hence the pressure exerted by the water at this depth is the same in all directions.

82

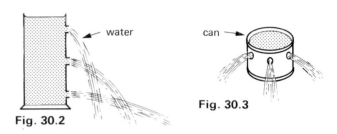

water

can

Fig. 30.2

Fig. 30.3

3. *A liquid finds its own level.* In the U-tube of Fig. 30.4*a* the liquid pressure at the foot of P is greater than at the foot of Q because the left-hand column is higher than the right-hand one. When the clip is opened the liquid flows from P to Q until the pressure and the levels are the same, i.e. the liquid finds its own level. Although the weight of liquid in Q is now greater than in P, it acts over a greater area since tube Q is wider.

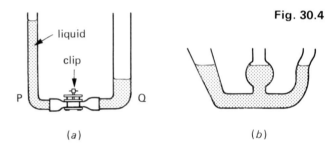

Fig. 30.4

liquid

clip

P   Q

(*a*)   (*b*)

In Fig. 30.4*b* the liquid is at the same level in each tube and confirms that the pressure at the foot of a liquid column depends only on the *vertical* depth of the liquid and not on the tube width or shape.

4. *Pressure depends on the density of the liquid.* The denser the liquid, the greater is the pressure at any given depth.

Water supply system

A town's water supply often comes from a reservoir on high ground. Water flows from it through pipes to any tap or storage tank that is below the level of water in the reservoir, Fig. 30.5. The lower the place supplied the greater is the water pressure at it. In very tall buildings it may be necessary first to pump the water to a large tank in the roof. Why?

Reservoirs for water supply or for hydroelectric power stations (p. 198) are often made in mountainous regions by building a dam at one end of a valley. The dam must be thicker at the bottom than at the top due to the large water pressure at the bottom, Fig. 30.6.

Fig. 30.6

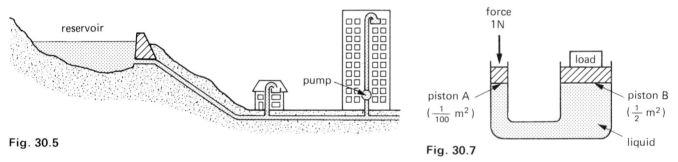

reservoir

pump

Fig. 30.5

force
1 N

load

piston A
($\frac{1}{100}$ m²)

piston B
($\frac{1}{2}$ m²)

liquid

Fig. 30.7

Hydraulic machines

Liquids are almost incompressible (i.e. their volume cannot be reduced by squeezing) and they pass on any pressure applied to them. Use is made of these facts in hydraulic machines. Fig. 30.7 shows the principle on which they work. Suppose a downwards force of 1 N acts on a piston A of area 1/100 m². The pressure transmitted through the liquid is

$$\text{pressure} = \frac{\text{force}}{\text{area}} = \frac{1}{1/100} = 100 \text{ Pa}$$

This pressure acts on piston B of area $\frac{1}{2}$ m². The total upwards force or thrust on B is given by

$$\text{force} = \text{pressure} \times \text{area} = 100 \times \tfrac{1}{2} = 50 \text{ N}$$

A force of 1 N thus produces a force of 50 N.

A *hydraulic jack*, Fig. 30.8, has a platform on top of piston B and is used in garages to lift cars. Both

valves open only to the right and they allow B to be raised a long way when A moves up and down repeatedly. A *hydraulic fork lift truck* works similarly. In a *hydraulic press* there is a fixed plate above B, and one like that in Fig. 30.9 shapes car bodies from sheets of steel (by permanently deforming them, p. 65) placed between B and the plate. *Hydraulic car brakes* are shown in Fig. 30.10. When the brake pedal is pushed the piston in the master cylinder

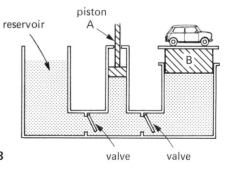

reservoir

piston
A

B

Fig. 30.8

valve   valve

Fig. 30.9

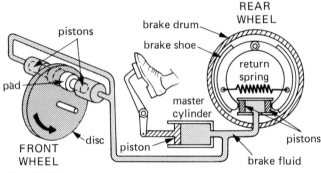

Fig. 30.10

exerts a force on the brake fluid and the resulting pressure is transmitted equally to eight other pistons (four are shown). These force the brake shoes or pads against the wheels and stop the car.

Expression for liquid pressure

In designing a dam an engineer has to calculate the pressure at various depths below the water surface.

An expression for the pressure at a depth h in a liquid of density d can be found by considering a horizontal area A, Fig. 30.11. The force acting vertically

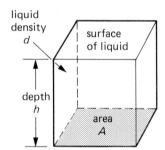

Fig. 30.11

downwards on A equals the weight of a liquid column of height h and cross-sectional area A above it. Then

volume of liquid column $= hA$

Since mass $=$ volume $\times$ density we can say

mass of liquid column $= hAd$

Taking a mass of 1 kg to have weight 10 N,

check list p.284

weight of liquid column $= 10hAd$

$\therefore$ force on area $A = 10\,hAd$

$\therefore$ pressure $=$ force/area $= 10\,hAd/A$

$\qquad\qquad\qquad\qquad = 10\,hd$

This pressure acts equally in all directions at depth h and depends only on h and d. Its value will be in Pa if h is in m and d in kg/m³.

Q Questions

1. A girl in stiletto heels is more likely to damage a wooden floor than an elephant is. Why?

2. **(a)** What is the pressure on a surface when a force of 50 N acts on an area of **(i)** 2.0 m², **(ii)** 100 m², **(iii)** 0.50 m²?
   (b) A pressure of 10 Pa acts on an area of 3.0 m². What is the force acting on the area?

3. A block of concrete weighs 900 N and its base is a square of side 2.0 m. What pressure does the block exert on the ground? *(E.A.)*

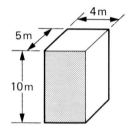

Fig. 30.12

4. **(a)** What is the volume of the block in Fig. 30.12?
   (b) What is the mass of the block if its density is 2000 kg/m³?
   (c) What is the weight of the block? (Assume a mass of 1 kg has weight 10 N.)
   (d) What pressure is exerted on the ground by the block?
   (e) If the shaded side of the block is on the ground, what effect, if any, will this have on **(i)** the force exerted by the block on the ground, **(ii)** the pressure exerted by the block on the ground?

5. In a hydraulic press a force of 20 N is applied to a piston of area 0.20 m². The area of the other piston is 2.0 m². What is **(a)** the pressure transmitted through the liquid, **(b)** the force on the other piston?

6. **(a)** Why must a liquid and not a gas be used as the 'fluid' in a hydraulic machine?
   (b) On what other important property of a liquid do hydraulic machines depend?

7. What is the pressure 100 m below the surface of sea water of density 1150 kg/m³?

8. The pressure in a water pipe in the ground floor of a building is 4×10^5 Pa but three floors up it is only 2×10^5 Pa. What is the height between the ground floor and the third floor? (The water in the pipe may be assumed to be stationary; density of water $= 1 \times 10^3$ kg/m³.) *(N.I.)*

31 Atmospheric pressure

The air forming the earth's atmosphere stretches upwards a long way. Air has weight and in a normal room weighs about the same as you do, e.g. 500 N. Owing to its weight the atmosphere exerts a large pressure at sea level, about 100 000 N/m² $= 10^5$Pa $= 100$ kPa. This pressure acts equally in all directions.

We do not normally feel atmospheric pressure because the pressure inside our bodies is almost the same as that outside. A similar balance exists with external objects, unless air is removed from one place. The effect of atmospheric pressure is then noticeable. A space from which all the air has been withdrawn is a *vacuum*.

Air pressure demonstrations

(a) **Collapsing can.** If air is removed from a can, Fig. 31.1a, by a vacuum pump, the can collapses because the air pressure inside becomes less than that outside, Fig. 31.1b.

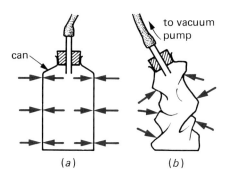

Fig. 31.1

(b) **Magdeburg hemispheres.** The vacuum pump was invented by von Guericke, the Mayor of Magdeburg. About 1650 he used it to remove the air from two large hollow metal hemispheres, fitted together

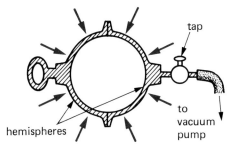

Fig. 31.2

to give an airtight sphere. So good was his pump that it took two teams, each of eight horses, to separate the hemispheres. A similar experiment can be done by two people with small brass hemispheres, Fig. 31.2.

Using air pressure

(a) **Drinking straw,** Fig. 31.3a. When you suck, your lungs expand and air passes into them from the straw. Atmospheric pressure pushing down on the surface of the liquid in the bottle is now greater than the pressure of the air in the straw and so forces the liquid up to your mouth.

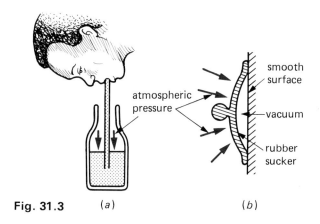

Fig. 31.3 (a) (b)

(b) **Rubber sucker,** Fig. 31.3b. When the sucker is moistened and pressed on a smooth flat surface, the air inside is pushed out. Atmospheric pressure then holds it firmly against the surface. Suckers are used as holders for towels in the home, for attaching car licences to windscreens and in industry for lifting metal sheets.

Pressure gauges

These measure the pressure exerted by a fluid, i.e. by a liquid or a gas.

(a) **Bourdon gauge.** It works like the toy in Fig. 31.4a. The harder you blow into the paper tube, the more does it uncurl. In a Bourdon gauge, Fig. 31.4b, when a fluid pressure is applied, the curved metal tube tries to straighten out and rotates a pointer over a scale. Car oil pressure gauges and the gauges on gas cylinders are of this type.

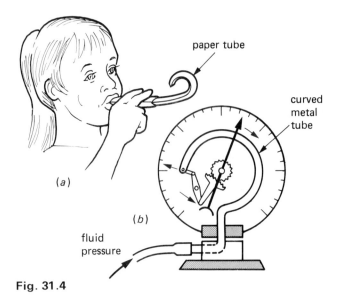

Fig. 31.4

(b) U-tube manometer. In Fig. 31.5a each surface of the liquid is acted on equally by atmospheric pressure and the levels are the same. If one side is connected to, for example, the gas supply, Fig. 31.5b, the gas exerts a pressure on surface A and level B rises until

pressure of gas = atmospheric pressure + pressure due to liquid column BC

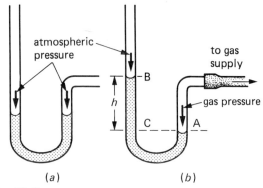

Fig. 31.5

The pressure of the liquid column BC therefore equals the amount by which the gas pressure *exceeds* atmospheric pressure. It equals 10 hd (in Pa) where h is the vertical height of BC (in m) and d is the density of the liquid (in kg/m^3). The height h is called the *head of liquid* and sometimes, instead of stating a pressure in Pa, we say that it is so many cm of water (or mercury for higher pressures).

Mercury barometer

A barometer is a manometer which measures atmospheric pressure. A simple barometer can be made by using a small funnel to nearly fill with mercury a

thick-walled glass tube about 1 m long. If the tube is slowly inverted several times, with a finger over the open end, the large air bubble runs up and down collecting any small air bubbles trapped in the mercury.

The tube is then filled with mercury, closed with the finger and inverted into a bowl of mercury. When the finger is removed the mercury falls until it is about 760 mm above the level in the bowl, Fig. 31.6a. The pressure at X due to the weight of the column of mercury XY equals the atmospheric pressure on the surface of the mercury in the bowl. XY measures the atmospheric pressure in mm of mercury (mmHg).

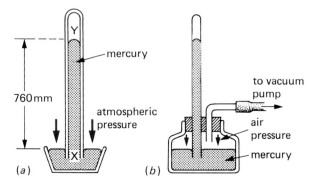

Fig. 31.6

The *vertical* height of the column is unchanged if the tube is tilted. Would it be different with a wider tube? Why? The space above the mercury in the tube is a vacuum (except for a little mercury vapour). How could you test this?

The apparatus in Fig. 31.6b may be used to show that it is atmospheric pressure which holds up the column. When the air above the mercury in the bottle is pumped out, the column falls.

Aneroid barometer

An aneroid (no liquid) barometer consists of a partially evacuated, thin metal box with corrugated sides to increase its strength, Fig. 31.7. The box is prevented from collapsing by a strong spring. If the atmospheric pressure increases, the box caves in

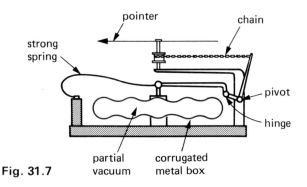

Fig. 31.7

slightly, if it decreases the spring pulls it out. A system of levers magnifies this movement and causes a chain to move a pointer over a scale.

Aneroid barometers are used as weather glasses, high pressure being associated with fine weather. They are also used as altimeters to measure the height of an aircraft since the pressure decreases the higher it goes.

Pressure, aviation and diving

Our bodies are designed to work at normal atmospheric pressure. At high altitudes breathing is difficult and modern aircraft have pressurized cabins in which the air pressure is increased sufficiently above that outside to safeguard the crew and passengers.

Ear 'popping' occurs when there are pressure changes, e.g. at aircraft take-off. It is due to the pressure difference between the air in the middle part of the ear and that in the outer ear. The eardrum becomes distorted. Swallowing helps to equalize the pressures.

If astronauts are not in pressurized space capsules they must rely on space suits for supplying a suitable atmosphere, Fig. 31.8. Otherwise blood and water in the body would boil and possibly explode (see p. 138).

A diver meets high pressures and different dangers. If he breathed air, extra nitrogen might dissolve in his blood and on returning to the surface he would suffer the painful and sometimes fatal condition called

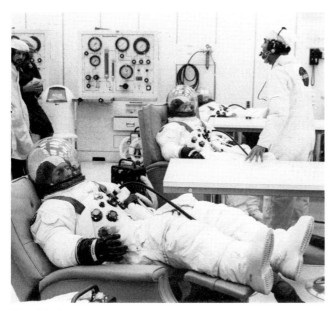

Fig. 31.8

'bends'. This is caused by the dissolved nitrogen forming bubbles in the blood. To prevent this he breathes a mixture of oxygen and helium. If he has been working in deep waters (e.g. 500 m) he is brought to the surface in a diving bell and kept in a decompression chamber for up to several days while the pressure is slowly reduced.

Q Questions

1. **(a)** Figs. 31.9*a* and *b* show an open ended and a closed tube manometer connected (at different times) to the same gas cylinder.

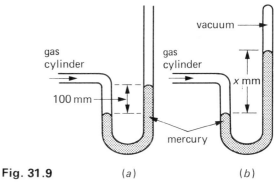

Fig. 31.9   (*a*)   (*b*)

The mercury barometer reads 760 mm. Assuming no loss in pressure from the gas cylinder, calculate **(i)** the pressure of gas in the cylinder, **(ii)** the height of the mercury in Fig. 31.9*b* (i.e. *x*).
(b) Draw a diagram of a manometer similar to that in **(i)** Fig. 31.9*a*, **(ii)** Fig. 31.9*b*, when it is connected to a gas supply at less than atmospheric pressure (760 mm of mercury).   (*W.M.*)

2. Fig. 31.10 shows a simple barometer.

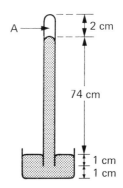

Fig. 31.10

(a) What is the region A?
(b) What keeps the mercury in the tube?
(c) What is the value of the atmospheric pressure being shown by the barometer?
(d) What would happen to this reading if the barometer were taken up a high mountain? Give a reason.
                                                        (*E.M.*)

3. What would be the height of a water barometer if atmospheric pressure is 1.0×10^5 Pa and the density of water is 1.0×10^3 kg/m³? (*Hint.* Use $p = 10\,hd$.)

check list
p.284

32 Pumps and pressure

Syringe

The syringe is used by doctors to give injections and by gardeners to spray plants. It consists of a tight-fitting piston in a barrel, Fig. 32.1, and may be filled by putting the nozzle under the liquid and drawing

Fig. 32.1

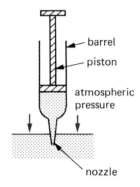

barrel
piston
atmospheric pressure
nozzle

back the piston. This reduces the air pressure in the barrel and atmospheric pressure forces the liquid up into it. Pushing down the piston drives liquid out of the nozzle.

Bicycle pump

When the piston is pushed in, the air between it and the tyre valve is compressed. This pushes the rim of the plastic washer against the wall of the barrel to form an airtight seal, Fig. 32.2. Air is forced past the tyre valve into the tyre when the pressure of the air between the plastic washer and the valve exceeds the pressure of the air in the tyre.

Fig. 32.2

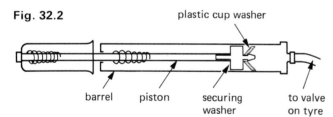

plastic cup washer

barrel   piston   securing washer   to valve on tyre

When the piston is withdrawn, the tyre valve is closed by the greater pressure in the tyre. Atmospheric pressure then forces air past the plastic washer, no longer pressed hard against the wall, into the barrel.

Boyle's law

(a) Experiment. Changes in the volume of a gas due to pressure changes can be studied using the apparatus in Fig. 32.3. The volume V of air trapped in the

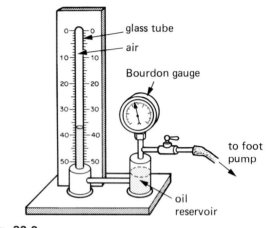

glass tube
air
Bourdon gauge
to foot pump
oil reservoir

Fig. 32.3

glass tube is read off on the scale behind. The pressure is altered by pumping air from a foot pump into the space above the oil reservoir. This forces more oil into the glass tube and increases the pressure p on the air in it; p is measured by the Bourdon gauge.

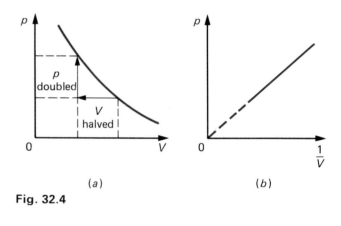

p doubled
V halved

(a)   (b)

Fig. 32.4

(b) Results. If a graph of pressure against volume is plotted using the results, a curve like that in Fig. 32.4a is obtained. Close examination of it shows that if p is doubled, V is halved. That is, p is *inversely proportional* to V. In symbols

$$p \propto \frac{1}{V} \quad \text{or} \quad p = \text{constant} \times \frac{1}{V}$$

$$\therefore pV = \text{constant}$$

If several pairs of readings p_1V_1, p_2V_2, etc. are taken, then $p_1V_1 = p_2V_2 = \text{constant}$. This is *Boyle's law* which is stated as follows.

The pressure of a fixed mass of gas is inversely proportional to its volume if its temperature is kept constant.

Since p is inversely proportional to V, then p is directly proportional to $1/V$. A graph of p against $1/V$ is therefore a straight line through the origin, Fig. 32.4b.

Aerosol sprays and pressure

An aerosol can uses gas under pressure to produce a fine spray. A simplified version is shown in Fig. 32.5. The active material to be sprayed, e.g. paint, hair spray, garden pesticide, is in liquid form with the propellent. The pressurized vapour of the propellent above the liquid forces the latter up the inner tube when the valve is opened by pressing the button on the top of the sealed can.

Fig. 32.5

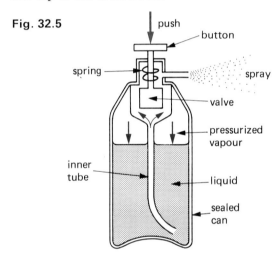

Aerosols should be used in short bursts and held about 30 cm away from whatever is being sprayed. The can should be protected from sunlight, stored away from heat and not pierced or burnt when empty but disposed of intact.

Q Questions

1. Fig. 32.6a shows air enclosed in a cylinder by an airtight piston. In Fig. 32.6b the piston has been pushed in

Fig. 32.6

so that the air occupies one-third of the length of the cylinder that it previously occupied. What, if anything, has happened to (a) the number of molecules of air, (b) the volume of air, (c) the pressure of the air? (E.A.)

2. If a certain quantity of gas has a volume of 30 cm^3 at a pressure of 1×10^5 Pa, what is its volume when the pressure is (a) 2×10^5 Pa, (b) 5×10^5 Pa?

3. An enclosed mass of air occupies $4.0 \times 10^{-3} \text{ m}^3$ at a pressure of 100 kPa; when the pressure is changed to 80 kPa (at constant temperature) the volume of the air, in m^3, will be

    A 2.2×10^{-3} B 3.2×10^{-3} C 4.0×10^{-3}
    D 5.0×10^{-3} E 9.0×10^{-3} (N.I.)

4. A diver at the bottom of a lake releases an air bubble of volume 2 cm^3. As the bubble rises its volume increases until at the surface it is 4 cm^3. How deep is the lake if atmospheric pressure equals the pressure due to a column of water 10 m high?

5. A sheet of plastic previously softened by heating can be permanently deformed to make a washing-up bowl by vacuum-forming techniques. Using Fig. 32.7 explain how this is done.

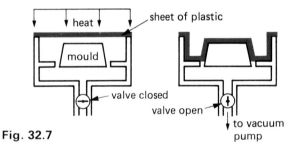

Fig. 32.7

check list p.284

33 Floating, sinking and flying

A ship floats because it gets support from the water. Any object in a liquid, whether floating or submerged, is acted on by an upward force or *upthrust*. This makes it seem to weigh less than normal.

Archimedes' principle

In Fig. 33.1*a* the block hanging from the spring balance weighs 10 N in air. When it is completely immersed in water the reading becomes 6 N, Fig. 33.1*b*. The loss in weight of the block is $(10 - 6)$ N = 4 N: the upthrust of the water on it is therefore 4 N.

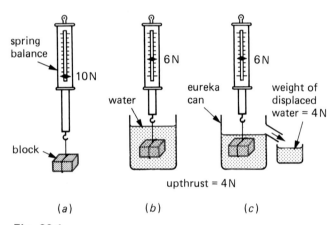

Fig. 33.1

If a can like that in Fig. 33.1*c* is used, full of water to the level of the spout, the water displaced (the overflow) can be collected and weighed. Its weight here is 4 N, the same as the upthrust. (Its volume is exactly equal to the volume of the block.) Experiments with other liquids and also with gases lead to the general conclusion called *Archimedes' principle*.

When a body is wholly or partly submerged in a fluid the upthrust equals the weight of fluid displaced (i.e. pushed aside).

A fluid means either a liquid or a gas. The case of gases will be dealt with later.

Floating

A stone held below the surface of water sinks when released, a cork rises. The weight of the stone is greater than the upthrust on it (i.e. the weight of water displaced) and there is a net or resultant downward force on it, Fig. 33.2*a*. If the cork has the

same volume as the stone, it will displace the same weight (and volume) of water. The upthrust on it is therefore the same as for the stone but it is greater than the weight of the cork. The resultant upward force on the cork makes it rise, Fig. 33.2*b*.

Fig. 33.2

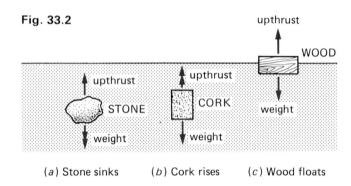

(*a*) Stone sinks   (*b*) Cork rises   (*c*) Wood floats

When a body floats in water the upthrust equals the weight of the body. The net force on the body is zero, Fig. 33.2*c*. This is a case of the *principle of flotation*.

A floating body displaces its own weight of fluid.

For example, a block of wood of weight 10 N displaces an amount of water (or any other liquid in which it floats) having weight 10 N (i.e. mass 1 kg).

Ships, submarines, balloons

(a) Ships. A floating ship displaces a weight of water equal to its own weight including that of the cargo. The load lines (called the Plimsoll mark) on the side of a ship show the levels to which it can legally be loaded under different conditions, Fig. 33.3. Why is it allowed to take a greater load in summer than in winter?

Fig. 33.3

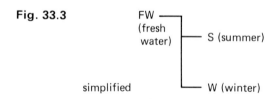

simplified

(b) Submarines. A submarine sinks by taking water into its buoyancy tanks. Once submerged, the upthrust is unchanged but the weight of the submarine increases with the inflow of water and it sinks faster. To surface, compressed air is used to blow the water out of the tanks.

In the Cartesian diver, Fig. 33.4, pressure on the cork forces more water into the bulb. The diver's weight increases and he sinks. Decreasing the pressure causes him to rise.

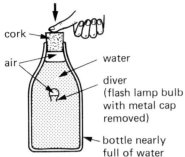

cork

air

water

diver
(flash lamp bulb
with metal cap
removed)

bottle nearly
full of water

Fig. 33.4

The Swedish warship *Wasa*, which sank in a squall in Stockholm Harbour over three hundred years ago, was refloated in 1961. Fig. 33.5*a* shows the ship emerging from the water. Fig. 33.5*b* shows it in dock

Fig. 33.6

with three of its four buoyancy tanks. The tanks, full of water, were attached to the partly submerged ship which was then raised by pumping the water out of the tanks. Can you explain why this happened?

(c) Balloons. A balloon filled with hydrogen or hot air weighs less than the weight of air it displaces. The upthrust is therefore greater than its weight and the resultant upwards force on the balloon causes it to rise.

Meteorological balloons carrying scientific instruments called *radiosondes* are sent into the upper atmosphere. A small radio transmitter sends signals back to earth which contain information about the temperature, pressure and humidity, Fig. 33.6. They are tracked by radar to give data on wind direction and speed.

Fig. 33.5*a*

Fig. 33.5*b*

Fig. 33.7

(d) Airships. The airship, *Skyship 600*, Fig. 33.7, has a plastic gas bag filled with non-flammable helium (a gas less dense than air). It is powered by two car engines which drive swivelling propellers that provide vertical thrust for take-off and landing and horizontal thrust for forward motion. It can climb at about 700 metres a minute, cruise sedately at 80 km per hour (50 mph) and travel 1600 km (1000 miles) on just 550 litres (120 gallons) of fuel.

It is used for aerial surveys, photographic, advertising and surveillance work as well as for cargo and passenger carrying.

Hydrometer

This is an instrument for measuring rapidly the density of a liquid, Fig. 33.8a. It is placed in the liquid and the scale read at the level of the liquid surface. The denser the liquid the higher it floats; the numbers on the scale increase downwards.

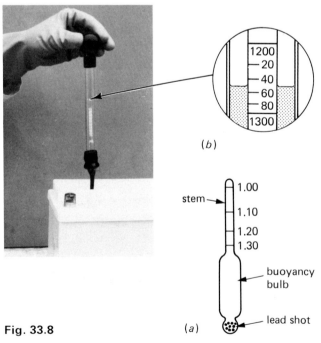

Fig. 33.8

92

The large bulb gives the instrument buoyancy and the small weighted bulb makes it float upright. The narrow stem gives greater sensitivity, i.e. a small density change causes the instrument to float much higher or lower.

Hydrometers enable the state of a car battery to be checked, Fig. 33.8b. In a fully charged one the density of the acid should be about 1.25. Recharging is required when the reading is less than 1.18. Hydrometers are also used in breweries to measure the density of beer. The 'watering down' of milk and wine can be tested similarly.

Bernoulli's principle

The pressure is the same at all points on the same level in a fluid at rest; this is not so when the fluid is moving.

(a) Liquids. When a liquid flows through a uniform tube the pressure falls steadily, as shown by the three manometer tubes in Fig. 33.9a. In Fig. 33.9b the

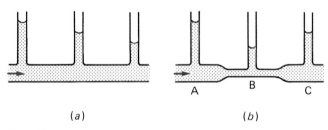

Fig. 33.9

pressure falls in the narrow part B but rises again in the wider part C. Since the same volume of liquid passes through B in a certain time as enters A, the liquid must be moving faster in B than in A. Therefore a decrease of pressure occurs when the speed of the liquid increases. Conversely an increase of pressure accompanies a fall in speed. This effect, called Bernoulli's principle, will be explained later (p. 111) and is stated as follows.

When the speed of a fluid in smooth flow increases, the pressure in the fluid decreases and vice versa.

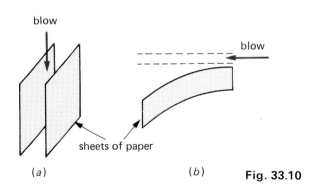

Fig. 33.10

(b) Gases. Bernoulli effects in air streams can be shown as in Fig. 33.10a and b. In (a) the two sheets of paper come together when you blow *between* them and in (b) the paper rises when you blow *over* it. In both cases the pressure falls in the moving air stream.

(c) Applications. As a fluid comes out of a jet its speed increases and its pressure decreases. This fact is used in a Bunsen burner, Fig. 33.11a, and in a carburettor. A spinning ball takes a curved path because it drags air round with it, thereby increasing the speed of the air flow on one side and reducing it on the other, Fig. 33.11b.

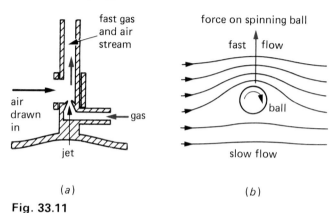

(a)   (b)

Fig. 33.11

An aircraft wing, called an *aerofoil*, is shaped so that the air has to travel farther and so faster over the top surface than underneath, Fig. 33.12a. The resultant upward force on the wing provides 'lift' for the aircraft.

The sail of a yacht tacking into the wind is another example of an aerofoil. The air flow over the sail produces a pressure increase on the windward side and a decrease on the leeward side. The resultant

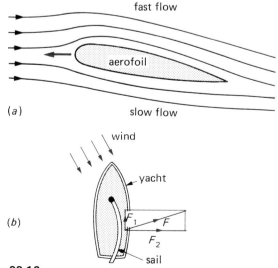

Fig. 33.12

driving force is roughly at right angles to the sail and can be resolved into a component F_1 producing forward motion and a greater component F_2 acting sideways, Fig. 33.12b. The keel produces a force to balance F_2. The drag force on the yacht acts backwards along the tangent to the sail and increases until it equals F_1 when, as we will see in Topic 38, the yacht stops accelerating and moves with constant speed.

Q Questions

1. A metal block is weighed (a) in air, (b) half-submerged in water, (c) fully submerged in water, (d) fully submerged in a strong salt solution.

The readings obtained, though not necessarily in the correct order, were 5 N, 8 N, 10 N and 6 N. Which reading was obtained for each weighing?

2. A block of density 2400 kg/m³ has a volume of 0.20 m³. What is (a) its mass, (b) its weight and (c) its apparent weight when completely immersed in a liquid of density 800 kg/m³?

3. A hot air balloon and its basket together weigh 3000 N and contains hot air weighing 17 000 N. If it displaces cold air of weight 25 000 N what is the maximum load it can lift?

4. A block of wood of volume 50 cm³ and density 0.60 g/cm³ floats on water. What is (a) the mass of the block, (b) the mass of water displaced, (c) the volume immersed in the water? (Density of water = 1.0 g/cm³.)

5. A swimmer dives off a raft in a pool. Does the raft rise or sink in the water? What happens to the water level in the pool? Give reasons for your answer.

6. (a) A test tube of mass 5 g, length 10 cm and uniform cross-sectional area 2 cm² is partly filled with lead shot and floats vertically in water with 5 cm of its length submerged, Fig. 33.13. Calculate (i) the volume of water displaced, (ii) the mass of water displaced, (iii) the combined mass of the test tube and lead shot, (iv) the mass of the lead shot, (v) the length of the test tube that would be submerged in a liquid of density 0.8 g/cm³.

Fig. 33.13

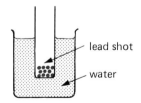

(b) (i) State one reason why it would be better to have chosen a tube of smaller cross-section if this method is to be used to measure the densities of liquids. **(ii)** What name is given to the commercial form of this apparatus? **(iii)** Give one example of the use of the commercial form. **(iv)** Why is a definite temperature usually marked on the instrument?

(E.M.)

check list
p.284

34 Additional questions

Core level (all students)

Moments and levers

1. Fig. 34.1 shows a nail being removed from a piece of wood by a claw hammer.
   (a) (i) What physical principle or law is involved in the withdrawal of the nail? (ii) The nail begins to move when a force of 100 N is applied at the end of the handle. What is the frictional resistance exerted on the nail by the wood?
   (b) State *one* more example of a lever of the same type as the claw hammer. (*N.W.*)

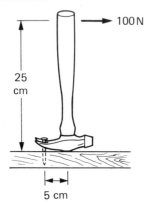

Fig. 34.1 5 cm

2. The uniform plank in Fig. 34.2 weighs 200 N, rests on two trestles A and B, and supports a boy of weight 500 N in the position shown. *P* and *Q* are the reaction forces at A and B.
   (a) Write down the moment of each force about A.
   (b) Use the principle of moments to find *Q*.
   (c) What is the total upward force $P+Q$?
   (d) What is the value of *P*?

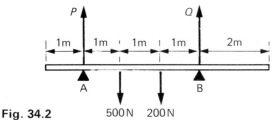

Fig. 34.2 500 N 200 N

3. Two uniform strip lights each of weight 30 N are supported from a ceiling as in Fig. 34.3*a*, *b*. Find the values of T_1, T_2, T_3 and T_4.

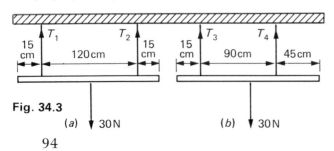

Fig. 34.3
(*a*) ↓ 30 N (*b*) ↓ 30 N

94

Centres of gravity

4. (a) (i) Define the *moment of a force about a point*.
      (ii) State the *principle of moments*.
   (b) A uniform plank PQ weighing 300 N and of length 3.0 m rests on two narrow supports C and D. C is 0.50 m from P and D is 1.0 m from Q.
      (i) Draw a neat diagram of the arrangement; on this diagram draw arrows at the points where the forces act ON THE PLANK, clearly marking the distances. The arrows should show the directions of the forces. Label the known force with its value.
      (ii) A weight *W* causes the plank to tip over when placed at one end, but does not do so when placed at the other end. At which end is it placed when it causes tipping? Give the reason for your answer.
      (iii) What is the least value of *W* which will cause tipping when placed (a) at end P, (b) at end Q?
      (iv) By hanging a bucket at Q and pouring water into it the plank can be made to balance about support D. How could this arrangement be used to check the value given for the weight of the plank?
      (*J.M.B./A.L./N.W.16+*)

Adding forces

5. The resultant of the two forces shown in Fig. 34.4 makes an angle of 30° with the force of 40 N. Determine, graphically or otherwise, the values of the resultant and of the force P. (*L.*)

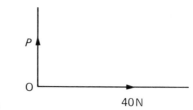

Fig. 34.4 40 N

Work, energy, power

6. Compressed air, a wound-up watch spring, a fly-wheel rotating on a fixed axle and a clock pendulum at maximum displacement.

The number of systems named above which possess p.e. is

A 0   B 1   C 2   D 3   E 4   (*J.M.B.*)

7. A boy is lifting sandbags from the floor onto a shelf 1.6 m high. The pull of the earth on each sandbag is 10 N. The boy lifts 80 sandbags in 100 s.

(a) How much useful work was done in lifting the sandbags?

(b) What is the total p.e. when all the bags are on the shelf?

(c) What was the boy's useful power output?

(d) The actual power output was greater than answer (c). Suggest two possible reasons for this difference.

One sandbag fell from the shelf.

(e) What is its p.e. when it is halfway to the ground?

(f) What is its k.e. when it is halfway to the ground?

(g) What happens to its k.e. when it hits the ground? Give two possible answers.

Fig. 34.5 shows the boy's arm when he is measuring the pull of the earth on a sandbag. He holds the balance still to take a reading.

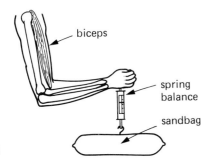

Fig. 34.5

(h) How much work is he doing as he checks the reading?

(i) Give a rough estimate (from the diagram) of the tension in the boy's biceps as he holds the balance. Explain how you arrived at this value.   (E.M.)

8. An escalator carries 60 people of average mass 70 kg to a height of 5 m in one minute. Find the power needed to do this.

Atmospheric pressure

9. (a) A mercury barometer is shown in Fig. 34.6. Copy it and mark the mercury levels when the tube is tilted from position A to positions B and C.

(b) How would the mercury height be affected if air got into the tube in position A?

(c) Why is mercury suitable as a liquid for use in a barometer?

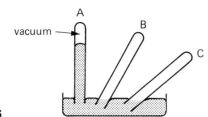

Fig. 34.6

Pumps and pressure

10. Explain the action of a syringe with the help of a labelled diagram.

11. The diagram, Fig. 34.7, shows the essential features of a bicycle pump connected to a tube.

(a) When the handle has moved halfway outwards, will the valves be open or closed? Explain.

(b) Under what circumstances will valve 1 open?

(c) Explain why the 'dead space' sets a limit to the pressure the pump can cause.   (S.R.)

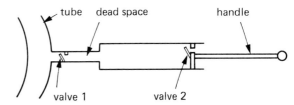

Fig. 34.7

12. Boyle's law refers to the product of the volume and pressure being constant.

(a) (i) To what does it apply? (ii) Under what condition?

(b) With the aid of a diagram describe an experiment to show that Boyle's law is obeyed.

(c) If the piston in Fig. 34.8 moves down 8 cm, what is the new pressure?   (A.L.)

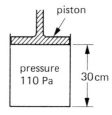

Fig. 34.8

Floating, sinking and flying

13. A body of mass 2 kg is suspended from a spring balance, calibrated in newtons, which reads 17 N when the body is completely submerged in water. The intensity of the earth's gravitational field is 10 N/kg (i.e. the acceleration due to gravity is 10 m/s^2).

(a) What is the upthrust (in newtons) of the water on the body?

(b) What is the mass (in kg) of water displaced by the body?

(c) If the density of water is 1000 kg/m^3, what is the volume (in m^3) of water displaced?
          (J.M.B./A.L./N.W.16+part qn.)

14. Describe the features of a common hydrometer. If it is to be used over the range of densities 0.70 to 1.00 indicate the appearance of its scale. State and explain what would happen if it was placed in a strong salt solution.

Further level (higher grade students)

Machines

15. (a) A load of 40 N is raised 0.50 m by a pulley system when the effort of 10 N moves 2.5 m. Calculate the efficiency.

(b) Using a pulley system operating at an efficiency of 75% a man lifts a load of weight 1800 N using an effort of 400 N. Calculate

(i) the work done by the effort when it moves through 6 m,

(ii) the distance the load moves when the effort moves 6 m.

Centres of gravity

16. (a) What is meant by (i) the centre of gravity of a body, (ii) a couple?

A metal sheet of irregular shape is suspended so that it can swing freely about a fixed horizontal needle which passes through a hole P near its edge. Draw a diagram showing the sheet at an instant during the swing when the sheet is not at its rest position. Show on the diagram the direction and point of application of each force acting on the metal sheet. Draw a second diagram showing the forces on the sheet when it has stopped swinging. Explain why the sheet comes to rest in this position.

(b) In order to 'weigh' a boy in the laboratory, a uniform plank of wood AB = 3.0 m long, having a mass of 8.0 kg, is pivoted about a point 0.50 m from A. The boy stands 0.30 m from A and a mass of 2.0 kg is placed 0.50 m from B in order to balance the plank horizontally.

Sketch the arrangement, representing each force acting on the plank by an arrow showing the direction of the force. Indicate the value of each force. Calculate the mass of the boy. (Take the weight of 1 kg as 10 N.) *(J.M.B.)*

Adding forces

17. A string attached to two hooks A, B, supports a weight W attached at a point C. CA makes an angle with the vertical of 30° and CB an angle of 45°, Fig. 34.9.

If the tension in BC is 100 N, find, by drawing or calculation, the tension in AC and the weight W.

(O. and C.)

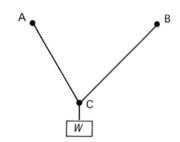

Fig. 34.9

96

18. What is meant by a *resultant* of two forces and by the *components* of a force?

What is the resultant of two equal forces, each of 20 N, if the direction of the two forces are at an angle of 60° to each other?

What is the component of a force of 20 N in a direction of 30° to the direction of the force if the two components are at right angles to each other? *(C.)*

Pressure in liquids

19. What is meant by the term *pressure?* Explain the fact that when someone uses his thumb to push a drawing pin into a block of wood, the pressure on the wood is greater than the pressure on the thumb.

Calculate the pressure on the thumb when the force exerted is 20 N and the top of the pin is square, the length of each side being 10 mm. *(C.)*

20. The pressure of the liquid column XY in Fig. 34.10 supports the mass M on the light piston.

(a) If the density of the liquid is 800 kg/m³ and a mass of 1 kg has a weight of 10 N, calculate the pressure at Y (in Pa) due to the column XY.

(b) What is (i) the pressure, (ii) the force on the lower surface of the piston if its area is 0.1 m²?

(c) Calculate the value of M.

(d) Would the value of M supported be affected if (i) tube XY were wider but with the same height of liquid, (ii) water of density 1000 kg/m³ replaced the liquid?

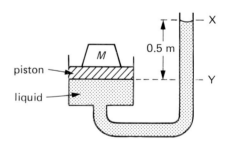

Fig. 34.10

21. Two jars, A and B, each of uniform cross-section, are filled to the brim with liquids X and Y respectively. A has twice the area of cross-section of B, but is the same height as B. The density of liquid X is twice the density of liquid Y.

(a) Compare the mass of liquid X in A with that of liquid Y in B.

(b) Compare the pressure at the bottom of A with that at the bottom of B due to the liquids X and Y respectively. *(W.)*

Atmospheric pressure

22. Describe how you would set up a simple mercury barometer, pointing out precautions you would take to obtain a reasonably accurate form of instrument.

How would you test whether there is a vacuum above the mercury?

The air pressure at the base of a mountain is 75.0 cm of mercury and at the top is 60.0 cm of mercury. Given that the average density of air is 1.25 kg/m^3 and the density of mercury is 1.36×10^4 kg/m^3, calculate the height of the mountain. (W.)

Pumps and pressure

23. Fig. 34.11 illustrates a simple experiment in which a column of air is enclosed in a very narrow (capillary) bore tube by a thread of mercury 1 cm in length. The capillary bore is uniform in cross-section. With the tube in the vertical position illustrated in diagram A the length of the

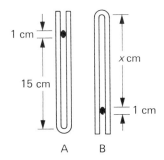

Fig. 34.11   A   B

air column is 15 cm as indicated. When the tube is inverted, as in B, the length of the air column changes to x cm. The height of the column of mercury in a mercury barometer at the time of the experiment is 76 cm and the temperature remains constant.

(a) What is the pressure (in cm of mercury) in the tube when it is as shown in diagram A?
(b) What is the pressure in the tube when it is as shown in diagram B?
(c) Calculate the value of x.

(J.M.B./A.L./N.W.16+)

24. The level of mercury in a perfect barometer stands at 75 cm above the level in the reservoir and is reduced to 70 cm by the introduction of a small quantity of air. Determine the volume of this air at atmospheric pressure if the tube is of uniform bore of cross-sectional area 2.0 cm^2 and its upper end is at a constant height of 80 cm above the external mercury level. You may assume that the external mercury level and the external temperature do not change. (L.)

Floating, sinking and flying

25. A rectangular block of wood, mass 0.32 kg, measuring 10 cm × 10 cm × 5 cm, floats in a liquid of density 1600 kg/m^3 with its large face horizontal. What is the height in cm of the upper face of the block above the surface of the liquid?

A 0.4   B 2.0   C 2.5   D 3.0   E 4.8
(N.I.)

Motion and energy

35 Velocity and acceleration

Speed

If a car travels 300 km from Liverpool to London in 5 hours, its *average speed* is 300 km/5 h = 60 km/h. The speedometer would certainly not read 60 km/h for the whole journey but might vary considerably from this value. That is why we state the average speed. If a car could travel at a constant speed of 60 km/h for 5 hours, the distance covered would still be 300 km. It is *always* true that

$$\text{AVERAGE SPEED} = \frac{\text{DISTANCE MOVED}}{\text{TIME TAKEN}}$$

To find the *actual speed* at any instant we would need to know the distance moved in a very short interval of time. This can be done by multiflash photography. In Fig. 35.1 the golfer is photographed whilst a flashing lamp illuminates him 100 times a second. The speed of the club-head as it hits the ball is about 190 km/h.

Velocity

Speed is the distance travelled in unit time, *velocity is the distance travelled in unit time in a stated direction*. If two cars travel due north at 20 m/s, they have the same speed of 20 m/s and the same velocity of 20 m/s *due north*. If one travels north and the other south, their speeds are the same but not their velocities since their directions of motion are different. Speed is a scalar and velocity a vector quantity (see p. 75).

$$\text{VELOCITY} = \frac{\text{DISTANCE MOVED}}{\text{TIME TAKEN}} \text{ in a stated direction}$$

The *velocity* of a body is *uniform* or *constant* if it moves with a steady speed in a straight line. It is not uniform if it moves in a curved path. Why?

The units of speed and velocity are the same, e.g. km/h, m/s, and

60 km/h = 60 000 m/3600 s = 17 m/s

Distance moved in a stated direction is called the *displacement*. It is a vector, unlike distance which is a scalar. Velocity may also be defined as

$$\text{VELOCITY} = \frac{\text{DISPLACEMENT}}{\text{TIME TAKEN}}$$

Acceleration

When the velocity of a body changes we say it accelerates. If a car starting from rest and moving due north has velocity 2 m/s after 1 second, its velocity has increased by 2 m/s in 1 s and its acceleration is 2 m/s per second due north. We write this as 2 m/s^2. *Acceleration is the change of velocity in unit time.*

$$\text{ACCELERATION} = \frac{\text{CHANGE OF VELOCITY}}{\text{TIME TAKEN FOR CHANGE}}$$

For a steady increase of velocity from 20 m/s to 50 m/s in 5 s

$$\text{acceleration} = \frac{(50-20) \text{ m/s}}{5 \text{ s}} = 6 \text{ m/s per second}$$

Acceleration is also a vector and both its magnitude and direction should be stated. However, at present we will consider only motion in a straight line and so the magnitude of the velocity will equal the speed, and the magnitude of the acceleration will equal the change of speed in unit time.

Fig. 35.1

98

The speeds of a car accelerating on a straight road are shown below.

Time (s)   0 1 2 3 4 5 6
Speed (m/s) 0 5 10 15 20 25 30

The speed increases by 5 m/s every second and the acceleration of 5 m/s² is said to be *uniform*.

An acceleration is positive if the velocity increases and negative if it decreases. A negative acceleration is also called a *deceleration* or *retardation*.

Tickertape timer: tape charts

A tickertape timer, Fig. 35.2*a*, enables us to measure speeds and hence accelerations. One type has a steel strip which vibrates 50 times a second and makes dots at 1/50 s intervals on the paper tape being pulled through it. 1/50 s is called a 'tick'. The type in Fig. 35.2*b* needs no carbon paper disc.

Fig. 35.2

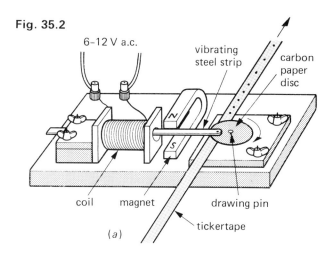

(a)

(b)

The distance between successive dots equals the average speed of whatever is pulling the tape in, say, cm per 1/50 s, i.e., cm per tick. The 'tentick' ($\frac{1}{5}$ s) is also used as a unit of time. Since ticks and tenticks are small we drop the 'average' and just refer to the 'speed'.

Tape charts are made by sticking successive strips of tape, usually tentick lengths, side by side. That in Fig. 35.3 represents a body moving with *uniform velocity* since equal distances have been moved in each tentick interval.

Fig. 35.3

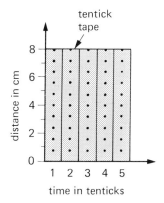

The chart in Fig. 35.4 is for *uniform acceleration*: the 'steps' are of equal size showing that the speed increased by the same amount in every tentick ($\frac{1}{5}$ s). The acceleration (average) can be found from the chart as follows.

Fig. 35.4

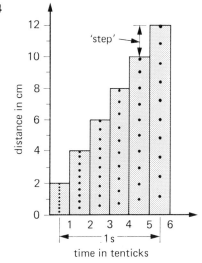

The speed during the *first* tentick is 2 cm/$\frac{1}{5}$ s or 10 cm/s. During the *sixth* tentick it is 12 cm/$\frac{1}{5}$ s or 60 cm/s. And so during this interval of 5 tenticks, i.e. 1 second, the change of speed is (60−10) cm/s = 50 cm/s.

$$\text{acceleration} = \frac{\text{change of speed}}{\text{time taken}}$$

$$= \frac{50 \text{ cm per s}}{1 \text{ s}}$$

$$= 50 \text{ cm per s per s}$$

$$= 50 \text{ cm/s}^2$$

ᴀᴀᴀ Experiment: analysing motion

(a) Your own motion. Pull a 2 m length of tape through a tickertape timer as you walk away from it quickly, then slowly, then speeding up again and finally stopping, Fig. 35.5a.

Cut the tape into tentick lengths and make a tape chart. Write labels on it to show where you speeded up, slowed down, etc.

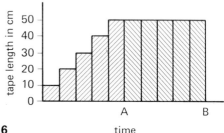

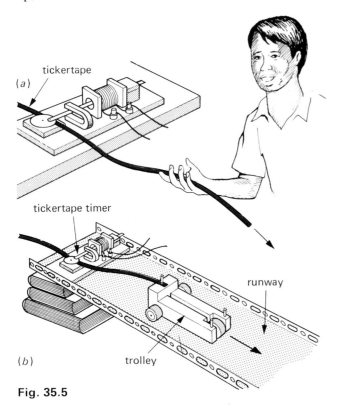

Fig. 35.5

(b) Trolley on a sloping runway. Attach a length of tape to a trolley and release it at the top of a runway, Fig. 35.5b. The dots will be very crowded at the start—ignore them; but beyond cut the tape into tentick lengths.

Make a tape chart. Is the acceleration uniform? What is its average value?

Q Questions

1. What is the average speed of **(a)** a car which travels 400 m in 20 s, **(b)** an athlete who runs 1500 m in 4 minutes?

2. A train increases its speed *steadily* from 10 m/s to 20 m/s in 1 minute.
    (a) What is its average speed during this time in m/s?
    (b) How far does it travel while increasing its speed?

3. A motor cyclist starts from rest and reaches a speed of 6 m/s after travelling with uniform acceleration for 3 s. What is his acceleration?

4. Each strip in the tape chart of Fig. 35.6 is for a time interval of 1 tentick.

Fig. 35.6

    (a) If the timer makes 50 dots per second, what time intervals are represented by OA and AB?
    (b) What is the acceleration between O and A in **(i)** cm/tentick², **(ii)** cm/s per tentick, **(iii)** cm/s²?
    (c) What is the acceleration between A and B?

5. An aircraft travelling at 600 km/h accelerates steadily at 10 km/h per second. Taking the speed of sound as 1100 km/h at the aircraft's altitude, how long will it take to reach the 'sound barrier'?

6. A vehicle moving with a uniform acceleration of 2 m/s² has a velocity of 4 m/s at a certain time. What will its velocity be **(a)** 1 s later, **(b)** 5 s later?

7. If a bus travelling at 20 m/s is subject to a steady deceleration of 5 m/s², how long will it take to come to rest?

8. The tape in Fig. 35.7 was pulled through a timer by a trolley travelling down a runway. It was marked off in tentick lengths.
    (a) What can you say about the trolley's motion?
    (b) Find its acceleration in cm/s².

Fig. 35.7

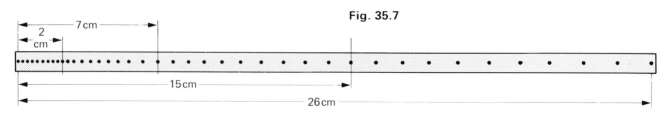

100

36 Equations of motion

Equations for uniform acceleration

Problems on bodies moving with *uniform acceleration* can often be solved quickly using the equations of motion.

First equation. If a body is moving with uniform acceleration a and its velocity increases from u to v in time t, then

$$a = \frac{\text{change of velocity}}{\text{time taken}} = \frac{v-u}{t}$$

$$\therefore \quad at = v-u$$
$$\text{or} \quad v = u+at \qquad (1)$$

Note that the initial velocity u and the final velocity v refer to the start and the finish of the *timing* and do not necessarily mean the start and finish of the motion.

Second equation. The velocity of a body moving with uniform acceleration increases steadily. Its average velocity therefore equals half the sum of its initial and final velocities, that is

$$\text{average velocity} = \frac{u+v}{2}$$

From (1) $\qquad\qquad v = u+at$

$$\therefore \quad \text{average velocity} = \frac{u+u+at}{2} = \frac{2u+at}{2}$$

$$= u+\tfrac{1}{2}at$$

If s is the distance moved in time t, then since average velocity = distance/time = s/t,

$$\frac{s}{t} = u+\tfrac{1}{2}at$$

$$\therefore \quad s = ut+\tfrac{1}{2}at^2 \qquad (2)$$

Third equation. This is obtained by eliminating t from equations (1) and (2). We have

$$v = u+at$$
$$\therefore \quad v^2 = u^2+2uat+a^2t^2$$
$$= u^2+2a(ut+\tfrac{1}{2}at^2)$$

But $\qquad s = ut+\tfrac{1}{2}at^2$
$$\therefore \quad v^2 = u^2+2as \qquad (3)$$

If we know any *three* of u, v, a, s and t, the others can be found from the equations.

✎ Worked example

A cyclist starts from rest and accelerates at 1 m/s² for 20 seconds. He then travels at a constant speed for 1 minute and finally decelerates at 2 m/s² until he stops. Find his maximum speed in km/h and the total distance covered in metres.

First stage

$$u = 0 \quad a = 1 \text{ m/s}^2 \quad t = 20 \text{ s}$$

We have $v = u+at = 0+1\times20 = 20$ m/s

$$= \frac{20}{1000}\times60\times60 = 72 \text{ km/h}$$

The distance s moved in the first stage is given by

$$s = ut+\tfrac{1}{2}at^2 = 0\times20+\tfrac{1}{2}\times1\times20^2 = 200 \text{ m}$$

Second stage

$$u = 20 \text{ m/s (constant)} \quad t = 60 \text{ s}$$

Distance moved = speed × time = $20\times60 = 1200$ m

Third stage

$$u = 20 \text{ m/s} \quad v = 0 \quad a = -2 \text{ m/s}^2 \text{ (a deceleration)}$$

We have $\quad v^2 = u^2+2as$

$$\therefore \quad s = \frac{v^2-u^2}{2a} = \frac{0-(20)^2}{2\times(-2)} = \frac{-400}{-4} = 100 \text{ m}$$

Answers

Maximum speed = 72 km/h
Total distance moved = $200+1200+100 = 1500$ m

Velocity–time graphs

If the velocity of a body is plotted against the time, the graph obtained is a velocity-time graph. It provides another way of solving motion problems. Tape charts are crude velocity-time graphs which show the velocity changing in jumps rather than smoothly, as occurs in practice.

(a) *The area under a velocity-time graph measures the distance travelled.*

In Fig. 36.1 AB is the velocity-time graph for a body moving with a *uniform velocity* of 20 m/s. Since distance = average velocity × time, after 5 s it will have moved 20 m/s × 5 s = 100 m. This is the shaded area under the graph, i.e. rectangle OABC.

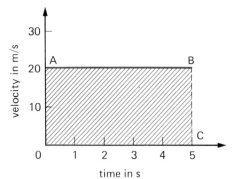

Fig. 36.1

In Fig. 36.2, PQ is the velocity-time graph for a body moving with *uniform acceleration*. At the start of the timing the velocity is 20 m/s but it increases steadily to 40 m/s after 5 s. If the distance covered equals the area under PQ, i.e. the shaded area OPQS, then

distance = area of rectangle OPRS +
area of triangle PQR
= OP × OS + $\frac{1}{2}$ × PR × QR
(area of a triangle = $\frac{1}{2}$ base × height)
= 20 m/s × 5 s + $\frac{1}{2}$ × 5 s × 20 m/s
= 100 + 50 = 150 m

We can check this using $s = ut + \frac{1}{2}at^2$. We have $u = 20$ m/s, $t = 5$ s, $a = (v-u)/t = (40-20)/5 = 4$ m/s². Hence

$s = 20 \times 5 + \frac{1}{2} \times 4 \times 5^2 = 100 + 50 = 150$ m

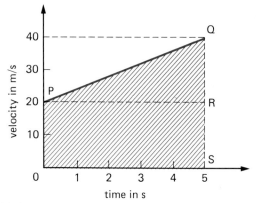

Fig. 36.2

Notes. 1. When calculating the area from the graph the unit of time must be the same on both axes.
2. This rule for finding distances travelled is true even if the acceleration is not uniform.

(b) *The slope or gradient of a velocity-time graph represents the acceleration of the body.*

In Fig. 36.1 the slope of AB is zero, as is the acceleration. In Fig. 36.2 the slope of PQ is QR/PR = 20/5 = 4: the acceleration is 4 m/s².

102

Distance–time graphs

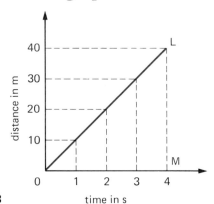

Fig. 36.3

A body travelling with uniform velocity covers equal distances in equal times. Its distance-time graph is a straight line, like OL in Fig. 36.3 for a velocity of 10 m/s. The slope of the graph = LM/OM = 40/4 = 10, which is the value of the velocity. The following statement is true in general.

The slope or gradient of a distance-time graph represents the velocity of the body.

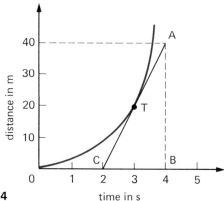

Fig. 36.4

When the velocity of the body is changing, the slope of the distance-time graph varies, Fig. 36.4, and at any point equals the slope of the tangent. For example, the slope of the tangent at T is AB/BC = 40/2 = 20. The velocity at the instant corresponding to T is therefore 20 m/s.

Ⓠ Questions

1. A body starts from rest and reaches a speed of 5 m/s after travelling with uniform acceleration in a straight line for 2 s. Calculate the acceleration of the body. (*E.A.*)

2. A body starts from rest and moves with a uniform acceleration of 2 m/s² in a straight line.
    (a) What is its velocity after 5 s?
    (b) How far has it travelled in this time?
    (c) After how long will the body be 100 m from its starting point? (*J.M.B./A.L./N.W.16+*)

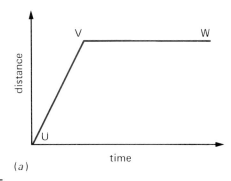

(a)

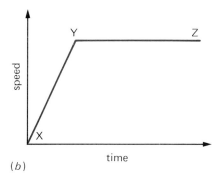

(b)

Fig. 36.5

3. A car accelerates from 4 m/s to 20 m/s in 8 s. How far does it travel in this time?

4. A motor cyclist travelling at 12 m/s decelerates at 3 m/s².
   (a) How long does he take to come to rest?
   (b) How far does he travel in coming to rest?

5. Fig. 36.5a shows the distance-time graph for a moving object and Fig. 36.5b shows the speed-time graph for another moving object.

Describe the motion, if any, of the objects in the regions (a) UV, (b) VW, (c) XY, (d) YZ. (E.A.)

6. The graph in Fig. 36.6 represents the distance travelled by a car plotted against time.

(a) How far has the car travelled at the end of 5 seconds?
(b) What is the speed of the car during the first 5 seconds?
(c) What has happened to the car after A?
(d) Plot a graph showing the speed of the car plotted against time during the first 5 seconds.

7. Fig. 36.7 shows an uncompleted velocity-time graph for a boy running a distance of 100 m.
(a) What is his acceleration during the first 4 seconds?
(b) How far does the boy travel during (i) the first 4 seconds, (ii) the next 9 seconds?
(c) Copy and complete the graph showing clearly at what time he has covered the distance of 100 m. Assume his speed remains constant at the value shown by the horizontal portion of the graph.

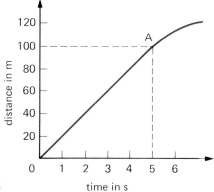

Fig. 36.6

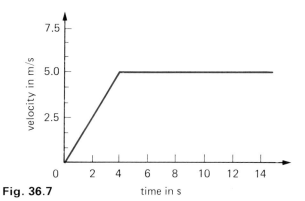

Fig. 36.7

check list P.285

37 Falling bodies

Fig. 37.2

In air, a coin falls faster than a bit of paper. In a vacuum they fall at the same rate as may be shown with the apparatus of Fig. 37.1. The difference, in air, is due to *air resistance* having a greater effect on light bodies than on heavy bodies. The air resistance to a light body is big when compared with the body's weight. With a dense piece of metal the resistance is negligible at low speeds.

Fig. 37.1

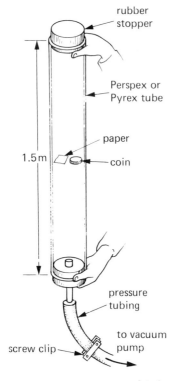

rubber stopper

Perspex or Pyrex tube

paper

coin

1.5 m

pressure tubing

screw clip

to vacuum pump

There is a story, untrue we now think, that in the sixteenth century the Italian scientist Galileo dropped a small iron ball and a large cannon ball ten times heavier, from the top of the Leaning Tower of Pisa, Fig. 37.2. And we are told, to the surprise of onlookers who expected the cannon ball to arrive first, they reached the ground almost simultaneously.

You will learn more about air resistance in the next topic.

⋀⋀⋀ Experiment: motion of a falling body

Arrange things as in Fig. 37.3 and investigate the motion of a 100 g mass falling from a height of about 2 m.

Construct a tape chart using *one tick* lengths. The dots at the start will be too close together but choose

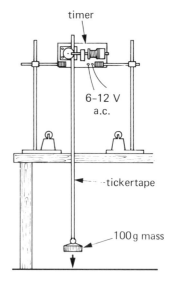

timer

6–12 V a.c.

tickertape

100 g mass

Fig. 37.3

as dot '0' the first one you can see clearly. What does the tape chart tell you about the motion of the falling mass?

Acceleration due to gravity

All bodies falling freely under the force of gravity do so with *uniform acceleration* if air resistance is negligible (i.e. the 'steps' in the previous tape chart should all be equal).

104

This acceleration, called the *acceleration due to gravity*, is denoted by the italic letter g. Its value varies over the earth. In Britain it is about 9.8 m/s^2 or near enough 10 m/s^2. The velocity of a free-falling body therefore increases by 10 m/s every second. A ball shot straight upwards with a velocity of 30 m/s decelerates by 10 m/s every second and reaches its highest point after 3 s.

In calculations using the equations of motion, g replaces a. It is given a positive sign for falling bodies (i.e. $a = +10$ m/s^2) and a negative sign for rising bodies since they are decelerating (i.e. $a = -10$ m/s^2).

Measuring g

Using the arrangement in Fig. 37.4 the time for a steel ball-bearing to fall a known distance is measured by an electric stop clock.

Fig. 37.4

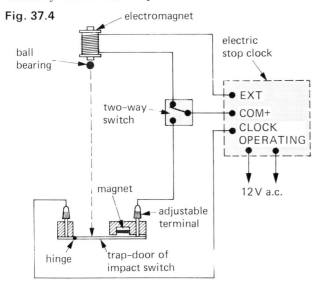

When the two-way switch is changed to the 'down' position, the electromagnet releases the ball and simultaneously the clock starts. At the end of its fall the ball opens the 'trap-door' on the impact switch and the clock stops.

The result is found from $s = ut + \frac{1}{2}at^2$, where s is the distance fallen (in m), t is the time taken (in s), $u = 0$ (the ball starts from rest) and $a = g$ (in m/s^2). Hence

$$s = \frac{1}{2}gt^2 \quad \text{or} \quad g = 2s/t^2$$

Air resistance is negligible for a dense object such as a steel ball-bearing falling a short distance.

✎ Worked example

A ball is projected vertically upwards with an initial velocity of 30 m/s. Find (a) its maximum height, (b) the time taken to return to its starting point. Neglect air resistance and take $g = 10$ m/s^2.

(a) We have $u = 30$ m/s, $a = -10$ m/s^2 (a deceleration) and $v = 0$ since the ball is momentarily at rest at its highest point. Substituting in

$$v^2 = u^2 + 2as$$

$$\therefore \quad 0 = 30^2 + 2(-10)s \quad \text{or} \quad -900 = -20s$$

$$\therefore \quad s = \frac{-900}{-20} = 45 \text{ m}$$

(b) If t is the time to reach the highest point, we have from

$$v = u + at$$

$$\therefore \quad 0 = 30 + (-10)t \quad \text{or} \quad -30 = -10t$$

$$\therefore \quad t = \frac{-30}{-10} = 3 \text{ s}$$

The downward trip takes exactly the same time as the upward one and so the answer is 6 s.

Distance–time graphs

For a body falling freely from rest we have

$$s = \frac{1}{2}gt^2$$

A graph of s against t is shown in Fig. 37.5a and for s against t^2 in Fig. 37.5b. The latter is a straight line through the origin since $s \propto t^2$ (g being constant at one place).

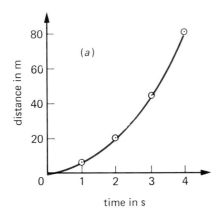

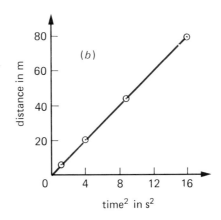

Fig. 37.5

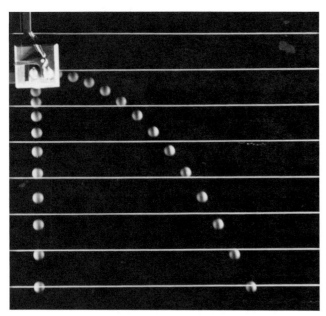

Fig. 37.6

Projectiles

The photograph in Fig. 37.6 was taken whilst a lamp emitted regular flashes of light. One ball was *dropped* from rest and the other, a 'projectile', was *thrown sideways* at the same time. Their vertical accelerations (due to gravity) are equal, showing that a projectile falls like a body which is dropped from rest. Its horizontal velocity does not affect its vertical motion.

The horizontal and vertical motions of a body are independent and can be treated separately.

For example if a ball is *thrown* horizontally from the top of a cliff and takes 3 s to reach the beach below we can calculate the height of the cliff by considering the vertical motion only. We have $u = 0$ (since the ball has no vertical velocity initially), $a = g = +10 \text{ m/s}^2$ and $t = 3$ s. The height s of the cliff is given by $s = ut + \frac{1}{2}at^2 = 0 \times 3 + \frac{1}{2}(+10)3^2 = 45$ m.

Q Questions

1. A stone falls from rest from the top of a high tower. Ignoring air resistance and taking $g = 10 \text{ m/s}^2$
   (a) what is its velocity after (i) 1 s, (ii) 2 s, (iii) 3 s, (iv) 5 s?
   (b) how far has it fallen after (i) 1 s, (ii) 2 s, (iii) 3 s, (iv) 5 s?

2. A body falls from a cliff 125 m high. Find the time of fall. How far did the body fall in half this time? (Take the acceleration due to gravity as 10 m/s^2.) *(E.A.)*

3. A mass is projected upwards with a velocity of 10 m/s. If the acceleration due to gravity is 10 m/s^2, what is the maximum height reached in metres?

   A 1   B 5   C 10   D 20   E 100   *(N.I.)*

4. An object is dropped from a helicopter at a height of 45 m above the ground.
   (a) If the helicopter is at rest, how long does the object take to reach the ground and what is its velocity on arrival?
   (b) If the helicopter is falling with a velocity of 1 m/s when the object is released, what will be the final velocity of the object? ($g = 10 \text{ m/s}^2$.)

5. An object is released from an aircraft travelling horizontally with a constant velocity of 200 m/s at a height of 500 m. Ignoring air resistance and taking $g = 10 \text{ m/s}^2$ find
   (a) how long it takes the object to reach the ground,
   (b) the horizontal distance covered by the object between leaving the aircraft and reaching the ground.

6. A gun pointing vertically upwards is fired from an open car B moving with uniform velocity. When the bullet returns to the level of the gun, car B has travelled to C and another car A has reached the position occupied by B when the gun was fired, Fig. 37.7. Are the occupants of A or B in danger?

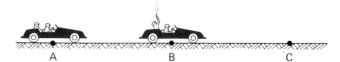

A   B   C

Fig. 37.7

check list p.285

38 Newton's laws of motion

First law

Friction and air resistance cause a car to come to rest when the engine is switched off. If these forces were absent we believe that a body, once set in motion, would go on moving forever with a constant speed in a straight line. That is, force is not needed to keep a body moving with uniform velocity so long as no opposing forces act on it.

This idea was proposed by Galileo and is summed up in Newton's first law.

A body stays at rest, or if moving it continues to move with uniform velocity, unless an external force makes it behave differently.

It seems that the question we should ask about a moving body is not 'what keeps it moving' but 'what changes or stops its motion'. The smaller the external forces opposing a moving body, the smaller is the force needed to keep it moving with uniform velocity. Friction is much reduced for a hovercraft floating on a cushion of air, Fig. 38.1.

Fig. 38.1

Mass and inertia

The first law is another way of saying that all matter has a built-in opposition to being moved if it is at rest or, if it is moving, to having its motion changed. This property of matter is called *inertia* (from the Latin word for laziness).

Its effect is evident on the occupants of a car which stops suddenly; they lurch forward in an attempt to continue moving, and this is why seat belts are needed. The reluctance of a stationary object to move can be shown by placing a large coin on a piece of card on your finger, Fig. 38.2. If the card is flicked *sharply* the coin stays where it is while the card flies off.

Fig. 38.2

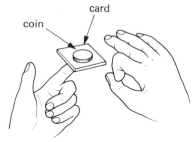

The larger the mass of a body the greater is its inertia, i.e. the more difficult is it to move when at rest and to stop when in motion. Because of this we consider that *the mass of a body measures its inertia*. This is a better definition of mass than the one given earlier (p. 56) in which it was stated to be the 'amount of matter' in a body.

◿◿ Experiment: effect of force and mass on acceleration

The apparatus consists of a trolley to which a force is applied by a stretched length of elastic, Fig. 38.3. The velocity of the trolley is found from a timer.

First compensate the runway for friction by raising one end until the trolley runs down with uniform velocity when given a push. The dots on the tape

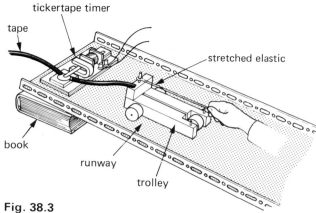

Fig. 38.3

108

should be equally spaced. There is now no resultant force on the trolley and any acceleration produced later will be due only to the force caused by the stretched elastic.

(a) Force and acceleration (mass constant). Fix one end of an elastic to the rod at the back of the trolley and stretch it until the other end is level with the front of the trolley. Before attaching a tape, practise pulling the trolley down the runway, keeping the same stretch on the elastic. After a few trials you should be able to produce a steady accelerating force. Now do it with a tape.

Repeat with fresh tapes using first two and then three *identical* elastics, stretched side by side by the same amount, to give two and three units of force.

Make a tape chart for each force and use it to find the acceleration produced in cm/tentick^2 (see p. 99). Ignore the start of the tape (where the dots are too close) and the end (where the force may not be steady). Does a steady force cause a steady acceleration? Put the results in a table. Do they suggest any relationship between a and F?

Force (*F*) (no. of elastics)	1	2	3
Acceleration (*a*) (cm/tentick^2)			

(b) Mass and acceleration (force constant). Do the experiment as in (*a*) using two elastics (i.e. constant *F*) to accelerate first one trolley, then two (stacked one above the other) and finally three. Check the friction compensation of the runway each time.

Find the accelerations from the tape charts and tabulate the results. Do they suggest any relationship between a and m?

Mass (*m*) (no. of trolleys)	1	2	3
Acceleration (*a*) (cm/tentick^2)			

Second law

The previous experiment should show roughly that the acceleration a is

(i) directly proportional to the applied force F for a fixed mass, i.e. $a \propto F$, and

(ii) inversely proportional to the mass m for a fixed force, i.e. $a \propto 1/m$.

Combining the results into one equation, we get

$$a \propto F/m \quad \text{or} \quad F \propto ma.$$

Therefore $F = kma$

where k is the constant of proportionality.

One newton is defined as the force which gives a mass of 1 kg an acceleration of 1 m/s².

Hence if $m = 1$ kg and $a = 1$ m/s² then $F = 1$ N. Substituting in $F = kma$ we get $k = 1$ and so we can write

$$F = ma$$

This is Newton's second law of motion. When using it two points should be noted. First, F is the resultant (or unbalanced) force causing the acceleration a. Second, F must be in newtons, m in kilograms and a in metres per second squared, otherwise k is not 1.

You should now appreciate that when the forces acting on a body do not balance, i.e. there is a net (resultant) force, they cause a *change* of motion, i.e. the body accelerates or decelerates. However, if they balance, there is no change of motion but there may be a change of shape. In that case internal forces in the body (i.e. forces between neighbouring atoms) balance the external forces.

 Worked examples

1. A block of mass 2 kg is pushed along a table with a constant velocity by a force of 5 N. When the push is increased to 9 N what is (a) the resultant force, (b) the acceleration?

When the block moves with constant velocity the forces acting on it are balanced. The force of friction opposing its motion must therefore be 5 N.

(a) When the push is increased to 9 N the resultant (unbalanced) force F on the block is $(9-5)\text{N} = 4$ N (since the frictional force is still 5 N).

(b) The acceleration a is obtained from $F = ma$ where $F = 4$ N and $m = 2$ kg.

$$\therefore \quad a = F/m = 4/2 = 2 \text{ m/s}^2$$

2. A car of mass 1200 kg travelling at 72 km/h is brought to rest in 4 s. Find the (a) average deceleration, (b) average braking force, (c) distance moved during the deceleration.

(a) The deceleration is found from $v = u + at$

where $v = 0$

$$u = 72 \text{ km/h} = \frac{72 \times 1000}{60 \times 60} = 20 \text{ m/s}$$

and $t = 4$ s.

Hence $0 = 20 + a \times 4$ or $-20 = 4a$

$$\therefore \quad a = -20/4 = -5 \text{ m/s}^2$$

(b) The average braking force F is given by $F = ma$ where $m = 1200$ kg and $a = -5$ m/s². Therefore

$$F = 1200 \times (-5) = -6000 \text{ N}$$

(c) To find the distance moved s we use $s = ut + \frac{1}{2}at^2$

$$\therefore \quad s = 20 \times 4 + \frac{1}{2} \times (-5) \times (4^2)$$
$$= 80 - 40 = 40 \text{ m}$$

Weight and gravity

The weight W of a body is the force of gravity acting on it which gives it an acceleration g when it is falling freely near the earth's surface. If the body has mass m, then W can be calculated from $F = ma$ if we put $F = W$ and $a = g$ to give

$$W = mg$$

Taking $g = 10$ m/s² and $m = 1$ kg, gives $W = 10$ N, i.e. a body of mass 1 kg has weight 10 N. Similarly a body of mass 2 kg has weight 20 N and so on. While the mass of a body is always the same, its weight varies depending on the value of g. On the moon the acceleration due to gravity is only about 1.6 m/s², and so a mass of 1 kg has a weight of just 1.6 N there.

The weight of a body is directly proportional to its mass, which explains why g is the same for all bodies. The greater the mass of a body, the greater is the force of gravity on it but it does not accelerate faster when falling because of its greater inertia (i.e. its greater resistance to acceleration).

Gravitational field

The force of gravity acts through space and can cause a body, not in contact with the earth, to fall to the ground. It is an invisible, action-at-a-distance force. We try to 'explain' its existence by saying that the earth is surrounded by a *gravitational field* which exerts a force on any body in the field. Later, magnetic and electric fields will be considered.

The strength of a gravitational field is defined as the force acting on unit mass in the field.

Measurement shows that on the earth's surface a mass of 1 kg experiences a force of 9.8 N, i.e. its weight is 9.8 N. The strength of the earth's field is therefore 9.8 N/kg (near enough 10 N/kg). It is denoted by g, the letter also used to denote the acceleration due to gravity. Hence

$$g = 9.8 \text{ N/kg} = 9.8 \text{ m/s}^2$$

109

We now have two ways of regarding *g*. When considering bodies *falling freely* we can think of it as an acceleration of 9.8 m/s², but when a body of known mass is *at rest* and we wish to know the force of gravity (in N) acting on it we think of *g* as the earth's gravitational field strength of 9.8 N/kg.

Third law

If a body A exerts a force on body B, then body B exerts an equal but opposite force on body A.

The law states that forces never occur singly but always in pairs as a result of the action between two bodies. For example, when you step forward from rest your foot pushes backwards on the earth and the earth exerts an equal and opposite force forward on you. Two bodies and two forces are involved. The small force you exert on the large mass of the earth gives no noticeable acceleration to the earth but the equal force it exerts on your very much smaller mass causes you to accelerate.

Note that the equal and opposite forces *do not act on the same body*; if they did, there could never be any resultant forces and acceleration would be impossible.

An appreciation of the third law and the effect of friction is desirable when stepping from a rowing boat, Fig. 38.4. You push backwards on the boat and, although the boat pushes you forwards with an equal force, it is itself now moving backwards (because friction with the water is slight), and this reduces your forwards motion by the same amount—and you might fall in!

Fig. 38.4

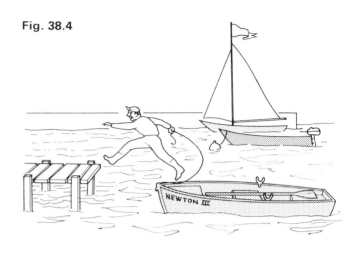

Air resistance: terminal velocity

When an object falls in air, the air resistance (fluid friction) opposing its motion *increases as its speed rises*, so reducing its acceleration. Eventually, air resistance acting upwards equals the weight of the object acting downwards. The resultant force on the object is then zero since the two opposing forces balance. The object falls at a constant velocity, called its *terminal velocity*, whose value depends on the size, shape and weight of the object.

A small dense object, e.g. a steel ball bearing, has a high terminal velocity and falls a considerable distance with a constant acceleration of 9.8 m/s² before air resistance equals its weight. A light object, e.g. a raindrop, or one with a large surface area, e.g. a parachute, has a low terminal velocity and only accelerates over a comparatively short distance before air resistance equals its weight. A sky diver,

Fig. 38.5

Fig. 38.5, has a terminal velocity of more than 50 m/s (100 m.p.h.).

Objects moving in viscous liquids, e.g. motor oil, behave similarly.

Explanation of Bernoulli's principle

In Fig. 33.9*b* (p. 92) the liquid speeds up going from the wide part A of the tube to the narrower part B, i.e. it is accelerated. Therefore, since $F = ma$, the force at A, and so also the pressure at A, must be greater than the force and pressure at B. Between B and C the liquid slows down due to the pressure at C being greater than that at B.

Ⓠ Questions

1. Which one of the diagrams in Fig. 38.6 shows the arrangement of forces which gives the block M the greatest acceleration?

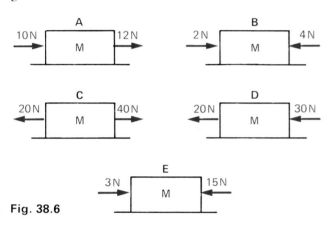

A
10N M 12N

B
2N M 4N

C
20N M 40N

D
20N M 30N

E
3N M 15N

Fig. 38.6

2. In Fig. 38.7 if *P* is a force of 20 N and the object moves with *constant velocity* what is the value of the opposing force *F*?

P   OBJECT   F

Fig. 38.7

3. (a) What resultant force produces an acceleration of 5 m/s² in a car of mass 1000 kg?
   (b) What acceleration is produced in a mass of 2 kg by a resultant force of 30 N?

4. A car of mass 500 kg accelerates steadily from rest to 40 m/s in 20 s.
   (a) What is its acceleration?
   (b) What resultant force produces this acceleration?
   (c) The actual force will be greater. Why?

5. A block of mass 500 g is pulled from rest on a horizontal frictionless bench by a steady force *F* and travels 8 m in 2 s. Find (a) the acceleration, (b) the value of *F*.

6. Starting from rest on a level road a boy can reach a speed of 5 m/s in 10 s on his bicycle. Find (i) the acceleration, (ii) the average speed during the 10 s and (iii) the distance he travels in 10 s.

Eventually, even though he is still pedalling as fast as he can, he stops accelerating and his speed reaches a maximum value. Explain in terms of the forces acting why this happens.

7. A trailer of mass 1000 kg is towed by means of a rope attached to a car moving at a steady speed along a level road. The tension in the rope is 400 N. Why is it not zero?

The car starts to accelerate steadily. If the tension in the rope is now 1650 N, with what acceleration is the trailer moving?   (O. and C.)

8. What does an astronaut of mass 100 kg weigh (a) on earth where the gravitational field strength is 10 N/kg, (b) on the moon where the gravitational field strength is 1.6 N/kg?

9. A rocket has a mass of 500 kg.
   (a) What is its weight on earth where $g = 10$ N/kg?
   (b) At lift-off the rocket engine exerts an upward force of 25 000 N. What is the resultant force on the rocket? What is its initial acceleration?

10. Fig. 38.8 shows the forces acting on a raindrop which is falling to the ground.

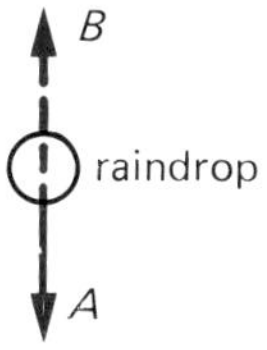

B

⬆ raindrop

A

Fig. 38.8

   (a) (i) *A* is the force which causes the raindrop to fall. What is this force called? (ii) *B* is the total force opposing the motion of the drop. State *one* possible cause of this force.
   (b) What happens to the raindrop when force *A* = force *B*?

111

39 Momentum

Momentum is a useful quantity to consider when bodies are involved in collisions and explosions. It is defined as the mass of the body multiplied by its velocity and is measured in *kilogram metre per second* (kg m/s).

MOMENTUM = MASS × VELOCITY

A 2 kg mass moving at 10 m/s has momentum 20 kg m/s, the same as the momentum of a 5 kg mass moving at 4 m/s.

〰 Experiment: collisions and momentum

Friction-compensate a runway as before (p. 108). Place one trolley at rest halfway down the runway and another at the top with a length of tickertape from it passing through a timer, Fig. 39.1. Each trolley should have a strip of Velcro (from haberdashery departments) fitted so that it 'sticks' to the other on collision.

Give the top trolley a good push. It will move forward with uniform velocity and should hit the second trolley so that they travel on as one.

From the tape find the velocity of the moving trolley before the collision and the common velocity of both trolleys after the collision (in cm/tentick say).

Repeat the experiment with another trolley stacked on top of the one to be pushed so that two are moving before the collision and three after.

Copy and complete the tables of results.

BEFORE COLLISION (m_2 at rest)

Mass m_1 (no. of trolleys)	Velocity v (cm/tentick)	Momentum $m_1 v$
1 2		

AFTER COLLISION (m_1 and m_2 together)

Mass $m_1 + m_2$ (no. of trolleys)	Velocity v_1 (cm/tentick)	Momentum $(m_1 + m_2)v_1$
2 3		

Do the results suggest any connection between the momentum before the collision and after it?

Conservation of momentum

When two or more bodies act on one another, as in a collision, the total momentum of the bodies remains constant, provided no external forces act (e.g. friction).

This statement is called the *principle of conservation of momentum*. Experiments like those above show that it is true for all types of collisions.

As an example, suppose a truck of mass 60 kg moving with velocity 3 m/s collides and couples with a stationary truck of mass 30 kg, Fig. 39.2a. The two move off together with the same velocity v which we can find as follows, Fig. 39.2b.

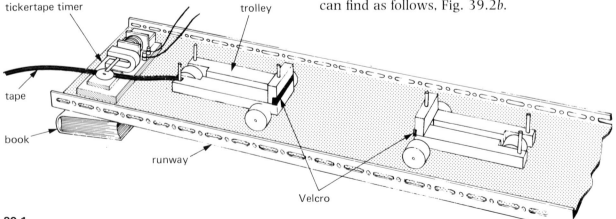

tickertape timer   trolley

tape

book

runway

Velcro

Fig. 39.1

112

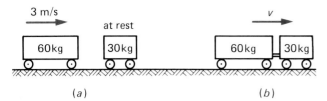

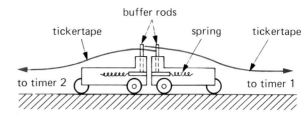

Fig. 39.2

Fig. 39.3

Total momentum before = $(60 \times 3 + 30 \times 0)$ kg m/s
$\qquad\qquad\qquad\qquad = 180$ kg m/s

Total momentum after $\;= (60 + 30)v$ kg m/s
$\qquad\qquad\qquad\qquad = 90v$ kg m/s

Since momentum is not lost

$90v = 180 \quad$ or $\quad v = 2$ m/s

Explosions

Momentum, like velocity, is a vector since it has both magnitude and direction. Vectors cannot be added by ordinary addition unless they act in the same direction. If they act in exactly opposite directions, e.g. east and west, the smaller subtracts from the greater, or if the same they cancel out.

Momentum is conserved in an explosion such as occurs when a rifle is fired. Before firing, the total momentum is zero since both rifle and bullet are at rest. During the firing the rifle and bullet receive *equal* but *opposite* amounts of momentum so that the *total* momentum after firing is zero.

For example, if a rifle fires a bullet of mass 0.01 kg with a velocity of 300 m/s then

forward momentum of bullet = 0.01 kg $\times$ 300 m/s
$\qquad\qquad\qquad\qquad\qquad = 3$ kg m/s

$\therefore \quad$ backward momentum of rifle = 3 kg m/s

If the rifle has mass m, it recoils (kicks back) with a velocity v such that

$mv = 3$ kg m/s

Taking $m = 6$ kg gives $v = 3/6 = 0.5$ m/s.

Experiment: explosions and momentum

The principle of conservation of momentum can be tested experimentally for 'explosions' with the apparatus in Fig. 39.3 arranged as shown. One tape from each trolley goes to a tickertape timer.

Tap one of the buffer rods to release the spring inside. The trolleys fly apart. Work out the velocities v_1 and v_2 of each from the tickertapes.

Since the trolleys are initially at rest

$\qquad$ total momentum before explosion = 0

If the trolleys have masses m_1 and m_2 then

$\qquad$ total momentum after explosion = $m_1v_1 - m_2v_2$

If the principle holds for explosions, $m_1v_1 - m_2v_2$ should be zero. (For trolleys of equal mass it simplifies matters to take '1 trolley' as the unit of mass (as before), then $m_1 = m_2 = 1$.)

Repeat the experiment with another trolley stacked on top of one of the trolleys so that, for example, $m_1 = 1$ and $m_2 = 2$.

Rockets and jets

If you release an inflated balloon with its neck open, it flies off in the opposite direction to that of the escaping air. In Fig. 39.4 the air has momentum to the left and the balloon moves to the right with equal momentum.

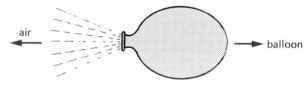

Fig. 39.4

This is the principle of rockets and jet engines. In both, a high-velocity stream of hot gas is produced by burning fuel and leaves the exhaust with large momentum. The rocket or jet engine itself acquires an equal forward momentum. Space rockets carry their own oxygen supply; jet engines use the surrounding air.

Force and momentum

If a steady force F acting on a body of mass m, increases its velocity from u to v in time t, the acceleration a is given by

$a = (v - u)/t \quad$ (from $v = u + at$)

Substituting for a in $F = ma$,

$$F = \frac{m(v-u)}{t} = \frac{mv - mu}{t}$$

$$\therefore \quad Ft = mv - mu$$

mv is the final momentum, mu the initial momentum and Ft is called the *impulse*. Therefore

$$force = \frac{change\ of\ momentum}{time} = \frac{rate\ of\ change\ of}{momentum}$$

This is another version of the second law of motion. For some problems it is more useful than $F = ma$.

Sport and momentum

The good cricketer or tennis player 'follows through' with the bat or racket when striking the ball, Figs. 39.5a, b. The force applied then acts for a longer time, the impulse is greater and so also is the gain of momentum (and velocity) of the ball.

(a)

Fig. 39.5

(b)

When we want to stop a moving ball such as a cricket ball, however, its momentum has to be reduced to zero. An impulse is then required in the form of an opposing force acting for a certain time. While any number of combinations of force and time will give a particular impulse, the 'sting' can be removed from the catch by drawing back the hands as the ball is caught. A smaller average force is then applied for a longer time.

[Q] Questions

1. What is the momentum in kg m/s of a 10 kg truck travelling at **(a)** 5 m/s, **(b)** 20 cm/s, **(c)** 36 km/h?

2. A ball X mass 1 kg travelling at 2 m/s has a head-on collision with an identical ball Y at rest. X stops and Y moves off. What is Y's velocity?

3. A boy with mass 50 kg running at 5 m/s jumps on to a 20 kg trolley travelling in the same direction at 1.5 m/s. What is their common velocity?

4. A 50 kg girl jumps out of a rowing boat of mass 300 kg on to the bank with a horizontal velocity of 3 m/s. With what velocity does the boat begin to move backwards?

5. A truck of mass 500 kg moving at 4 m/s collides with another truck of mass 1500 kg moving in the same direction at 2 m/s. What is their common velocity just after the collision if they move off together?

6. The velocity of a body of mass 10 kg increases from 4 m/s to 8 m/s when a force acts on it for 2 s.
    (a) What is the momentum before the force acts?
    (b) What is the momentum after the force acts?
    (c) What is the momentum gain per second?
    (d) What is the value of the force?

7. A rocket launched vertically sends out 50 kg of exhaust gases every second with a velocity of 200 m/s.
    (a) What is the upward force on the rocket?
    (b) If the mass of the rocket is 500 kg, what is its initial upward acceleration?

8. Explain why the crumple zones at the front and back of a car are designed to crumple up gradually when hit in a collision, rather than stay stiff and rigid.

check list p.285

40 K.E. and P.E.

Energy and its different forms were discussed earlier (p. 76). Here we will consider kinetic energy (k.e.) and potential energy (p.e.) in more detail.

Kinetic energy

Kinetic energy is the energy a body has because of its motion.

We can obtain an expression for k.e. Suppose a body of mass m starts from rest and is acted on by a steady force F which gives it a uniform acceleration a. If the velocity of the body is v when it has travelled a distance s, then, using $v^2 = u^2 + 2as$, we get, since $u = 0$,

$$v^2 = 2as \quad \text{or} \quad a = v^2/2s$$

Substituting in $F = ma$

$$F = m\left(\frac{v^2}{2s}\right) \qquad \text{or} \qquad Fs = \tfrac{1}{2}mv^2$$

Fs is the work done on the body to give it velocity v and therefore equals its k.e. Hence

KINETIC ENERGY $= E_k = \tfrac{1}{2}mv^2$

If m is in kg and v in m/s, then k.e. is in J. For example, a cricket ball of mass 0.2 kg (200 g) moving with velocity 20 m/s has k.e. $= \tfrac{1}{2}mv^2 = \tfrac{1}{2} \times 0.2 \times (20)^2 = 0.1 \times 400 = 40$ J.

Since k.e. depends on v^2, a vehicle, such as a High Speed Train, Fig. 40.1, travelling at 200 km/h (125 m.p.h.), has four times the k.e. it has at 100 km/h.

Fig. 40.1

Potential energy

Potential energy is the energy a body has because of its position or condition.

A body above the earth's surface is considered to have an amount of gravitational p.e. equal to the work that has been done against gravity by the force used to raise it. To lift a body of mass m through a *vertical* height h at a place where the earth's gravitational field strength is g, needs a force equal and opposite to the weight mg of the body. Hence

$$\text{work done by force} = \text{force} \times \text{vertical height}$$
$$= mg \times h$$
$$\therefore \quad \text{POTENTIAL ENERGY} = E_p = mgh$$

When m is in kg, g in N/kg (or m/s²) and h in m, the p.e. is in J. For example, if $g = 10$ N/kg, the p.e. gained by a 0.1 kg (100 g) mass raised vertically by 1 m is $0.1 \text{ kg} \times 10 \text{ N/kg} \times 1 \text{ m} = 1 \text{ N m} = 1 \text{ J}$.

Note. Strictly speaking we are concerned with changes in p.e. from the p.e. that a body has at the earth's surface, rather than with actual values. The expression for p.e. is therefore more correctly written

$$\Delta E_p = mgh$$

where Δ (pronounced delta) is the Greek capital D and stands for 'a change in'.

⩗ Experiment: change of p.e. to k.e.

Friction-compensate a runway and arrange the apparatus as in Fig. 40.2 with the bottom of the 0.1 kg (100 g) mass 0.5 m from the floor.

Start the timer and release the trolley. It will accelerate until the falling mass reaches the floor; after that it moves with *constant* velocity v.

From the tickertape measure v in m/s (50 ticks = 1 s). Find the mass of the trolley in kg. Work out:

k.e. gained by trolley and 0.1 kg mass =   J
p.e. lost by 0.1 kg mass =   J

Compare and comment on the results.

Conservation of energy

A mass m at height h above the ground has p.e. $= mgh$, Fig. 40.3. When it falls, its velocity increases and it gains k.e. at the expense of its p.e. If it

Fig. 40.3

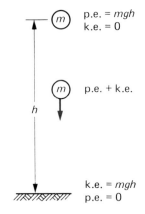

p.e. = mgh
k.e. = 0

p.e. + k.e.

k.e. = mgh
p.e. = 0

starts from rest and air resistance is negligible, its velocity v on reaching the ground is given by

$$v^2 = u^2 + 2as = 0 + 2gh = 2gh$$

Also, as it reaches the ground, its

$$\text{k.e.} = \tfrac{1}{2}mv^2 = \tfrac{1}{2}m \times 2gh = mgh$$
$$\therefore \quad \text{loss of p.e.} = \text{gain of k.e.}$$

This is an example of the *principle of conservation of energy*.

Energy may be changed from one form to another but it cannot be destroyed.

Fig. 40.2

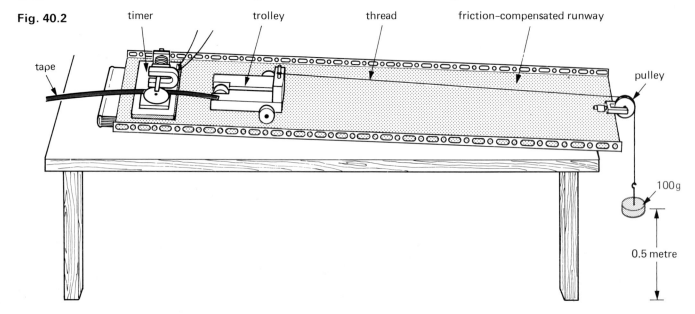

Fig. 40.4

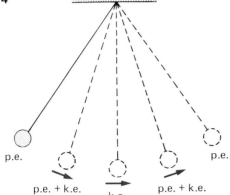

In a pendulum k.e. and p.e. are interchanged continually. The energy of the bob is all p.e. at the end of the swing and all k.e. as it passes through its central position. In other positions it has both p.e. and k.e., Fig. 40.4. Eventually all the energy is changed to heat as a result of overcoming air resistance.

Potential energy stored by a stretched spring

Energy has to be supplied to stretch a spring and is stored in the spring as p.e. So long as the elastic limit (p. 61) is not exceeded, the energy can usually be recovered completely. If the force–extension graph is a straight line, i.e. Hooke's law is obeyed, Fig. 40.5, it can be shown that when a force F (in newtons) caused an extension e (in metres), the p.e. stored (in joules) is given by

$$\text{p.e.} = \text{area } \Delta OAB = \tfrac{1}{2} AB \times OA = \tfrac{1}{2}F \times e$$

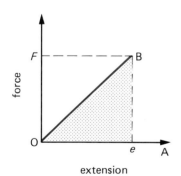

Fig. 40.5   extension

Worked example

A boulder of mass 4 kg rolls over a cliff and reaches the beach below with a velocity of 20 m/s.
  (a) What is the k.e. of the boulder just before it lands?
  (b) What is its p.e. on the cliff?
  (c) How high is the cliff?

(a) Mass of boulder $= m = 4$ kg
    Velocity of boulder as it lands $= v = 20$ m/s
$\therefore$ k.e. of boulder as it lands $= E_k = \tfrac{1}{2}mv^2$
$$= \tfrac{1}{2} \times 4 \times (20)^2$$
$$= 800 \text{ J}$$

(b) Applying the principle of conservation of energy (and neglecting energy lost in overcoming air resistance),

    p.e. of boulder on cliff = k.e. as it lands
$\therefore$ $\Delta E_p = E_k = 800$ J

(c) If h is the height of the cliff,

$\Delta E_p = mgh$

$$\therefore \quad h = \frac{\Delta E_p}{mg} = \frac{800}{4 \times 10} = 20 \text{ m}$$

Q Questions

1. Calculate the k.e. of
    (a) a 1 kg trolley travelling at 2 m/s,
    (b) a 2 g (0.002 kg) bullet travelling at 400 m/s,
    (c) a 500 kg minicar travelling at 72 km/h.

2. **(a)** What is the velocity of an object of mass 1 kg which has 200 J of k.e.?
    (b) Calculate the p.e. of a 5 kg mass when it is **(i)** 3 m, **(ii)** 6 m, above the ground. ($g = 10$ N/kg)

3. A 100 g steel ball falls from a height of 1.8 m on to a plate and rebounds to a height of 1.25 m. Find
    (a) the p.e. of the ball before the fall ($g = 10$ m/s²),
    (b) its k.e. as it hits the plate,
    (c) its velocity on hitting the plate,
    (d) its k.e. as it leaves the plate on the rebound,
    (e) its velocity of rebound.

4. A body of mass 5 kg falls from rest and has a k.e. of 1000 J just before touching the ground. Assuming there is no friction and using a value of 10 m/s² for the acceleration due to gravity, calculate
    (a) **(i)** the loss in p.e. during the fall, **(ii)** the height from which the body has fallen.
    (b) Name an important principle which applies in this situation.   (*J.M.B./A.L./N.W.16+*)

5. At what height above the ground must a mass of 5 kg be to have a p.e. equal in value to the k.e. possessed by a mass of 5 kg moving with a velocity of 10 m/s? (Assume $g = 10$ m/s².)

    A 1 m   **B** 5 m   **C** 10 m   **D** 50 m   **E** 100 m
                                                        (*J.M.B.*)

6. It is estimated that 7×10^6 kg of water pours over the Niagara Falls every second. If the Falls are 50 m high, and if all the energy of the falling water could be harnessed, what power would be available? ($g = 10$ N/kg)

check list   p.285

41 Circular motion

Fig. 41.1

Circular motion

There are many examples of bodies moving in circular paths—chair-o-planes at a fun fair, clothes in a spin dryer, the planets going round the sun and the moon circling the earth. 'Throwing the hammer' is a sport practised at Highland Games in Scotland, Fig. 41.1, in which the hammer is whirled round and round before it is released. When a car turns a corner it may follow an arc of a circle.

Centripetal force

In Fig. 41.2 a ball attached to a string is being whirled round in a horizontal circle. Its direction of motion is constantly changing. At A it is along the tangent at A; shortly afterwards, at B, it is along the tangent at B and so on.

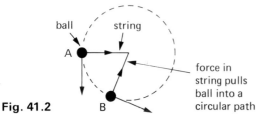

Fig. 41.2

ball   string

force in string pulls ball into a circular path

118

Velocity has both size and direction; speed has only size. Velocity is speed in a stated direction and if the direction of a moving body changes, even if its speed does not, then its velocity has changed. A change of velocity is an acceleration and so during its whirling motion the ball is accelerating.

It follows from Newton's first law of motion that *if we consider a body moving in a circle to be accelerating* then there must be a force acting on it to cause the acceleration. In the case of the whirling ball it is reasonable to say the force is provided by the string pulling inwards on the ball. Like the acceleration, the force acts towards the centre of the circle and keeps the body at a fixed distance from the centre.

A larger force is needed if

(a) the speed of the ball is increased,
(b) the radius of the circle decreases,
(c) the mass of the ball is increased.

The rate of change of direction, i.e. the acceleration, is increased by (a) and (b) and from $F = ma$, if m increases, F must. Should the force be greater than the string can bear, the string breaks and the ball flies off with steady speed in a straight line along the *tangent*, i.e. in the direction of travel when the string broke (as the first law of motion predicts). It is not thrown outwards.

The force which acts *towards the centre* and keeps a body in a circular path is called the *centripetal force* (centre-seeking force). Whenever a body moves in a circle (or circular arc) there must be a centripetal force acting on it. In throwing the hammer it is the pull of the athlete's arms acting on the hammer towards the centre of the whirling path. When a car rounds a bend a frictional force is exerted inwards by the road on the car's tyres.

Universal gravitation

Newton proposed that all objects in the universe attract each other with a force he called *gravitation.* He suggested that the gravitational attraction of the sun for the planets was the centripetal force which kept the planets in near-circular orbits round the sun. The gravitational attraction between two ordinary objects, e.g. two 1 kg masses 1 m apart, is extremely small and difficult to detect. The greater the masses of the objects and the smaller their separation, the more do they attract.

Newton regarded *gravity* as the gravitational attraction of the earth for nearby objects. He saw it as being responsible for holding the moon in its orbit round the earth.

Satellite motion

To put an artificial satellite in orbit at a certain height above the earth it must enter the orbit at the correct speed. If it does not, the force of gravity, which decreases with height, will not be equal to the centripetal force needed for the orbit.

This can be seen by imagining a shell fired horizontally from the top of a very high mountain, Fig. 41.3. If gravity did not pull it towards the centre of the earth it would continue to travel horizontally, taking

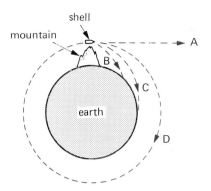

Fig. 41.3

path A. In practice it might take path B. A second shell fired faster might take path C and travel farther. If a third shell is fired even faster, it might never catch up with the rate at which the earth's surface is falling away. It would remain at the same height above the earth (path D) and return to the mountain top, behaving like a satellite.

How might a fourth shell behave if fired faster than the speed needed for orbit D?

Q Questions

1. An apple is whirled round in a horizontal circle on the end of a string which is tied to the stalk. It is whirled faster and faster and at a certain speed the apple is torn from the stalk. Why?

2. A car rounding a bend travels in an arc of a circle.
 (a) What provides the centripetal force?
 (b) Is a larger or a smaller centripetal force required if:
  (i) the car travels faster,
  (ii) the bend is less curved,
  (iii) the car has more passengers?

3. A space shuttle is in orbit round the earth at a certain height.
 (a) What keeps it in orbit?
 (b) If the shuttle reduces its mass by launching a communications satellite, how does this affect the centripetal force?
 (c) If the shuttle moves to a higher orbit does there have to be an increase or a decrease in (i) the centripetal force, (ii) the orbit speed?

check list
p.285

42 Additional questions

Core level (all students)

Velocity and acceleration: equations of motion

1. **(a)** A body travelling with uniform velocity covers a distance of 840 m in 1 minute. What is its velocity in m/s?
(b) If the body begins to accelerate uniformly at 0.2 m/s², what will be its velocity after a *further* minute?
(c) How far will the body have travelled from the instant at which it began to accelerate?
(d) If the body is then retarded uniformly and comes to rest in a *further* distance of 338 m, what is the value of the retardation?

(*J.M.B./A.L./N.W.16+*)

Falling bodies

2. A man drops, without throwing, an iron ball from the top of a tower and it takes 1.5 s to reach the ground. Which one of the following actions will result in a similar ball taking more than 1.5 s to reach the ground?

A The ball is thrown downwards from the top of a shorter tower.
B The ball is thrown horizontally with a speed of 10 m/s from the top of a shorter tower.
C The ball is thrown horizontally with a speed of 10 m/s from the top of the same tower.
D The ball is thrown upwards at an angle of 10° to the horizontal from the top of the same tower.
E The ball is thrown downwards at an angle of 10° to the horizontal from the top of the same tower.

(*O.L.E./S.R.16+*)

Newton's laws of motion

3. If the forces acting on a moving body cancel each other out (i.e. are in equilibrium) the body will

A move in a straight line at a steady speed
B slow down to a steady slower speed
C speed up to a steady faster speed
D be brought to a state of rest. (*W.M.*)

Momentum

4. Fig. 42.1 shows a truck of mass 3 kg moving at 10 m/s along a horizontal track about to collide with another stationary truck of mass 2 kg. After the collision, the trucks link and move together. The speed of the trucks immediately after the collision is

A 1 m/s B 2 m/s C 6 m/s D 6.67 m/s
E 10 m/s (*O.L.E./S.R.16+*)

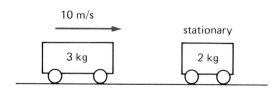

Fig. 42.1

Further level (higher grade students)

Velocity and acceleration: equations of motion

5. The distance from the starting point of a steel ball which starts from rest and rolls with uniform acceleration down an inclined plane is noted at one second intervals. The data are shown in the table.

Time in s	0	1	2	3	4
Distance in cm	0	2	8	18	32

The speed at the end of the fourth second in cm/s is
A 8   B 14   C 16   D 32   E 64   (*N.I.*)

6. **(a)** Define the terms 'velocity' and 'acceleration'. Choose *one* of these terms and explain what is meant when the quantity is said to be 'uniform'.
(b) A car runs at a constant speed of 15 m/s for 300 s and then accelerates uniformly to a speed of 25 m/s over a period of 20 s. This speed is maintained for 300 s before the car is brought to rest with uniform deceleration in 30 s.
Draw a velocity-time graph to represent the journey described above.
From the graph find
(i) the acceleration while the velocity changes from 15 m/s to 25 m/s,
(ii) the total distance travelled in the time described,
(iii) the average speed over the time described.

(*J.M.B.*)

Falling bodies

7. A stone is projected vertically upwards with an initial velocity of 90 m/s and at the same instant another stone is allowed to fall from rest from a height of 405 m. At what distance from the ground will the stones pass? ($g = 10$ m/s²: air resistance may be neglected.) (*N.I.*)

8. A stone is projected vertically upwards from the ground and is observed to pass a point situated at a height *H* above ground level at 4 s and again at 5 s after projection. Assuming that air resistance is negligible, determine:

(i) the velocity of projection of the stone
(ii) the height *H*
(iii) the velocity of the stone when at height *H*. Take $g = 10$ m/s^2. (*L.*)

Newton's laws of motion

9. (a) Define the newton.
Describe an experiment to investigate the relationship between the force applied to a trolley and the acceleration produced in the trolley.
(b) The reading on the speedometer of a car of mass 500 kg is 18 km/h when the brakes are applied. The car is brought to rest in 10 m. Find **(i)** the average retardation, **(ii)** the average braking force. (*J.M.B.*)

10. What is meant by *acceleration*? What evidence have you for believing that a moving body will proceed at a constant speed in a straight line indefinitely unless it is acted on by a resultant force?

A cyclist of mass 50 kg riding a cycle of mass 10 kg raises his speed from 2 m/s to 5 m/s by accelerating uniformly for 6 s while travelling due north along a horizontal road. What horizontal force is exerted on the machine by the road, and in which direction? Does the cyclist exert any force on the machine and, if so, in what direction?

How far does he travel in the 6 s? (*O. and C.*)

Momentum: K.E. and P.E.

11. A body of mass 20 kg, moving with uniform acceleration, has an initial momentum of 200 kg m/s and after 10 s the momentum is 300 kg m/s. What is the acceleration of the body?

    A 0.5 m/s^2 **B** 5 m/s^2 **C** 25 m/s^2
    D 50 m/s^2 **E** 100 m/s^2 (*J.M.B.*)

12. A small steel ball of mass 80 g is released from rest at a height of 1.25 m above a rigid horizontal metal plate. After the rebound the ball rises vertically to a height of 1.00 m above the plate. Calculate **(a)** the velocity of the ball just before impact, **(b)** the momentum of the ball just before impact, **(c)** the kinetic energy of the ball just before impact, **(d)** the loss of energy on impact. Give reasons for this loss of energy. ($g = 10$ m/s^2) (*S.*)

13. Fig. 42.2 shows a trolley A, of mass 3 kg with light frictionless wheels, which is held at rest on a smooth inclined plane. When it is released its centre of mass is lowered through a *vertical* distance of 1.25 m while it is accelerating down the slope. It then collides with a second similar trolley B of mass 2 kg which is at rest on a horizontal plane. After the collision the two trolleys travel forward as a single body.

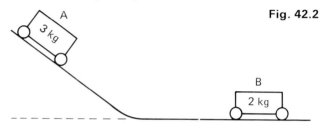

Fig. 42.2

(a) Calculate the speed of A immediately before the collision.
(b) Use the principle of conservation of momentum to find the common speed of the two trolleys after the collision.
(c) Calculate the kinetic energy of A just *before* the collision and the kinetic energy of the two trolleys *after* the collision.
(d) Explain why the kinetic energies before and after the collision are different. ($g = 10$ m/s^2) (*O.L.E.*)

Circular motion

14. A stone of mass 0.6 kg is whirled round in a horizontal circle at a constant speed.

(a) In what direction is the stone being accelerated at each point of its path?
(b) If the value of this acceleration of the stone is 10 m/s^2, what is the value of the horizontal force acting on the stone?

Heat and energy

43 Thermometers

The temperature of a body tells us how hot it is and is measured by a thermometer, usually in *degrees Celsius* (°C). The kinetic theory (Topic 22) regards temperature as a measure of the average k.e. of the molecules of the body. The greater this is, the faster do the molecules move and the higher is the temperature of the object.

There are different kinds of thermometer, each type being more suitable than another for a certain job. The one in Fig. 43.1 is being used to find the temperature of a furnace.

Liquid-in-glass thermometer

In this type the liquid in a glass bulb expands up a capillary tube when the bulb is heated. The liquid must be easily seen, expand (or contract) rapidly and by a large amount over a wide range of temperature. It must not stick to the inside of the tube or the reading will be low when the temperature is falling.

Mercury and coloured alcohol are in common use. Mercury freezes at −39 °C and boils at 357 °C; alcohol freezes at −115 °C and boils at 78 °C and is therefore more suitable for low temperatures.

Scale of temperature

A scale and unit of temperature are obtained by choosing two temperatures, called the *fixed points*, and dividing the range between them into a number of equal divisions or *degrees*.

Fig. 43.1

On the Celsius scale (named after the Swedish scientist who suggested it), *the lower fixed point is the temperature of pure melting ice* and is taken as 0 °C. Impurities in the ice would lower its melting point (p. 138).

The upper fixed point is the temperature of the steam above water boiling at normal atmospheric pressure of 760 mmHg and is taken as 100°C. The temperature of the boiling water itself is not used because any impurities in the water raise its boiling point and there is a variation in temperature below the surface increasing with depth; the temperature of the steam is not affected, however (p. 138).

Fig. 43.2

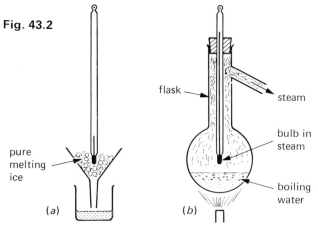

Methods of finding the fixed points are shown in Fig. 43.2a,b. When they have been marked on the thermometer, the distance between them is divided into 100 equal degrees, Fig. 43.3. The thermometer now has a scale, i.e. it has been calibrated or graduated.

Fig. 43.3

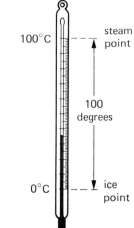

Clinical thermometer

This is a special type of mercury-in-glass thermometer used by doctors and nurses. Its scale only extends over a few degrees on either side of the normal body temperature of 37 °C, Fig. 43.4.

The tube has a constriction (i.e. a narrower part) just beyond the bulb. When the thermometer is placed under the tongue the mercury expands, forcing its

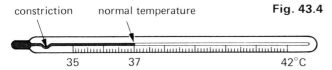

constriction   normal temperature   **Fig. 43.4**

35   37   42°C

way past the constriction. When the thermometer is removed (after 1 minute) from the mouth, the mercury in the bulb cools and contracts, breaking the mercury thread at the constriction. The mercury beyond the constriction stays in the tube and shows the body temperature. After use the mercury is returned to the bulb—by a flick of the wrist.

Thermocouple thermometer

A thermocouple consists of two wires of different materials, e.g. copper and iron, joined together, Fig. 43.5. When one junction is at a higher temperature than the other an electric current flows and produces a deflection on a galvanometer (a sensitive ammeter) which depends on the temperature difference.

Fig. 43.5

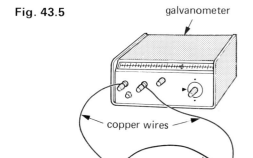

galvanometer

copper wires

hot   iron   cold
junction   wire   junction

Thermocouples are used in industry to measure a wide range of temperatures up to about 1 500 °C (Fig. 43.1), especially changing ones.

Heat and temperature

It is important not to confuse the temperature of a body with the heat energy that can be obtained from it. For example, a red-hot spark from a fire is at a higher temperature than the boiling water in a saucepan. If the spark landed in the water, heat would pass from it to the water even though much more heat-energy could be supplied by the water.

Heat is also called *thermal* or *internal* energy since it is the energy a body has because of the kinetic energy *and* the potential energy of its molecules. Increasing the temperature of a body increases its heat energy due to the k.e. of its molecules increasing. But as we will see later (Topic 47), the heat energy of a body can also be increased by increasing the p.e. of its molecules. In a pan of boiling water the molecules have less kinetic energy per molecule than in a spark but since there are many more water molecules, their total energy is greater.

Heat passes from a body at a higher temperature to one at a lower temperature. This is due to the average k.e. (and speed) of the molecules in the 'hot' body falling as a result of having collisions with molecules of the 'cold' body whose average k.e., and therefore temperature, increases. When the average k.e. of the molecules is the same in both bodies, they are at the same temperature.

In a solid, k.e. and p.e. are present in roughly equal amounts. In a gas, where the intermolecular forces are weak, the molecules have more k.e. than p.e.

Ⓠ Questions

1. 1 530 °C   120 °C   55 °C   37 °C   19 °C   0 °C
   − 12 °C   − 50 °C

From the above list of temperatures choose the most likely value for *each* of the following:
   (a) melting point of iron, **(b)** the temperature of a room that is comfortably warm, **(c)** the melting point of pure ice at normal pressure, **(d)** the lowest outdoor temperature recorded in London in winter, **(e)** the normal body temperature of a healthy person.

2. In order to make a mercury thermometer which will measure small changes in temperature accurately

   A decrease the volume of the mercury bulb
   B put the degree markings farther apart
   C decrease the diameter of the capillary tube
   D put the degree markings closer together
   E leave the capillary tube open to the air.

3. A simple thermometer has a stem and a bulb containing a liquid. Which one of the following will help the thermometer to rapidly register a new temperature?

   A a large bulb
   B a long stem
   C a thick-walled bulb
   D a liquid of high density
   E a thin-walled bulb   (O.L.E./S.R.16 +)

4. Why does a clinical thermometer **(a)** have a constriction just above the bulb, **(b)** cover only a narrow range of temperature, **(c)** have a very fine bore?   (E.A.)

123

44 Expansion of solids and liquids

In general, when matter is heated it expands and when cooled it contracts. If the changes are resisted large forces are created which are sometimes useful but at other times are a nuisance.

According to the kinetic theory (p. 62) the molecules of solids and liquids are in constant vibration. When heated they vibrate faster and force each other a little farther apart. Expansion results.

Uses of expansion

(a) Shrink fitting. In Fig. 44.1 the axle is being shrunk by cooling in liquid nitrogen at $-196\,°C$ until the gear wheel can be slipped on to it. On regaining normal temperature the axle expands to give a very tight fit.

(b) Riveting metal plates. A white-hot rivet is placed in the rivet hole and its end hammered flat. On

Fig. 44.2

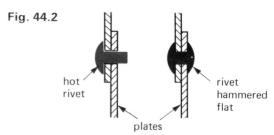

hot rivet

rivet hammered flat

plates

cooling it contracts and pulls the plates together, Fig. 44.2. Steel plates are riveted in shipbuilding.

Precautions against expansion

(a) Railway lines. Previously gaps were left between the lengths of rail to allow for expansion in summer. They caused a 'clickety-click' sound as the train passed over them.

Today rails are welded into lengths of about 1 km and are held by concrete sleepers that can withstand

Fig. 44.1

124

Fig. 44.3

the large forces created, without buckling. Also, at the joints the ends are tapered and overlap, Fig. 44.3. This gives a smoother journey and allows some expansion near the ends of each length of rail.

(b) Bridges. One end is fixed and the other rests on rollers which permit movement.

Bimetallic strip

If equal lengths of two different metals, e.g. copper and iron, are riveted together so that they cannot move separately, they form a bimetallic strip, Fig. 44.4a. When heated, copper expands more than iron and to allow this, the strip bends with copper on the outside, Fig. 44.4b. If they had expanded equally the strip would have stayed straight.

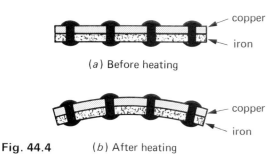

(a) Before heating

copper
iron

copper
iron

Fig. 44.4 (b) After heating

Bimetallic strips have many uses.

(a) Fire alarm, Fig. 44.5a. Heat from the fire makes the bimetallic strip bend and complete the electrical circuit, so ringing the alarm bell.

It is also used in this way to work the flashing direction indicator lamps in a car, being warmed by an electric heating coil wound round it.

(b) Thermostat. A thermostat keeps the temperature of a room or an appliance constant. The one in Fig. 44.5b uses a bimetallic strip in the electrical heating circuit of, for example, an electric iron.

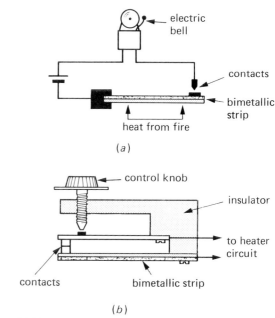

electric bell

contacts

bimetallic strip

heat from fire

(a)

control knob

insulator

to heater circuit

contacts

bimetallic strip

(b)

Fig. 44.5

When the iron reaches the required temperature the strip bends down, breaks the circuit at the contacts and switches off the heater. After cooling a little the strip remakes contact and turns the heater on again. A near-steady temperature results.

If the control knob is screwed down, the strip has to bend more to break the heating circuit and this needs a higher temperature.

125

Gas oven thermostat

Use is made of the fact that Invar, an alloy of steel and nickel, has a very small expansion.

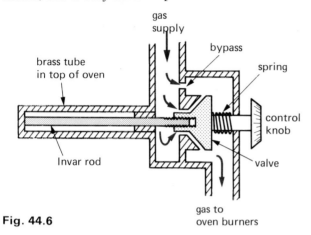

Fig. 44.6

The arrangement is shown in Fig. 44.6. As the temperature of the oven rises the brass tube expands to the left pulling the Invar rod with it. Since the expansion of the Invar is negligible, the gas supply through the valve to the burners is reduced. The control knob alters the original position of the valve and selects the steady temperature reached.

Linear expansivity

An engineer has to allow for the linear (lengthways) expansion of a bridge when designing it. The expansion can be calculated if he knows (i) the length of the bridge, (ii) the range of temperature it will experience and (iii) the *linear expansivity* of the material to be used.

The linear expansivity of a substance is the increase in length of unit length per degree rise in temperature.

The linear expansivity of a material is found by experiment. For steel it is 0.000 012 per °C. This means that 1 m will become 1.000 012 m for a temperature rise of 1 °C. A steel bridge 100 m long will expand by 0.000 012 × 100 m for each 1 °C rise in temperature. If the maximum temperature *change* expected is 60 °C (e.g. from −15 °C to +45 °C), the expansion will be 0.000 012 × 100 × 60 = 0.072 m = 7.2 cm. In general,

EXPANSION = LINEAR EXPANSIVITY × ORIGINAL
LENGTH × TEMPERATURE RISE

Measurement of linear expansivity

Rewriting the above expression we get

$$\frac{\text{linear}}{\text{expansivity}} = \frac{\text{expansion}}{(\text{original length}) \times (\text{temp. rise})}$$

126

To find the linear expansivity of a material we must therefore measure the expansion of a known length for a known temperature rise.

One form of apparatus is shown in Fig. 44.7. The original length of a rod of the material (about 50 cm) is measured with a metre rule. It is placed in the steam jacket with one end against the stop. The micrometer is adjusted until it just touches the other end. The reading is taken, as is the temperature of the rod.

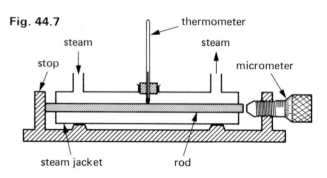

Fig. 44.7

The micrometer is unscrewed and steam passed through the jacket for several minutes. It is screwed up again and the new reading noted. This is repeated shortly afterwards to ensure it has not changed, showing that the rod is at the steam temperature— which is also taken. The difference in the micrometer readings equals the expansion of the rod. The linear expansivity can then be calculated.

Unusual expansion of water

As water is cooled to 4 °C it contracts, as we would expect. However between 4 °C and 0 °C it expands, surprisingly. *Water therefore has a maximum density at 4 °C.*

At 0 °C, when it freezes, a considerable expansion occurs and every 100 cm³ of water becomes 109 cm³ of ice—which accounts for the bursting of water pipes in very cold weather.

These changes are represented in Fig. 44.8.

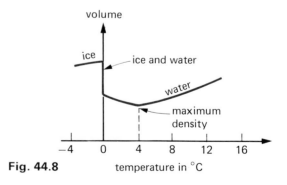

Fig. 44.8

The unusual expansion of water between 4 °C and 0 °C explains why fish survive in a frozen pond. The water at the top of the pond cools first, contracts and being denser sinks to the bottom. Warmer less dense water rises to the surface to be cooled. When all the water is at 4 °C the circulation stops. If the temperature of the surface water falls below 4 °C, it becomes less dense and *remains at the top*, eventually forming a layer of ice at 0 °C. Temperatures in the pond are then as in Fig. 44.9.

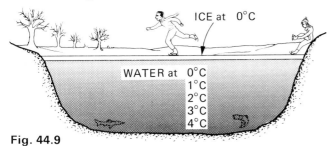

Fig. 44.9

The expansion of water between 4 °C and 0 °C is due to the breaking up below 4 °C of the groups which water molecules form above 4 °C. The new arrangement requires a larger volume and more than cancels out the contraction due to the fall in temperature.

Q Questions

1. Explain why (a) the metal lid on a glass jam jar can be unscrewed easily if the jar is inverted for a few seconds with the *lid* in very hot water, (b) furniture may creak at night after a warm day, (c) concrete roads are laid in sections with pitch between them.

2. A bimetallic strip is made from aluminium and copper. When heated it bends in the direction shown in Fig. 44.10.

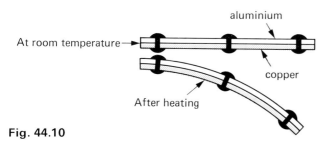

Fig. 44.10

Which metal expands more for the same rise in temperature?

Draw a diagram to show how the bimetallic strip would appear if it were cooled to below room temperature.

3. How does the bimetallic thermometer in Fig. 44.11 work?

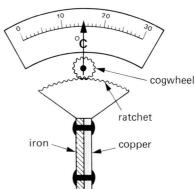

Fig. 44.11

4. The linear expansivities of some common substances are given below.

Aluminium	0.000 03 per °C	Glass	0.000 009 per °C
Concrete	0.000 01 per °C	Platinum	0.000 009 per °C
Copper	0.000 02 per °C	Steel	0.000 01 per °C

(a) Rods of aluminium, copper and steel are the same length at a temperature of 0 °C. Which will be the longest at 300 °C?
(b) If the aluminium rod is 1.000 m long at 0 °C, what will be its length at 300 °C?
(c) Why is steel a suitable material for reinforcing concrete?
(d) Which of the substances listed above would be most suitable for carrying a current of electricity through the walls of a glass vessel?
(e) How might a steel cylinder liner be fitted tightly inside a cylinder block made of aluminium?

5. Using the values for linear expansivities in Qn. 4 calculate the expansion of
(a) 100 m of copper pipe heated through 50 °C.
(b) 50 cm of steel pipe heated through 80 °C.
(c) 200 m of aluminium pipe heated from 10 °C to 60 °C.

6. When a particular substance at a certain temperature is heated, it expands. When the same substance at the same temperature is cooled, it also expands. (a) What is the substance? (b) What is the temperature? (E.M.)

check list
P.286

45 The gas laws

When a gas is heated, as air is in a jet engine, its *pressure* as well as its *volume* may change. To study the effect of temperature on these two quantities we must keep one fixed while the other is changed.

☒ Experiment: effect on volume of temperature (pressure constant)— Charles' law

Arrange the apparatus as in Fig. 45.1. The index of concentrated sulphuric acid traps the air column to be investigated and also dries it. Adjust the capillary tube so that the bottom of the air column is opposite a convenient mark on the ruler.

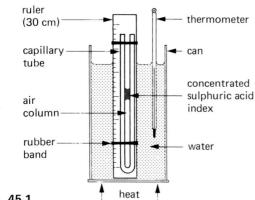

Fig. 45.1

Note the length of the air column (to the *lower* end of the index) at different temperatures. Put the results in a table but before taking a reading, stop heating and stir well to make sure that the air has reached the temperature of the water.

Plot a graph of volume (in cm, since the length of the air column is a measure of it) on the y-axis and temperature (in °C) on the x-axis.

The pressure of (and on) the air column is constant and equals atmospheric pressure plus the pressure of the acid index.

☒ Experiment: effect on pressure of temperature (volume constant)— the Pressure law

The apparatus is shown in Fig. 45.2. The rubber tubing from the flask to the pressure gauge should be as short as possible. The flask must be in water almost to the top of its neck and be securely clamped to keep it off the bottom of the can.

128

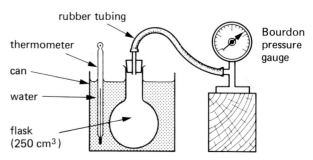

Fig. 45.2

Record the pressure over a wide range of temperatures. Tabulate the results but before taking a reading, stop heating, stir and allow time for the gauge reading to become steady; the air in the flask will then be at the temperature of the water.

Plot a graph of pressure on the y-axis and temperature on the x-axis.

Absolute zero

The volume-temperature and pressure-temperature graphs for a gas (obtained by experiments like the above) are straight lines, Fig. 45.3a,b. They show that gases expand *uniformly* with temperature as measured on a mercury thermometer, i.e. equal temperature increases cause equal volume or pressure increases.

The graphs do not pass through the Celsius temperature origin (0 °C). If they are produced backwards they cut the temperature axis at about −273 °C. This temperature is called *absolute zero* because we believe it is the lowest temperature possible. It is the zero of the *absolute* or *Kelvin scale of temperature*.

Degrees on this scale are called *kelvins* and are denoted by K. They are exactly the same size as Celsius degrees, i.e. 1 °C = 1 K. Since −273 °C = 0 K, conversions from °C to K are made by adding 273. For example

$$0\,°C = 273\ K$$
$$15\,°C = 273+15 = 288\ K$$
$$100\,°C = 273+100 = 373\ K$$

Kelvin or absolute temperatures are represented by the letter T and if t stands for a Celsius scale temperature then, in general

$$T = 273+t$$

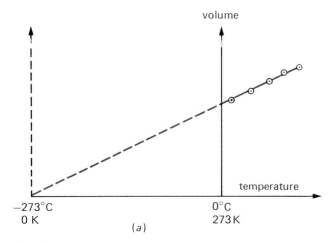

Fig. 45.3

Near absolute zero strange things occur. Liquid helium becomes a 'superfluid'. It cannot be kept in an open vessel because it flows up the inside of the vessel, over the edge and down the outside. Some metals and compounds become 'superconductors' of electricity and a current once started in them flows for ever without a battery. Electromagnets with coils of superconducting wires are being developed for use in industry and research, Fig. 45.4.

The gas laws

Using absolute temperatures the gas laws can be stated in a convenient form for calculations.

(a) Charles' law. In Fig. 45.3a the volume-temperature graph does pass through the origin if temperatures are measured on the Kelvin scale. That is, if we take 0 K as the origin. We can then say that the volume V is directly proportional to the absolute temperature T, i.e. doubling T doubles V, etc. Therefore

$$V \propto T \quad \text{or} \quad V = \text{constant} \times T$$
$$\text{or} \quad V/T = \text{constant} \tag{1}$$

Charles' law may be stated as follows.

The volume of a fixed mass of gas is directly proportional to its absolute temperature if the pressure is kept constant.

(b) Pressure law. From Fig. 45.3b we can say similarly for the pressure p that

$$p \propto T \quad \text{or} \quad p = \text{constant} \times T$$
$$\text{or} \quad p/T = \text{constant} \tag{2}$$

The Pressure law may be stated as follows.

The pressure of a fixed mass of gas is directly proportional to its absolute temperature if the volume is kept constant.

(c) Boyle's law. Previously (p. 88) we found for a fixed mass of gas at constant temperature, that

$$pV = \text{constant} \tag{3}$$

These three equations can be combined giving

$$\frac{pV}{T} = \text{constant}$$

It is useful for cases in which p, V and T all change from say, p_1, V_1 and T_1 to p_2, V_2 and T_2, then

$$\frac{p_1 V_1}{T_1} = \frac{p_2 V_2}{T_2} \tag{4}$$

Fig. 45.4

129

Worked example

A cycle pump contains 50 cm³ of air at 17 °C and a pressure of 1.0 atmosphere. Find the pressure when the air is compressed to 10 cm³ and its temperature rises to 27 °C.

We have,

$p_1 = 1.0$ atmosphere $\quad p_2 = ?$
$V_1 = 50$ cm³ $\qquad\qquad V_2 = 10$ cm³
$T_1 = 273 + 17 = 290$ K $\quad T_2 = 273 + 27 = 300$ K

From equation (4) above we get

$$p_2 = p_1 \times \frac{V_1}{V_2} \times \frac{T_2}{T_1}$$

Replacing

$$p_2 = 1 \times \frac{50}{10} \times \frac{300}{290} = 5.2 \text{ atmosphere}$$

Notes. 1. All temperatures must be in K.
2. Any units can be used for p and V so long as they are the same on both sides of the equation.
3. In some calculations the volume of the gas has to be found at s.t.p. (standard temperature and pressure). This is 0 °C and 760 mmHg pressure.

Gases and the kinetic theory

The kinetic theory can explain the behaviour of gases.

(a) Cause of gas pressure. All the molecules in a gas are in rapid motion, with a wide range of speeds and repeatedly hit the walls of the container in huge numbers per second. The average force and therefore the pressure they exert on the walls is constant since pressure is force on unit area.

(b) Boyle's law. If the volume of a fixed mass of gas is halved by halving the volume of the container, Fig. 45.5, the number of molecules per cm³ will be doubled. There will be twice as many collisions per second with the walls, i.e. the pressure is doubled. This is Boyle's law.

Fig. 45.5

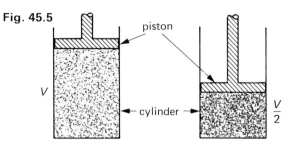

(c) Temperature. When a gas is heated and its temperature rises, the average speed of its molecules increases. If the volume of the gas is to remain constant, its pressure increases due to more frequent and more violent collisions of the molecules with the walls. If the pressure of the gas is to remain constant, the volume must increase so that the number of collisions does not.

(d) Absolute zero. This is the temperature at which the molecules have their lowest possible kinetic energy.

Q Questions

1. A gas of volume 2 m³ at 27 °C is **(a)** heated to 327 °C, **(b)** cooled to −123 °C, at constant pressure. What are its new volumes?

2. A container holds gas at 0 °C. To what temperature must it be heated for its pressure to double? Assume the container does not expand. *(E.A.)*

3. A mass of gas occupies a volume of 200 cm³ at a temperature of 27 °C and a pressure of 1 atmosphere. Calculate the volume when
 (i) the pressure is doubled at constant temperature,
 (ii) the absolute temperature is doubled at constant pressure,
 (iii) the pressure is $1\frac{1}{2}$ atmospheres and the Celsius temperature is 127 °C. *(E.A.)*

4. A gas is heated in a closed container so that its volume cannot change. Which of the following will NOT happen?

 A The average speed of the molecules will increase.
 B The molecules will move in all directions.
 C The number of molecules will increase.
 D The molecules will hit the walls of the container more often.
 E The pressure of the gas will increase.

check list p.286

46 Specific heat capacity

If 1 kg of water and 1 kg of paraffin are heated in turn for the same time by the same heater, the temperature rise of the paraffin is about *twice* that of the water. Since the heater gives equal amounts of heat energy to each liquid, it seems that different substances require different amounts of heat to cause the same temperature rise in the same mass, say 1 °C in 1 kg.

The 'thirst' of a substance for heat is measured by its *specific heat capacity* (symbol c).

The specific heat capacity of a substance is the heat required to produce unit temperature rise in unit mass.

Heat, like other forms of energy, is measured in joules (J) and the unit of specific heat capacity is the joule per kilogram °C, i.e. J/(kg °C).

In physics the word 'specific' means 'unit mass' is being considered.

The heat equation

If a substance has a specific heat capacity of 1000 J/(kg °C) then

$$1000 \text{ J raise the temp. of 1 kg by 1 °C}$$
$$\therefore \quad 2 \times 1000 \text{ J raise the temp. of 2 kg by 1 °C}$$
$$\therefore \quad 3 \times 2 \times 1000 \text{ J raise the temp. of 2 kg by 3 °C}$$

That is 6000 J will raise the temperature of 2 kg of this substance by 3 °C. We have obtained this answer by multiplying together:

(i) the *mass* in kg,
(ii) the *temperature rise* in °C, and
(iii) the *specific heat capacity* in J/(kg °C).

If the temperature of the substance fell by 3 °C, the heat given out would also be 6000 J. In general, we can write the 'heat equation' as

HEAT RECEIVED OR GIVEN OUT
= MASS × TEMP. CHANGE × SPEC. HEAT CAPACITY

In symbols

$$Q = m \times \Delta\theta \times c$$

For example, if the temperature of a 5 kg mass of copper of specific heat capacity 400 J/(kg °C) rises from 15 °C to 25 °C, the heat received Q

$$= 5 \text{ kg} \times (25-15)\text{°C} \times 400 \text{ J/(kg °C)}$$
$$= 5 \times 10 \times 400 = 20\,000 \text{ J}$$

Experiment: finding specific heat capacities

You need to know the power of the 12-volt electric immersion heater to be used. (N.B. *Do not use one with a cracked seal.*) A 40-watt heater converts 40 joules of electrical energy into heat energy per second. If the power is not marked on the heater ask about it.[1]

(a) Water. Weigh out 1 kg of water into a container, e.g. an aluminium saucepan. Note the temperature of the water, insert the heater, Fig. 46.1, and switch

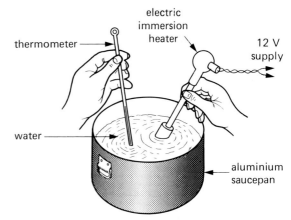

Fig. 46.1

on the 12 V supply. Stir the water and after 5 minutes switch off, but continue stirring and note the *highest* temperature reached.

Assuming that the heat supplied by the heater equals the heat received by the water, work out the specific heat capacity of water in J/(kg °C), as shown below.

Heat received by water (J)
  = power of heater (J/s) × time heater on (s)

Rearranging the 'heat equation' we get

$$\text{sp. heat cap. of water} = \frac{\text{heat received by water (J)}}{\text{mass (kg)} \times \text{temp. rise (°C)}}$$

Suggest causes of error in this experiment.

(b) Aluminium. A cylinder weighing 1 kg and having two holes drilled in it is used. Place the immersion heater in the central hole and a thermometer in the other hole, Fig. 46.2.

[1] The power is found by immersing the heater in water, connecting it to a 12-volt d.c. supply and measuring the current taken (usually 3–4 amperes). Then power in watts = volts × amperes.

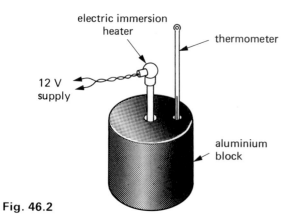

electric immersion heater

thermometer

12 V supply

aluminium block

Fig. 46.2

Note the temperature, connect the heater to a 12 V supply and switch it on for 5 minutes. When the temperature stops rising record its highest value.

Calculate the specific heat capacity as before.

Importance of high specific heat capacity of water

The specific heat capacity of water is 4200 J/(kg °C) and of soil it is about 800 J/(kg °C). As a result, the temperature of the sea rises and falls more slowly than that of the land. A certain mass of water needs five times more heat for its temperature to rise by 1 °C than does the same mass of soil. Water also has to give out more heat to fall 1 °C. Since islands are surrounded by water they experience much smaller changes of temperature from summer to winter than do large land masses such as Central Asia.

The high specific heat capacity of water (as well as its cheapness and availability) accounts for its use to cool car engines and in the radiators of central heating systems.

✎ Worked examples

1. A tank holding 60 kg of water is heated by a 3 kW electric immersion heater. If the specific heat capacity of water is 4200 J/(kg °C), estimate the time for the temperature to rise from 10 °C to 60 °C.

A 3 kW (3000 W) heater supplies 3000 J of heat energy per second.

Let t = time taken in seconds to raise the temperature of the water by $(60 - 10) = 50$ °C,

$\therefore$ heat supplied to water in time $t = 3000 \times t$ J

From the 'heat equation', we can say

heat received by water = $60 \times 4200 \times 50$ J

Assuming heat supplied = heat received

$3000 \times t = 60 \times 4200 \times 50$

132

$\therefore \quad t = \dfrac{60 \times 4200 \times 50}{3000} = 4200$ s (70 mins)

2. A piece of aluminium of mass 0.5 kg is heated to 100 °C and then placed in 0.4 kg of water at 10 °C. If the resulting temperature of the mixture is 30 °C, what is the specific heat capacity of aluminium if that of water is 4200 J/(kg °C)?

When two substances at different temperatures are mixed, heat flows from the one at the higher temperature to the one at the lower temperature until both are at the same temperature—the temperature of the mixture. If there is no loss of heat, then in this case

heat given out by aluminium
                = heat taken in by water

Using the 'heat equation' and letting c be the sp. heat cap. of aluminium in J/(kg °C), we have

heat given out = $0.5 \times c \times (100 - 30)$ J

heat taken in = $0.4 \times 4200 \times (30 - 10)$ J

$0.5 \times c \times 70 = 0.4 \times 4200 \times 20$

$\therefore \quad c = \dfrac{4200 \times 8}{35} = 960$ J/(kg °C)

World energy resources

Energy is a necessary 'raw material' in the modern world. At present the main *primary*, but *non-renewable* sources are fossil fuels, i.e. oil (which supplies about half the energy used), natural gas, coal, and nuclear fuels (uranium). Alternative, *renewable* but less convenient, primary sources are tidal, hydroelectric, solar, geothermal, biomass, wind and wave energy.

The primary sources supply us with *secondary* forms of energy such as electrical, chemical and mechanical. All of these end up as heat, often as a result of overcoming forces such as friction. Heat is the 'lowest grade' of energy because it is very difficult to change it into other, more useful, 'higher grade' forms, such as electrical and chemical energy.

(a) Fossil fuels. These were formed from the decayed remains of animals and plants which lived millions of years ago on earth. Unfortunately oil and gas are unlikely to last beyond the early part of the 21st century, though coal may not run out for another 200 or so years. As a result, there is concern to develop alternative sources. Burning fossil fuels pollutes the atmosphere, causing smog and acid rain.

(b) Nuclear energy. The physics of nuclear energy will be considered in Topic 75. Although it presents *safety* problems if radiation leakage occurs and *environmental* problems due to the disposal of dangerous waste materials with long half-lives (p. 215), there are risks in not using it as well as using it. Views differ on what is the best course to follow.

(c) Solar energy. The sun's energy can be harnessed directly using large curved mirrors to focus its rays on to a small area. The energy can then be used to turn water to steam for driving the turbine of an electrical generator in a power station.

Roof-top solar collectors for solar water heating systems are now common in hot countries (see Further Revision Question 92, p. 267).

(d) Geothermal energy. Scientists have known that huge amounts of heat are stored throughout the earth, deep in certain rocks, e.g. granite, due to radioactivity (p. 213). At a depth of 2 km the temperature is typically 80°C and at 6 km about 200°C. To tap these 'heat reservoirs' two holes are drilled and the cold water pumped down one comes up the other as hot water or steam. The latter can be used to generate electricity or heat buildings.

(e) Wind and wave energy. Electrical generators driven by giant windmills are being developed for use in windy situations. Wave energy is also being investigated using the rocking motion of a floating object to operate a generator. Winds, which cause waves, are due to the uneven heating of the earth's surface by the sun creating convection currents in the air (p. 142).

(f) Hydroelectric power. This is an important source of energy in some countries, e.g. Norway. It arises from the sun creating rain clouds, Fig. 28.4d.

(g) Tidal energy. This can be harnessed by building a barrage (barrier) containing water turbines (Fig. 28.4d) and sluice gates across the mouth of a river (e.g. the Rance in France). Water is trapped on the incoming tide and then used at low tide to drive the turbines which turn electrical generators.

Tides are due mainly to the gravitational pull of the moon on the oceans causing a high tide at places nearest to and farthest from the moon on the earth.

(h) Biomass (vegetable fuels). These include cultivated crops (e.g. oil-seed rape), crop residues (e.g. cereal straw), natural vegetation (e.g. gorse), trees (e.g. spruce) grown for their wood, animal dung and sewage. *Biofuels* such as methane gas and alcohol (ethanol) are obtained from them by fermentation using enzymes and decomposition by bacterial action in the absence of air.

Liquid biofuels can replace petrol and although they have up to 50% less energy per litre they are lead- and sulphur-free and so cleaner. *Biogas* is a mix of methane and carbon dioxide with an energy content about 2/3 that of natural gas. It is used for heating and cooking.

Ｑ Questions

1. How much heat is needed to raise the temperature by 10°C of 5 kg of a substance of specific heat capacity 300 J/(kg °C)?

2. 2000 J of energy is needed to heat 1 kg of paraffin through 1 °C. How much heat is needed to heat 2 kg of paraffin through 10 °C?

   A 4010 J   B 4020 J   C 10 000 J
   D 24 000 J   E 40 000 J

(J.M.B./A.L./N.W.16+)

3. The same quantity of heat was given to different masses of three substances A, B and C. The temperature rise in each case is shown in the table. Calculate the specific heat capacities of A, B and C.

Material	Mass (kg)	Heat given (J)	Temp. rise (°C)
A	1.0	2000	1.0
B	2.0	2000	5.0
C	0.5	2000	4.0

4. How much heat is given out when an iron ball of mass 2 kg and specific heat capacity 440 J/(kg °C) cools from 300 °C to 200 °C?

5. How many joules of heat energy are supplied by a 2 kW heater in (a) 10 s, (b) 1 minute?

6. 1 kg of water is contained in a vessel with a 50 W immersion heater. When the immersion heater is switched on the temperature of the water soon begins to rise at a rate of 1 °C every 2 minutes.
   (a) How much heat is supplied by the immersion heater every 2 minutes?
   (b) From these results what is the approximate specific heat capacity of water?   (*O.L.E./S.R.16+*)

7. An electric water heater raises the temperature of 0.5 kg of water by 30 °C every minute. The specific heat capacity of water is 4200 J/(kg °C). Assuming that no heat is lost, what is the power of the heater?

8. The jam in a hot 'roly-poly' pudding always seems hotter than the pastry. Why?

9. What mass of cold water at 10 °C must be added to 60 kg of hot water at 80 °C by someone who wants to have a bath at 50 °C? Neglect heat losses and take the specific heat capacity of water as 4200 J/(kg °C).

check list p.286

47 Latent heat

When a solid is heated, it may melt and change its state from solid to liquid. If ice is heated it becomes water. The opposite process of freezing occurs when a liquid solidifies.

A pure substance melts at a definite temperature, called the *melting point*; it solidifies at the same temperature—*the freezing point.*

〰 Experiment: cooling curve of ethanamide

Half-fill a test-tube with ethanamide (acetamide) and place it in a beaker of water, Fig. 47.1a. Heat the water until all the ethanamide has melted.

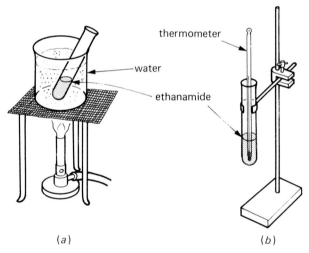

(a)   (b)

Fig. 47.1

Remove the test-tube and arrange it as in Fig. 47.1b with a thermometer in the liquid ethanamide. Record the temperature every minute until it has fallen to 70 °C.

Plot a cooling curve of temperature against time. What is the freezing (melting) point of ethanamide?

Latent heat of fusion

The previous experiment shows that the temperature of liquid ethanamide falls until it starts to solidify (at 82 °C) and remains constant till it has all solidified. The cooling curve in Fig. 47.2 is for a pure substance; the flat part AB occurs at the melting point when the substance is solidifying.

134

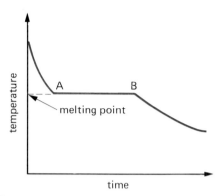

Fig. 47.2

During solidification a substance loses heat to its surroundings but its temperature does not fall. Conversely when a solid is melting, the heat supplied does not cause a temperature rise. For example, the temperature of a well-stirred ice-water mixture remains at 0 °C until all the ice is melted.

Heat which is *absorbed* by a solid during melting or *given out* by a liquid during solidification is called *latent heat of fusion*. Latent means hidden and fusion means melting. Latent heat does not cause a temperature change; it seems to disappear.

The specific latent heat of fusion (l_f) of a substance is the quantity of heat needed to change unit mass from solid to liquid without temperature change.

It is measured in J/kg or J/g. In general, the quantity of heat Q to change a mass m from solid to liquid is given by

$$Q = m \times l_f$$

〰 Experiment: specific latent heat of fusion of ice

Place a 12 V electric immersion heater of known power in a filter funnel and pack small pieces of ice round it, Fig. 47.3. (N.B. *Do not use one with a cracked seal.*) Switch on the heater for 3 minutes and find the mass of water which collects in a beaker. Arrange the results as shown.

	Power of immersion heater	=	W (J/s)
	Time heat supplied	=	s
∴	Heat supplied to ice	=	J
	Mass of beaker empty	=	g
	Mass of beaker + melted ice	=	g
∴	Mass of melted ice	=	g

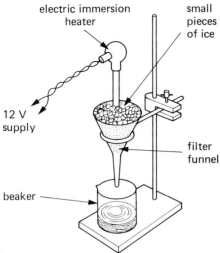

Fig. 47.3

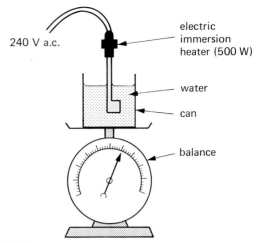

Fig. 47.4

Calculate the heat needed to melt 1 g of ice.

Suppose that a 40 W heater melts 20 g of ice in 3 minutes (180 s), then

heat supplied in 180 s $= 40\,\text{J/s} \times 180\,\text{s} = 7200\,\text{J}$

$\therefore$ latent heat of fusion $= 7200\,\text{J}/20\,\text{g} = 360\,\text{J/g}$

To obtain a more accurate result a control experiment can be set up at the same time using exactly the same apparatus but with the heater OFF. The mass of melted ice in the control beaker after 3 minutes should then be subtracted from that obtained in the beaker within the heater ON. The ice is melted by heat from the room and we can assume that the same mass is melted by room heat in the funnel with the heater on.

The accepted value is 340 J/g.

Latent heat of vaporization

Latent heat is also needed to change a liquid into a vapour. The reading of a thermometer in boiling water remains constant at 100 °C even though heat, called *latent heat of vaporization*, is still being absorbed by the water from whatever is heating it.

When steam condenses to form water, latent heat is given out. This is why a scald from steam may be more serious than one from boiling water.

The specific latent heat of vaporization (l_v) of a substance is the quantity of heat needed to change unit mass from liquid to vapour without change of temperature.

It is measured in J/kg or J/g. In general, the quantity of heat Q to change a mass m from liquid to vapour is given by

$Q = m \times l_v$

An estimate of its value for water can be made using the apparatus of Fig. 47.4. The mains-operated immersion heater is clamped so that it is well covered by the water in the can. When the water is *boiling briskly* the reading on the balance is noted and a stop clock started. The time for 50 g of water to be boiled off is found.

Suppose a 500 W heater is used and that the time required is 4 minutes (240 s), then

heat supplied in 240 s $= 500\,\text{J/s} \times 240\,\text{s}$
$= 120\,000\,\text{J}$

$\therefore$ latent heat of vaporization $= 120\,000\,\text{J}/50\,\text{g}$
$= 2400\,\text{J/g}$

The accepted value is 2300 J/g $= 2\,300\,000$ J/kg $= 2.3 \times 10^6$ J/kg $= 2.3$ MJ/kg. Errors arise because the can and water lose heat to the surroundings.

Latent heat and the kinetic theory

(a) **Fusion.** The kinetic theory explains latent heat of fusion as being the energy which enables the molecules of a solid to overcome the intermolecular forces that hold them in place and when it exceeds a certain value they break free. Their vibratory motion about fixed positions changes to the slightly greater range of movement they have as liquid molecules. Their p.e. increases but not their average k.e. as happens when the heat causes a temperature rise.

(b) **Vaporization.** If liquid molecules are to overcome the forces holding them together and gain the freedom to move around independently as gas molecules, they need a large amount of energy. They receive this as latent heat of vaporization which, like latent heat of fusion, increases the p.e. of the molecules but not their k.e. It also gives the mole-

135

cules the energy required to push back the surrounding atmosphere in the large expansion that occurs when a liquid vaporizes.

To change 1 kg of water at $100\,°C$ to steam at $100\,°C$ needs over *five* times as much heat as it does to raise the temperature of 1 kg of water at $0\,°C$ to water at $100\,°C$ (see *Worked example 1*).

 Worked examples

The following values are required.

	Water	Ice	Aluminium
Sp. ht. cap. $(J/(g\,°C))$	4.2	2.0	0.90
Sp. lat. ht. (J/g)	2300	340	

1. How much heat is needed to change 20 g of ice at $0\,°C$ to steam at $100\,°C$?

There are three stages in the change.

Heat to change 20 g ice at $0\,°C$ to *water at $0\,°C$*
    = mass of ice × sp. lat. ht. of ice
    = 20 g × 340 J/g = 6800 J

Heat to change 20 g *water at $0\,°C$* to *water at $100\,°C$*
    = mass of water × sp. ht. cap. of water × temp. rise
    = 20 g × 4.2 J/(g °C) × 100 °C = 8400 J

Heat to change 20 g *water at $100\,°C$* to *steam at $100\,°C$*
    = mass of water × sp. lat. ht. of steam
    = 20 g × 2300 J/g = 46 000 J

∴ Total heat supplied
    = 6800 + 8400 + 46 000 = 61 200 J

2. An aluminium can of mass 100 g contains 200 g of water. Both, initially at $15\,°C$, are placed in a refrigerator at $-5.0\,°C$. Calculate the quantity of heat that has to be removed from the water and the can for their temperatures to fall to $-5.0\,°C$.

Heat lost by *can* in falling from $15\,°C$ to $-5.0\,°C$
    = mass of can × sp. ht. cap. aluminium × temp. fall
    = $100 × 0.90 × [15-(-5)]$
    = $100 × 0.90 × 20$
    = 1800 J

Heat lost by *water* in falling from $15\,°C$ to $0\,°C$
    = mass of water × sp. ht. cap. water × temp. fall
    = $200 × 4.2 × 15 = 12\,600$ J

Heat lost by *water* at $0\,°C$ freezing to ice at $0\,°C$
    = mass of water × sp. lat. ht. of ice
    = $200 × 340 = 68\,000$ J

Heat lost by *ice* in falling from $0\,°C$ to $-5.0\,°C$
    = mass of ice × sp. ht. cap. of ice × temp. fall
    = $200 × 2.0 × 5.0 = 2000$ J

∴ Total heat removed
    = $1800 + 12\,600 + 68\,000 + 2000 = 84\,400$ J

Q Questions

Use values given in *Worked examples*.

1. (a) How much heat will change 10 g of ice at $0\,°C$ to water at $0\,°C$?
(b) What quantity of heat must be removed from 20 g of water at $0\,°C$ to change it to ice at $0\,°C$?

2. (a) How much heat is needed to change 5 g of ice at $0\,°C$ to water at $50\,°C$?
(b) If a refrigerator cools 200 g of water from $20\,°C$ to its freezing point in 10 minutes, how much heat is removed per minute from the water?

3. How long will it take a 50 W heater to melt 100 g of ice at $0\,°C$?

4. Some small aluminium rivets of total mass 170 g and at $100\,°C$ are emptied into a hole in a large block of ice at $0\,°C$.
(a) What will be the final temperature of the rivets?
(b) How much ice will melt?

5. (a) How much heat is needed to change 4 g of water at $100\,°C$ to steam at $100\,°C$?
(b) Find the heat given out when 10 g of steam at $100\,°C$ condenses and cools to water at $50\,°C$.

6. A 3 kW electric kettle is left on for 2 minutes after the water starts to boil. What mass of water is boiled off in this time?

7. 200 g of metal are heated in a flame to a temperature of $600\,°C$ and dropped into a boiling liquid. It is found that 20 g of liquid vaporizes. If the specific heat capacity of the metal is 0.50 J/(g °C) and the boiling point of the liquid is $100\,°C$, find the specific latent heat of the liquid. (*N.W.*)

8. (a) Why is ice good for cooling drinks?
(b) Why do engineers often use superheated steam (steam above 100°C) to transfer heat?

48 Melting and boiling

Effect of pressure on melting point

Increasing the pressure on ice lowers its melting point. This may be shown by the demonstration in Fig. 48.1 in which a weighted copper wire passes through a block of ice without cutting it in two.

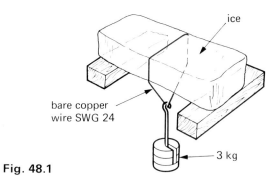

Fig. 48.1

A large pressure is exerted on the ice below the wire. The *melting point* is lowered and the ice, being at 0 °C, melts since its temperature is now above its new melting point. The wire sinks through the water which is no longer under pressure and refreezes above the wire because the melting point returns to 0 °C. In refreezing the water gives out latent heat of fusion and this is *conducted* down through the wire to enable the ice below it to melt.

The effect is called *regelation* (refreezing).

If an iron wire is used in the demonstration it passes through the ice more slowly. No effect is obtained if string is used. Why?

Regelation causes the snow to 'bind' together when a snowball is made. It may also account for the motion of glaciers like that in Fig. 48.2.

The actual depression of the melting point is only 0.0075 °C for ice if normal atmospheric pressure is doubled. Larger pressures cause proportionate decreases.

Fig. 48.2

Effect of impurities on melting point

The temperature of a well-stirred ice-water mixture is normally 0 °C but when an 'impurity' such as salt is added, it may fall to −20 °C. The freezing mixture so formed can be used for cooling purposes.

The fall in temperature is due to the salt lowering the *melting point* of the ice. The ice, however, is still at 0 °C, i.e. above its new melting point. It therefore melts, absorbing latent heat from the mixture whose temperature falls till it freezes at the new melting point.

This effect explains the use of anti-freeze in car radiators and also why brine (and sea sand) is spread on icy roads. The 'impurity' lowers the freezing point of the mixture and may prevent it freezing. When water does freeze it expands and exerts tremendous forces if resisted.

Effect of pressure on boiling point

The boiling point of water rises when the pressure above it is raised. Using the apparatus in Fig. 48.3*a* the pressure can be increased by pinching the rubber tube for *just long enough* to see that the thermometer reading rises.

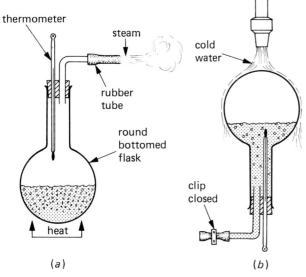

Fig. 48.3

To show that decreasing the pressure lowers the boiling point, the water should be boiled for a few minutes so that the steam sweeps out most of the air. The heating is then stopped and the clip closed. Cold water is run over the inverted flask, Fig. 48.3*b*, so condensing the water vapour inside it and reducing the pressure above the water. The water starts to boil and if the cooling in this way is continued, it may go on boiling until about 40 °C.

Fig. 48.4

In a pressure cooker, Fig. 48.4, food cooks more quickly because the pressure of the steam above the water in the cooker can rise to twice the normal atmospheric value. The water then boils at about 120 °C.

Effect of impurities on boiling point

An 'impurity' such as salt when added to water raises the boiling point.

Evaporation

In evaporation a liquid changes to a vapour without ever reaching its boiling point. A pool of water in the road evaporates and does so most rapidly when there is (i) a wind, (ii) sun and (iii) only a little water vapour present in the air.

Evaporation and boiling are compared below.

Evaporation	Boiling
1. Occurs at any temperature.	Occurs at a definite temperature—the boiling point.
2. Occurs at surface of liquid; no bubbles.	Occurs within liquid: bubbles appear.

Both processes need latent heat of vaporization. In evaporation this is obtained by the liquid from its surroundings, as may be shown by the demonstration in Fig. 48.5 (done in a fume cupboard).

Dichloromethane is a *volatile* liquid, that is, it has a low boiling point and evaporates readily at room temperature, especially when air is blown through it. Latent heat is taken first from the dichloromethane

Fig. 48.5

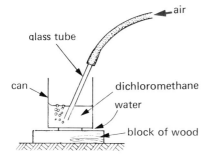

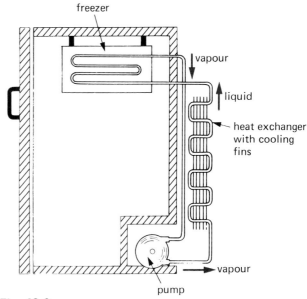

Fig. 48.6

and then from the water below the can. The water soon freezes causing the block and can to stick together.

Volatile liquids feel cold when spilt on the hand. They are used in perfumes.

Water evaporates from the skin when we sweat. This is the body's way of using unwanted heat and keeping a constant temperature. After vigorous exercise there is a risk of the body being overcooled, especially in a draught; it is then less able to resist infection.

Cooling by evaporation and the kinetic theory

The molecules of a liquid have an average speed at a particular temperature but some are moving faster than others. Evaporation occurs when faster molecules escape from the surface of the liquid. The average speed, and therefore the average k.e. of the molecules left behind, decreases, i.e. the temperature of the liquid falls.

Refrigerator

In a refrigerator heat is taken in at one place and given out at another by the refrigerating substance as it is pumped round a circuit, Fig. 48.6.

The coiled pipe round the *freezer* at the top of the refrigerator contains a volatile liquid. This evaporates and takes latent heat from its surroundings, so cooling the air near the top. This becomes more dense, falls to the bottom and causes warmer air to rise to the top to be cooled. The electrically-driven pump removes the vapour (so reducing the pressure, lowering the boiling point and encouraging evaporation or even boiling) and forces it into the *heat exchanger* (pipes with cooling fins outside the rear of

the refrigerator). Here the vapour is compressed and liquefies, giving out latent heat of vaporization to the surrounding air. The liquid returns to the coils round the freezer and the cycle is repeated.

An adjustable thermostat switches the pump on and off, controlling the rate of evaporation and so the temperature in the refrigerator.

Q Questions

1. Explain the following.
   (a) On a very cold day good snowballs cannot be made.
   (b) When walking on snow it 'cakes' and sticks to the sole of the shoe.
   (c) A good cup of tea cannot be brewed on a high mountain.

2. (a) Give two reasons why the boiling point of water is sometimes greater than 100 °C.
   (b) What would you expect the freezing point of the water to be if its boiling point were greater than 100 °C?
   (c) Name and explain the use of a practical device which makes use of one of the effects mentioned in (a).
   (*W.M.*)

3. Some water is stored in a bag of a porous material, e.g. canvas, which is hung where it is exposed to a draught of air. Explain why the temperature of the water is lower than that of the air.

4. Explain why a bottle of milk keeps better when it stands in water in a porous pot in a draught.

check list
p.286

49 Conduction and convection

To keep a building or a house at a comfortable temperature, in winter and summer, requires a knowledge of how heat travels, if it is to be done economically and efficiently.

Conduction

The handle of a metal spoon held in a hot drink soon gets warm. Heat passes along the spoon by *conduction*.

Conduction is the flow of heat through matter from places of higher to places of lower temperature without movement of the matter as a whole.

A simple demonstration of the different conducting powers of various metals is shown in Fig. 49.1. A matchstick is fixed to one end of each rod using a little melted wax. The other ends of the rods are

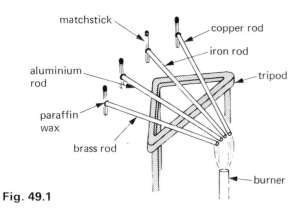

Fig. 49.1

heated by a burner. When the temperatures of the far ends reach the melting point of wax, the matches drop off. The match on copper falls first showing it is the best conductor, followed by aluminium, brass and perhaps iron.

Heat is conducted faster through a rod if it has a large cross-section area, is short and has a large temperature difference between its ends.

Most metals are good conductors of heat; materials such as wood, glass, cork, plastics and fabrics are bad conductors. The apparatus in Fig. 49.2 can be used to show the difference between brass and wood. If the rod is passed through a flame several times, the paper over the wood scorches but not over the brass. The brass conducts the heat away from the paper quickly and prevents it reaching the temperature at which it burns. The wood only conducts it away slowly.

140

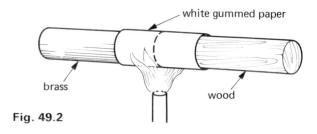

Fig. 49.2

Metal objects below body temperature *feel* colder than those made of bad conductors because they carry heat away faster from the hand—even though all the objects are at the same temperature.

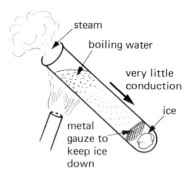

Fig. 49.3

Liquids and gases also conduct heat but only very slowly. Water is a very poor conductor as may be shown in Fig. 49.3. The water at the top of the tube can be boiled before the ice at the bottom melts.

Uses of conductors

(a) Good conductors. These are used whenever heat is required to travel quickly through something. Kettles, saucepans, boilers and radiators are made of metals such as aluminium, iron and copper.

(b) Bad conductors (insulators). The handles of teapots, kettles and saucepans are made of wood or plastic. Cork is used for table mats.

Air is one of the worst conductors, i.e. best insulators. This is why houses with cavity walls (i.e. two walls separated by an air space), Fig. 49.4a, and double-glazed windows, keep warmer in winter and cooler in summer.

Materials which trap air, e.g. wool, felt, fur, feathers, polystyrene, fibre glass, are also very bad conductors. Some are used as 'lagging' to insulate water pipes,

(a)

(b)

Fig. 49.4

hot water cylinders, ovens, refrigerators and the roofs (and walls) of houses, Fig. 49.4b. Others make warm winter clothes.

Wet suits are worn by water skiers to keep them warm. During skiing the suit gets wet and a layer of water gathers between the skier's body and the suit. The water is warmed by body heat and stays warm because the suit is made of an insulating fabric, e.g. neoprene.

Conduction and the kinetic theory

Two processes occur in metals. These have a large number of 'free' electrons (p. 166) which wander about inside them. When one part of a metal is heated, the electrons there move faster (i.e. their k.e. increases) and farther. As a result they 'jostle' atoms in cooler parts, so passing on their energy and raising the temperature of these parts. This process occurs quickly.

The second process is much slower. It is less important in metals but is the only way conduction occurs in non-metals since these do not have 'free' electrons. In it the atoms themselves at the hot part make 'colder' neighbouring atoms vibrate more vigorously.

Convection in liquids

Convection is the usual method by which heat travels through fluids, i.e. liquids and gases. It can be shown in water by dropping a few crystals of potassium permanganate down a tube to the bottom of a flask of water. When the tube is removed and the flask heated just below the crystals by a *small* flame,

Fig. 49.5, purple streaks of water rise upwards and fan outwards.

Streams of warm moving fluids are called *convection currents*. They arise when a fluid is heated because it expands, becomes less dense and is forced upwards

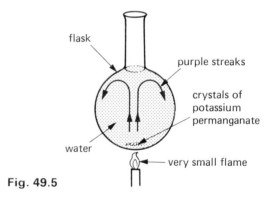

Fig. 49.5

by surrounding cooler, denser fluid which moves under it. We say 'hot water (or hot air) rises'. Warm fluid behaves like a cork released under water: being less dense it bobs up. In convection, however, a fluid floats in a fluid, not a solid in a fluid.

Convection is the flow of heat through a fluid from places of higher to places of lower temperature by movement of the fluid itself.

House hot water and central heating systems

(a) Hot water system. One is shown in Fig. 49.6. A convection current of hot water from the *top* of the boiler rises up pipe A to the *top* of the hot water cylinder and cold water flows down pipe B from the *bottom* of the cylinder to the *bottom* of the boiler. Hot

141

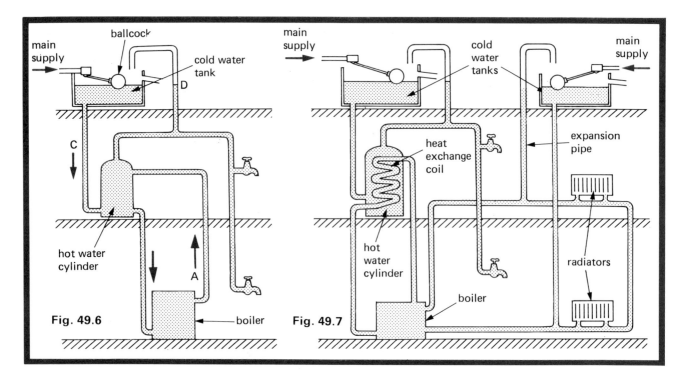

Fig. 49.6

Fig. 49.7

water is drawn off from the *top* of the cylinder and is replaced by cold water from the cold water tank which enters the *bottom* of the cylinder by pipe C.

Water is supplied from the mains to the cold tank through a ballcock. The expansion pipe D allows the escape of (i) dissolved air which comes out of the water when it is heated and (ii) steam if the water boils. The first might otherwise cause air-locks in the pipes and the second an explosion.

(b) Combined hot water and central heating system, Fig. 49.7. The water in the central heating part is quite separate from the water in the hot water part. The latter is similar to Fig. 49.6 except that the hot water cylinder is heated indirectly by a heat exchange coil in the cylinder. When hot water is drawn off, the temperature of the central heating water in the boiler and radiators is therefore hardly affected. A second small cold water tank keeps the boiler and radiators topped up.

Modern systems use narrow pipes and the water is pumped round them.

Convection in air

Black marks often appear on the wall or ceiling above a lamp or a radiator. They are caused by dust being carried upwards in air convection currents produced by the hot lamp or radiator.

A laboratory demonstration of convection currents in air can be given using the apparatus of Fig. 49.8. The direction of the convection current created by

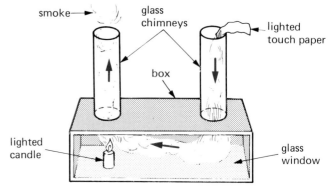

Fig. 49.8

the candle is made visible by the smoke from the touch paper (made by soaking brown paper in strong potassium nitrate solution and drying it).

Convection currents set up by electric, gas and oil heaters help to warm our homes. Many so-called 'radiators' are really convector heaters.

Natural convection currents

(a) Coastal breezes. During the day the temperature of the land increases more quickly than that of the sea (because the specific heat capacity of the land is much smaller). The hot air above the land rises and is replaced by colder air from the sea. A breeze from the sea results, Fig. 49.9a.

At night the opposite happens. The sea has more heat to lose and cools more slowly. The air above the sea is warmer than that over the land and a breeze blows from the land, Fig. 49.9b.

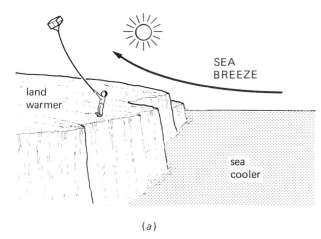

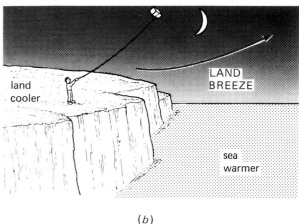

(a) (b)

Fig. 49.9

(b) Gliding. Gliders, including 'hang-gliders', Fig. 49.10, depend on hot air currents, called *thermals*. By flying from one thermal to another gliders can stay airborne for several hours.

Fig. 49.10

Ⓠ Questions

1. Explain why **(a)** newspaper wrapping keeps hot things hot, e.g. fish and chips, and cold things cold, e.g. ice-cream, **(b)** fur coats would keep their owners warmer if they were worn inside out, **(c)** a string vest keeps a person warm even though it is a collection of holes bounded by string.

2. Why is a glass milk bottle likely to crack if boiling water is poured into it?

3. As fuels become more expensive, more people are finding it worthwhile to reduce heat losses from their homes. Fig. 49.11 illustrates three ways of doing this.

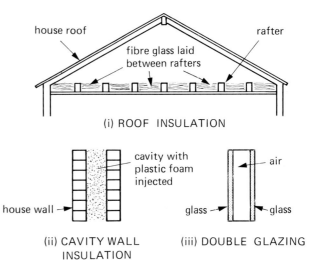

Fig. 49.11

(a) As far as you can, explain how each of the three methods reduces heat losses. Draw diagrams where they will help your explanations.
(b) Why are fibreglass and plastic foam good substances to use?
(c) Air is one of the worst conductors of heat. What is the point of replacing it by the plastic foam in (*ii*)?
(d) A vacuum is an even better heat insulator than air. Suggest one scientific reason why the double glazing should not have a vacuum between the sheets of glass.
(e) The manufacturers suggest that two layers of fibreglass are more effective than one. Suggest how you might set up an experiment in the laboratory to test whether this is true.

4. Convection takes place

A only in solids   B only in liquids   C only in gases
D in solids and liquids   E in liquids and gases.

(N.W.)

5. What is the advantage of placing an electric immersion heater **(a)** near the top, **(b)** near the bottom, of a tank of water?

check list p.287

50 Radiation

Radiation is a third way in which heat can travel but whereas conduction and convection both need matter to be present, radiation can occur in a vacuum. It is the way heat reaches us from the sun.

Radiation has all the properties of electromagnetic waves (p. 43), e.g. it travels at the speed of radio waves and gives interference effects. When it falls on an object, it is partly reflected, partly transmitted and partly absorbed; the absorbed part raises the temperature of the object.

Radiation is the flow of heat from one place to another by means of electromagnetic waves.

Radiation is emitted by all bodies above absolute zero and consists mostly of infrared radiation (p. 43) but light and ultraviolet are also present if the body is very hot (e.g. the sun).

Good and bad absorbers

Some surfaces absorb radiation better than others as may be shown using the apparatus in Fig. 50.1. The inside surface of one lid is shiny and of the other dull black. The coins are stuck on the outside of each lid with candle wax. If the heater is mid-way between the lids they each receive the same amount of radiation. After a few minutes the wax on the black lid melts and the coin falls off. The shiny lids stays cool and the wax unmelted.

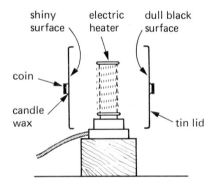

Fig. 50.1

Dull black surfaces are better absorbers of radiation than white shiny surfaces—the latter are good reflectors of radiation. This is why buildings in hot countries are often painted white and why light-coloured clothes are cooler in summer. Also, reflectors on electric fires are made of polished metal because of their good reflecting properties.

144

Good and bad emitters

Some surfaces also emit radiation better than others when they are hot. If you hold the backs of your hands on either side of a hot copper sheet which has one side polished and the other blackened, Fig. 50.2, it will be found that the *dull black surface is a better emitter of radiation than the shiny one.*

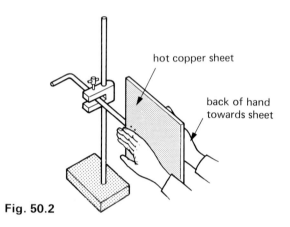

Fig. 50.2

The cooling fins on the heat exchanger of a refrigerator (p. 139, Fig. 48.6) are painted black so that they lose heat more quickly. By contrast teapots and kettles which are polished are poor emitters and keep their heat longer.

In general surfaces that are good absorbers of radiation are good emitters when hot.

Vacuum flask

A vacuum or Thermos flask keeps hot liquids hot or cold liquids cold. It is very difficult for heat to travel into or out of the flask.

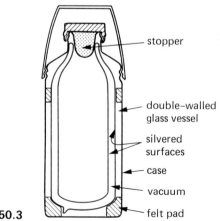

Fig. 50.3

Transfer by conduction and convection is minimized by making the flask a double-walled glass vessel with a vacuum between the walls, Fig. 50.3. Radiation is reduced by silvering both walls on the vacuum side. Then, if for example, a hot liquid is stored, the small amount of radiation from the hot inside wall is reflected back across the vacuum by the silvering on the outer wall. The slight heat loss which does occur is by conduction up the walls and through the stopper.

The greenhouse

The warmth from the sun is not cut off by a sheet of glass but the warmth from a red-hot fire is. The radiation from very hot bodies like the sun is mostly in the form of light and short wavelength infrared. The radiation from less hot objects, e.g. a fire, is largely long wavelength infrared which, unlike light and short wavelength infrared, cannot pass through glass.

Light and short wavelength infrared from the sun penetrate the glass of a greenhouse and are absorbed by the soil, plants, etc., raising their temperature. These in turn emit infrared but, because of their relatively low temperature, this has a long wavelength and is not transmitted by the glass. The greenhouse thus acts as a 'heat-trap' and its temperature rises.

Fig. 50.4

Rate of cooling of an object

The rate at which an object cools, i.e. at which its temperature falls, can be shown to be proportional to the ratio of its surface area A to its volume V.

For a cube of side l

$$A_1/V_1 = 6 \times l^2/l^3 = 6/l$$

For a cube of side $2l$

$$A_2/V_2 = 6 \times 4\, l^2/8\, l^3 = 3/l = \tfrac{1}{2} \times 6/l = \tfrac{1}{2} A_1 V_1$$

The larger cube has the smaller A/V ratio and so cools more slowly.

You could investigate this using two aluminium cubes, one having twice the length of side of the other. Each needs holes for a thermometer and an electric heater to raise them to the same starting temperature. Temperature against time graphs for both blocks can then be obtained.

Q Questions

1. We feel the heat from a coal fire by

   A convection   B conduction   C regelation
   D diffusion   E radiation

2. (a) Three beakers are of identical size and shape; one beaker is painted matt black, one is dull white and one is gloss white. The beakers are filled with boiling water. In which beaker will the water cool most quickly? Give a reason.
   (b) State a process, in addition to conduction, convection and radiation, by which heat energy will be lost from the beakers. (E.A.)

3. The door canopy in Fig. 50.4 shows in a striking way the difference between white and black surfaces when radiation falls on them. Explain why.

4. Fig. 50.5 shows an electric heater H placed mid-way between two flasks A and B. Flask A is shiny on the outside and flask B is blackened on the outside.

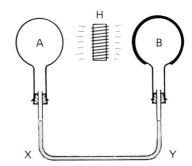

Fig. 50.5

(a) Name the process by which heat travels from the heater to the flasks.
(b) What happens to the liquid in the tube XY?
(c) Give a reason for your answer to (b). (E.M.)

5. (a) The earth has been warmed by the radiation from the sun for millions of years yet we think its average temperature has remained fairly steady. Why is this?
   (b) Why is frost less likely on a cloudy night than a clear one?

6. A Thermos flask has a silver layer on its thin glass walls to reduce loss of heat by

   A convection   B evaporation
   C conduction   D radiation   (W.M.)

check list p.287

51 Heat engines

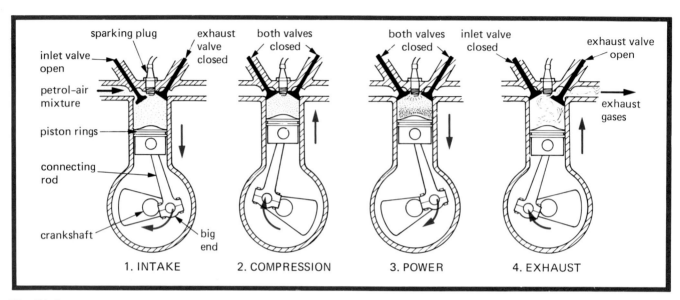

Fig. 51.1

A heat engine, called a *prime mover*, is a machine which extracts mechanical energy (k.e.) from a gas at a high temperature (obtained by burning a fuel) and then expelling it at a lower temperature into the atmosphere.

Oil is the main energy source for transport but other alternatives exist including ethanol, methane and electric batteries.

The same quantity of different fuels gives out different amounts of heat energy when burnt. For example, the values in kJ/g are, for crude oil 50, methane 55, ethanol 30, coal 25–33 and wood 17.

Petrol engines

(a) Four-stroke engine, Fig. 51.1 shows the action. On the *intake stroke*, the piston moves down (due to the starter motor in a car or the kickstart in a motor cycle turning the crankshaft) so reducing the pressure inside the cylinder. The inlet valve opens and the petrol-air mixture from the carburettor is forced into the cylinder by atmospheric pressure.

On the *compression stroke*, both valves are closed and the piston moves up, compressing the mixture.

On the *power stroke*, a spark jumps across the points of the sparking plug and the mixture burns rapidly, forcing the piston down.

On the *exhaust stroke*, the outlet valve opens and the

146

piston rises, pushing the exhaust gases out of the cylinder.

The crankshaft turns a flywheel (a heavy wheel) whose momentum keeps the piston moving between power strokes.

Most cars have at least four cylinders on the same crankshaft, Fig. 51.2. Each cylinder 'fires' in turn, in the order 1–3–4–2, giving a power stroke every half revolution of the crankshaft. Smoother running results.

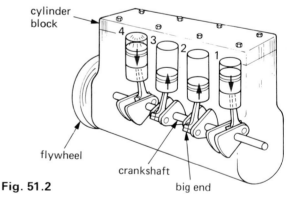

Fig. 51.2

(b) Two-stroke engine. This is used in mopeds, lawnmowers and small boats. Valves are replaced by ports on the side of the cylinder which are opened and closed by the piston as it moves.

In Fig. 51.3*a*, the piston is at the top of the cylinder and the mixture above it is compressed. The exhaust and transfer ports are closed but fresh mixture enters

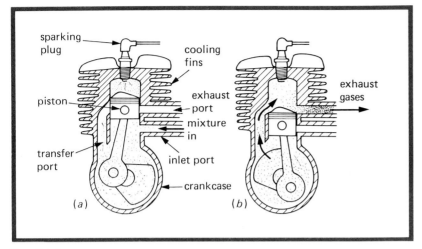

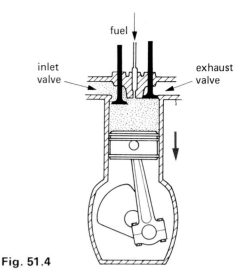

Fig. 51.3

Fig. 51.4

the crankcase by the inlet port. When the spark ignites the compressed mixture, the piston is driven down on its power stroke, closing the inlet port and compressing the mixture in the crankcase.

In Fig. 51.3*b*, the piston is near the bottom of the power stroke and the mixture below it passes through the transfer port into the cylinder above the piston. The exhaust port is now open and the burnt gases are pushed out as the piston moves up again. The shape of the piston helps to stop fresh fuel and burnt gases mixing. The cycle of operations is completed in two strokes.

The efficiency of petrol engines varies with the load but is about 30%, which means that only 30% of the heat energy supplied becomes kinetic energy; much of the rest is lost with the exhaust gases.

Diesel engines

The operation of two- and four-stroke Diesel engines is similar to that of the petrol varieties. However, fuel oil is used instead of petrol, there is no sparking plug and the carburettor is replaced by a fuel injector.

Air is drawn into the cylinder on the downstroke of the piston, Fig. 51.4, and on the upstroke it is compressed to about one-sixteenth of its original volume (which is twice the compression in a petrol engine). This very high compression increases the temperature of the air considerably and when, at the end of the compression stroke, fuel is pumped into the cylinder by the fuel injector, it ignites automatically. The resulting explosion drives the piston down on its power stroke. (You may have noticed that the air in a bicycle pump gets hot when it is squeezed.)

Diesel engines, sometimes called *compression ignition* (C.I.) engines, though heavier than petrol engines,

are reliable and economical. Their efficiency of about 40% is higher than that of any other heat engine.

Jet engines (gas turbines)

There are several kinds of jet engine; Fig. 51.5 is a simplified diagram of a turbo-jet.

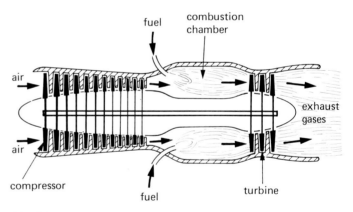

Fig. 51.5

To start the engine, an electric motor sets the compressor rotating. The compressor is like a fan, its blades draw in and compress air at the front of the engine. Compression raises the temperature of the air before it reaches the combustion chamber. Here, fuel (kerosene) is injected and burns to produce a high-speed stream of hot gas which escapes from the rear of the engine, so thrusting it forward (as explained on p. 113 when we considered momentum). The exhaust gas also drives a turbine (another fan) which, once the engine is started, turns the compressor, since both are on the same shaft.

Turbo-jet engines have a high power-to-weight ratio (i.e. produce large power for their weight) and are ideal for use in aircraft.

147

Rockets

Rockets, like jet engines, obtain their thrust from the hot gases they eject by burning a fuel. They can, however, travel where there is no air since they carry the oxygen needed for burning, instead of taking it from the atmosphere as does a jet engine. Space rockets use liquid oxygen (at $-183\,°C$). Common fuels are kerosene and liquid hydrogen (at $-253\,°C$), but solid fuels are also used. Fig. 51.6 is a simplified drawing of a rocket.

Steam turbines

Steam turbines are used in power stations and nuclear submarines, Fig. 51.7. They have efficiencies of about 30%.

The action of a steam turbine resembles that of a water wheel but moving steam not moving water causes the motion. Steam produced in a separate boiler enters the turbine and is directed by the *stator* or *diaphragm* (sets of fixed blades) on to the *rotor* (sets of blades on a shaft that can rotate). The rotor revolves and drives whatever is coupled to it, e.g. an electrical generator or a ship's propeller. The steam expands as it passes through the turbine and the size of the blades increases along the turbine to allow for this. Fig. 51.8 shows the rotor of a steam turbine.

Rotary engines like the steam turbine are smoother than piston (reciprocating) engines.

Fig. 51.6

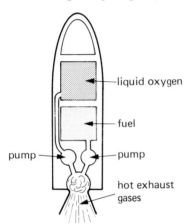

liquid oxygen

fuel

pump

pump

hot exhaust gases

Fig. 51.7

Fig. 51.8

Q Questions

1. Most car engines are four-stroke petrol engines.
   (a) Name the four strokes, in the correct sequence.
   (b) How many times does the crankshaft revolve during the four strokes? *(E.A.)*

2. (a) Fig. 51.9 shows the cylinder of a four-stroke petrol engine. If the spark plug has just fired, what is wrong with the diagram? Explain why the engine would not work very well with this fault.
   (b) What are the basic differences between four-stroke and two-stroke engines? *(E.A.)*

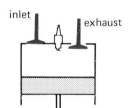

Fig. 51.9

3. In a motor-car engine chemical energy is converted into mechanical energy. What other forms of energy are produced? Some of the mechanical energy is not usefully employed. What happens to it? Describe the energy changes involved when the brakes are applied.

4. (a) Explain the actions of (i) a compressor, (ii) a turbine, in a jet engine.
   (b) Draw a labelled diagram of a jet engine, showing the positions of the compressor and the turbine.
   (c) The graphs in Fig. 51.10 compare the performances of a propeller-driven aircraft with that of a turbo-jet. Explain what the graphs show.

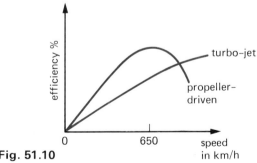

Fig. 51.10

5. (a) Explain how a rocket engine obtains its forward thrust.
   (b) Draw a labelled diagram of a typical rocket engine.
   (c) What fuels are used in modern rockets?
   (d) Why does a rocket work better outside the atmosphere?

149

check list
p.287

52 Additional questions

Core level (all students)

Thermometers

1. Fig. 52.1 shows three thermometers, A, B and C. The lowest reading on each thermometer is 0 °C. One thermometer reads up to 50 °C, another to 100 °C and the third to 200 °C. Which thermometer is which? Give the reasoning by which you obtain your answer.   (E.A.)

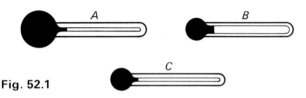

Fig. 52.1

Expansion of solids and liquids

2. When a metal bar is heated the increase in length is greater if
   1 the bar is long
   2 the temperature rise is large
   3 the bar has a large diameter

Which statement(s) is (are) correct?

   A 1, 2, 3   B 1, 2   C 2, 3   D 1   E 3

3. A bimetallic thermostat for use in an electric iron is shown in Fig. 52.2.

Fig. 52.2

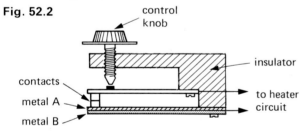

   1 It operates by the bimetallic strip bending away from the contact.
   2 Metal A has a greater expansivity than metal B.
   3 Screwing in the control knob raises the temperature at which the contacts open.

Which statement(s) is (are) correct?

   A 1, 2, 3   B 1, 2   C 2, 3   D 1   E 3

4. Which one of the graphs in Fig. 52.3 most nearly shows how the volume of water changes with temperature between −5 °C and +10 °C?
                    (J.M.B./A.L./N.W.16+)

The gas laws

5. The pressure of the air inside a car tyre increases if the car stands for some time in full sunlight. According to kinetic theory this is due to an increase inside the tyre of

   A the size of the molecules
   B the number of air molecules
   C the speed of the air molecules
   D the average distance between the air molecules
   E the total mass of the air molecules.
                    (O.L.E./S.R.16+)

Specific heat capacity

6. A certain liquid has a specific heat capacity of 3.0 J/(g °C). What mass of the liquid may be heated from 20 °C to 50 °C by 630 J of heat?

   A 4.2 g   B 7.0 g   C 10.5 g   D 11.4 g
   E 12.0 g   (J.M.B./A.L./N.W.16+)

7. 100 g of metal at 100 °C is dropped into 50 g of liquid at 20 °C and the final temperature of the mixture is 40 °C. If the specific heat capacity of the liquid is 3.0 J/(g °C), calculate the specific heat capacity of the metal.   (N.W.)

Latent heat

8. 3.000 kg of ice at 0 °C are supplied with 1 509 000 J of heat. The specific latent heat of ice is 335 000 J/kg and the specific heat capacity of water is 4200 J/(kg °C).

   (a) How much heat is used in melting the ice?
   (b) How much heat is used to warm the water?
   (c) What will be the final temperature of the water?
                    (J.M.B./A.L./N.W.16+)

Fig. 52.3

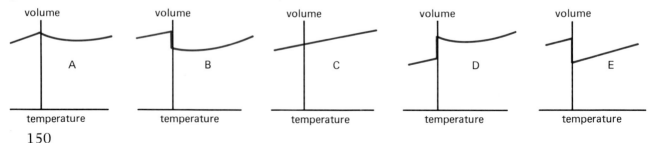

9. Fig. 52.4 shows a graph of temperature against time for a substance which is heated at a constant rate from a low to high temperature. Use the letters to answer the following questions.

    (a) Which part, or parts, of the graph correspond to the substance existing in two states at the same time?
    (b) Over which part is the substance increasing in temperature at the fastest rate?
    (c) Which point of the graph corresponds to the molecules of the substance having the greatest average kinetic energy?   (*W.*)

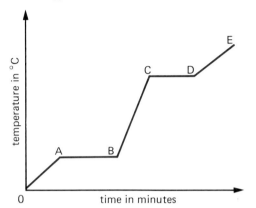

Fig. 52.4

Melting and boiling

10. (a) State *two* similarities and *two* differences between the processes of evaporation and boiling.
    State two ways in which the rate of evaporation of a liquid could be increased.
    (b) Describe experiments, one in each case, to show the effect on the boiling point of water of **(i)** dissolving common salt in the water, **(ii)** reducing the air pressure over the pure water. State the result you would expect in each case.   (*J.M.B.*)

Conduction and convection: radiation

11. On a frosty day the metal handlebars of a bicycle feel colder than the rubber grips because

  A the rubber is a better absorber of radiation than the metal
  B the metal is colder than the rubber
  C the rubber has a higher heat capacity than the metal
  D the metal is a better conductor of heat than the rubber
  E the metal is a better radiator of heat than the rubber.   (*O.L.E./S.R.16+*)

12. Explain in terms of heat transfer the function of the following features of a Thermos flask.
    (a) A tight fitting cork or plastic stopper.
    (b) The silver coating on the inside of the outer glass wall and the outside of the inner glass wall.
    (c) The evacuated enclosure between the walls.
    (*N.I.*)

Further level (higher grade students)

Thermometers

13. (a) What is meant by **(i)** scale of temperature and **(ii)** fixed points?
    State the advantages and disadvantages of mercury and alcohol as thermometric liquids.
    (b) Explain why a clinical thermometer is **(i)** short in length, **(ii)** fast-acting and **(iii)** sensitive.
    A clinical thermometer should not be sterilized in boiling water. Why?

14. A mercury-in-glass thermometer is graduated from −5 °C to 105 °C. Describe how you would test experimentally that the 0 and 100 marks are correct. Illustrate with diagrams where necessary.

Explain why, if water is used to check the 0 °C reading, the water must be pure, but for checking the 100 °C temperature the water need not be pure. Indicate the effect on the water at both these temperatures of the impurities normally considered.

State, giving a reason, a circumstance in which an alcohol thermometer could be used but a mercury one could not.
    (*L.*)

Expansion of solids and liquids

15. Fig. 52.5 shows a metal rod which is 0.50 m long at 15 °C. It is fixed at one end and the other end touches a lever arm pivoted at the fixed point P; the tip B of the free end moves over a circular scale graduated in mm. The rod can be heated or cooled by suitable means which are not shown. The linear expansivity of the metal is 4.0×10^{-5} per °C.

    (a) What is the expansion of the rod when it is heated from 15 °C to 85 °C?
    (b) The end of the pointer B reads zero on the scale at 15 °C. What is its reading when the rod is at 85 °C?
    (c) If the end of the pointer reads 25 mm on the scale, what is the temperature of the rod, supposing this is uniform?

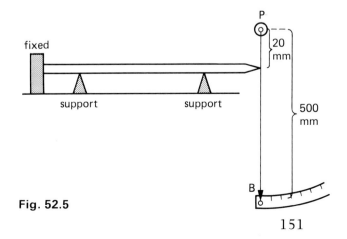

Fig. 52.5

The gas laws

16. Describe fully the apparatus to be used, the precautions which should be taken, and the observations to be made to establish the relationship at constant pressure between the volume of a fixed mass of gas and its temperature over the range approximately 0 °C to 100 °C.

Sketch the graph you would expect to obtain and indicate the expected relationship.

A thick-walled steel cylinder used for storing compressed air is fitted with a safety valve which lifts at a pressure of 1.0 Pa. It contains the air at 17 °C and 0.8 Pa. At what temperature will the valve lift? (L.)

17. A long capillary tube of uniform bore, and sealed at one end, has dry air trapped in it by a small pellet of mercury. The length of the air column is 145 mm at 17 °C. Find (i) the length of the air column at 91 °C, (ii) the temperature at which the air column is 132 mm long.

(W.)

Specific heat capacity

18. An immersion heater rated at 100 W supplies heat for 440 s to 2 kg of paraffin oil. Assuming that the specific heat capacity of paraffin oil is 2.2×10^3 J/(kg °C) and that all the heat from the heater is used to heat the paraffin, the rise in temperature of the paraffin in °C, is

  A 1   B 10   C 20   D 40   E 100   (N.I.)

19. If 5 kg of water at 290 K falls through a vertical height of 315 m what would be its temperature in K after the fall if there is no heat exchange with the air or the ground? (Specific heat capacity of water = 4200 J/(kg K); g = 10 m/s².)

  A 289.250   B 290.000   C 290.075
  D 290.750   E 293.750   (N.I.)

20. Explain the meaning of the term *specific heat capacity*.

Mention one consequence of the high specific heat capacity of water.

In order to reduce its diameter a wire is pulled through a small hole in a metal plate. The wire is made of metal whose specific heat capacity is 400 J/(kg °C) and, on emerging from the hole, has a mass of 5.0 g per m. A steady force of 600 N is required. If all the heat generated were retained in the wire, what would be the rise in temperature of the wire?

If, however, the temperature of the wire were kept constant by spraying it with cold water as it emerged at a speed of 8.4 m/s, what mass of water would be needed per second if the temperature rise of this water were to be 12 °C?

Take the specific heat capacity of water as 4200 J/(kg °C).

(O. and C.)

Latent heat

21. A mass of water was heated in a vessel with an immersion heater of power 40 W. The heater was used to boil the water for 100 s during which time the mass of water decreased by 0.002 kg. Assuming all the heat energy was given to the water, the specific latent heat of vaporization of water is

  A $40 \times 100 \times 0.002$ J/kg
  B $40/(100 \times 0.002)$ J/kg
  C $100/(40 \times 0.002)$ J/kg
  D $40 \times 100/0.002$ J/kg
  E 40×0.002 J/kg   (J.M.B.)

22. Describe an experiment to determine a value for the specific latent heat of fusion of ice. Your account should include the precautions taken to obtain as accurate a result as possible and show how the result is calculated from the readings taken.

An aluminium tray of mass 400 g containing 300 g of water is placed in a refrigerator. After 80 minutes the tray is removed and it is found that 60 g of water remain unfrozen at 0 °C. If the initial temperature of the tray and its contents was 20 °C, determine the average amount of heat removed per minute by the refrigerator. (Assume specific heat capacity of aluminium = 1 J/(g °C), specific heat capacity of water = 4 J/(g °C); specific latent heat of fusion of ice = 340 J/g.) (L.)

23. Explain the terms *specific heat capacity of water*, *specific latent heat of steam*.

Describe an experiment to determine the specific heat capacity of water, explaining carefully how the result is calculated.

When heat is supplied at the rate of 450 W to an electric kettle containing boiling water, steam escapes at such a rate that the loss of water is 0.15 g/s. When heat is supplied at the rate of 680 W the rate of water loss becomes 0.25 g/s. Calculate (a) the specific latent heat of steam, and (b) the rate of loss of heat from the kettle due to factors other than evaporation. (O. and C.)

Melting and boiling

24. (a) How would you show experimentally that evaporation produces cooling? Explain the effect by considering the kinetic energy of the molecules.
(b) Draw a labelled diagram of a refrigerator and describe the action of the refrigerator. (J.M.B.)

Conduction and convection: radiation

25. (a) Two rods, of identical dimensions but of *different* metals, are placed on a sheet of heat-sensitive paper with the same fraction of their lengths projecting from one of its edges, as in Fig. 52.6. The projecting ends are surrounded by water which is kept boiling.

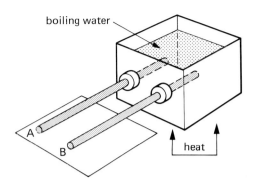

boiling water

A

B

heat

Fig. 52.6

As the rods become hot the paper beneath them changes colour at a certain temperature. When the steady state has been reached it is found that the discoloration of the paper around rod A extends farther from the edge than does that around rod B.
(i) Explain the process by which heat is transmitted along the rods.
(ii) Why does the final temperature vary with position along each of the rods?

(iii) What do you deduce from the fact that the point on A at a certain final temperature is seen to be farther from the edge of the paper than the point on B at this temperature?
(iv) It is observed that initially the discoloration spreads faster along the paper around B than around A. What can you deduce from this?

(b) What is meant by convection? Describe a simple experiment which illustrates the circulation of convection currents in a fluid.   (O.L.E.)

26. (a) Describe the differences in the manner in which heat is transmitted by *conduction* and *radiation*.

Describe an experiment which shows that a shiny or white surface is a poorer *absorber* of heat radiation than a dull or black surface. Give a brief explanation of one use made of this fact in everyday life.

(b) Describe a method of showing that heat radiation is to be found spread over an infrared region in the spectrum of electromagnetic radiation from the sun.

How does the action of a greenhouse depend on the difference in behaviour between longer and shorter wavelengths of infrared radiation?   (L.)

Electricity and magnetism

53 Permanent magnets

Permanent magnets do not readily lose their magnetism with normal treatment. The first permanent magnets were made of steel (an alloy of iron); modern magnets are much stronger and are of two types.

1. Alloy magnets contain metals, e.g. iron, nickel, copper, cobalt, aluminium. They have trade names such as Alnico and Alcomax.

Fig. 53.1 shows a magnet being used to remove an open safety pin someone has accidentally swallowed.

2. Ceramic magnets are made from powders called ferrites which consist of iron oxide and barium oxide. They are brittle. One has the trade name Magnadur.

Magnets are used in electric generators and motors, loudspeakers and telephones. Ferrite powder can be bonded with plastic and rubber to give a flexible magnet or one of any shape. Very fine powder, each particle of which can be magnetized, is used to coat tapes for tape recorders and for computer memories.

Fig. 53.1

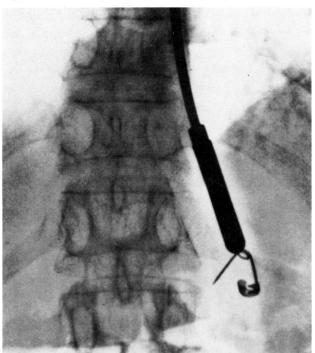

Properties of magnets

1. Magnetic materials. Magnets only attract strongly certain materials such as iron, steel, nickel, cobalt, which are called ferromagnetics.

2. Magnetic poles. These are the places in a magnet to which magnetic materials are attracted, e.g. iron filings. They are near the ends of a bar magnet and occur in pairs of equal strength.

3. North and south poles. If a magnet is supported so that it can swing in a horizontal plane it comes to rest with one pole, the North-seeking or N pole, always pointing roughly towards the earth's north pole. A magnet can therefore be used as a compass.

4. Law of magnetic poles. If the N pole of a magnet is brought near the N pole of a suspended magnet repulsion occurs, Fig. 53.2a. Two S poles also repel. By contrast, N and S poles always attract, Fig. 53.2b.

Fig. 53.2

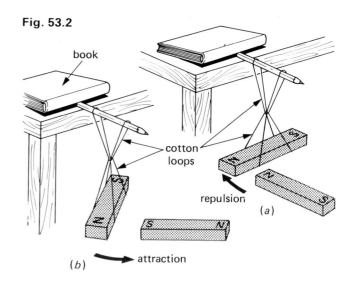

The law of magnetic poles summarizes these facts and states:

like poles repel, unlike poles attract.

The force between magnetic poles decreases as their separation increases.

Fig. 53.3

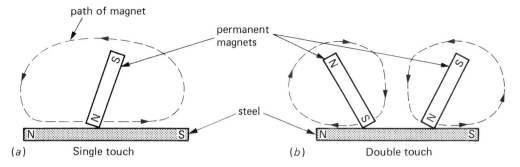

(a) Single touch   (b) Double touch

Test for a magnet

A permanent magnet causes repulsion with one pole when both poles are, in turn, brought near to a suspended magnet. An unmagnetized magnetic material would give attraction with both poles of the suspended magnet.

Repulsion is the only sure test for a magnet.

Making a magnet

(a) By stroking. The methods of single and double touch are shown in Fig. 53.3a,b; steel knitting needles, hair grips or pieces of clockspring can be magnetized. In single touch, the steel is stroked from end to end about 20 times in the same direction by the same pole of a permanent magnet. In the better method of double touch, stroking is done from the centre outwards with unlike poles of two magnets at the same time. The magnets must be lifted high above the steel at the end of each stroke in both methods.

The pole produced at the end of the steel where the stroke ends is of the opposite kind to that of the stroking pole.

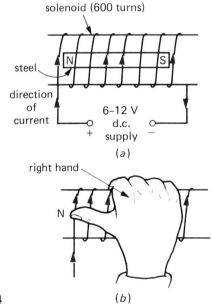

Fig. 53.4

(b) Electrically. The magnetic material is placed inside a cylindrical coil called a solenoid, having several hundred turns of insulated copper wire, which is connected to a 6–12 V *direct current* (d.c.) supply, Fig. 53.4a. If the current is switched on for a second and then off, the material is found to be a magnet when removed from the solenoid.

The polarity of the magnet depends on the direction of the current and is given by the *Right-hand grip rule.* It states that if the fingers of the right hand grip the solenoid in the direction of the current (i.e. from the positive of the supply), the thumb points to the N pole, Fig. 53.4b.

In practice magnets are made electrically using a very large current for a fraction of a second. Fig. 53.5 shows loudspeakers being magnetized.

Fig. 53.5

Demagnetizing a magnet

The magnet is placed inside a solenoid through which *alternating current* (a.c.) is flowing, Fig. 53.6. With the current still passing, the magnet is slowly removed to a distance from the solenoid.

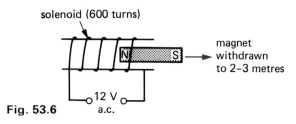

Fig. 53.6

Induced magnetism

When a piece of unmagnetized magnetic material touches or is brought near to the pole of a permanent magnet, it becomes a magnet itself. The material is said to have magnetism *induced* in it. Fig. 53.7 shows that a N pole induces a N pole in the far end.

Fig. 53.7

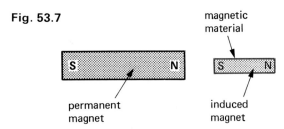

This can be checked by hanging two iron nails from the N pole of a magnet. Their lower ends repel each other, Fig. 53.8*a*, and both are repelled by the N pole of another magnet, Fig. 53.8*b*.

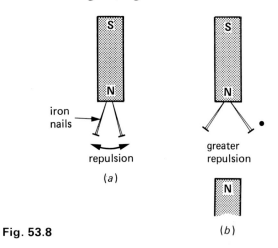

Fig. 53.8

Magnetic properties of iron and steel

Chains of small iron paper clips and steel pen nibs can be hung from a magnet, Fig. 53.9. Each clip or nib magnetizes the one below it by induction and the unlike poles so formed attract.

156

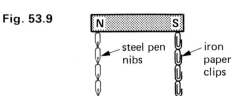
Fig. 53.9

If the iron chain is removed by pulling the top clip away from the magnet, the chain collapses, showing that *magnetism induced in iron is temporary*. When the same is done with the steel chain, it does not collapse; *magnetism induced in steel is permanent*.

Magnetic materials like iron which magnetize easily but do not keep their magnetism, are said to be 'soft'. Those like steel which are harder to magnetize than iron but stay magnetized, are 'hard'. Both types have their uses; very hard ones are used to make permanent magnets.

Theory of magnetism

If a magnetized piece of steel clockspring or thin rod is cut into smaller pieces, each piece is a magnet with a N and a S pole, Fig. 53.10. It is therefore reasonable

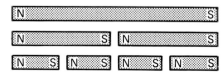

Fig. 53.10

to suppose that a magnet is made up of lots of 'tiny' magnets all lined up with their N poles pointing in the same direction, Fig. 53.11*a*. At the ends, the 'free' poles of the 'tiny' magnets repel each other and fan out so that the poles of the magnet are *round* the ends.

In an unmagnetized bar we can imagine the 'tiny' magnets pointing in all directions, the N pole of one being neutralized by the S pole of another. Their magnetic effects cancel out and there are no 'free' poles near the ends, Fig. 53.11*b*.

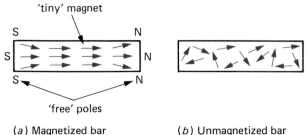

(*a*) Magnetized bar   (*b*) Unmagnetized bar
Fig. 53.11

As well as explaining the breaking of a magnet, this theory accounts for the following.

(i) Magnetic saturation. There is a limit to the strength of a magnet. It occurs when all the 'tiny' magnets are lined up.

(ii) Demagnetization by heating or hammering. Both processes cause the atoms of the magnet to vibrate more vigorously and disturb the alignment of the 'tiny' magnets.

There is evidence to show that the 'tiny' magnets are groups of millions of atoms, called *domains*. In a ferromagnetic material each atom is a magnet and the magnetic effect of every atom in a particular domain acts in the same direction.

Storing magnets

A magnet tends to become weaker with time due to the 'free' poles near the ends repelling each other and upsetting the alignment of the domains. To prevent this bar magnets are stored in pairs with unlike poles opposite and pieces of soft iron, called *keepers*, across the ends, Fig. 53.12.

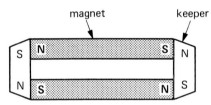

Fig. 53.12

The keepers become induced magnets and their poles neutralize the poles of the bar magnets. The domains in both magnets and keepers form closed chains with no 'free' poles.

Ⓠ Questions

1. A magnet attracts

    A plastics   B any metal   C iron and steel
    D aluminium   E carbon

2. To test a piece of metal to determine whether it was a magnet or not, one would see if it

    A attracted steel filings   B repelled a metal bar
    C attracted a magnet   D repelled a known magnet.
    (*W.M.*)

3. The north pole N of a magnet is stroked along a metal bar in the direction shown in Fig. 53.13.

    (a) Name a metal which would become permanently magnetized by stroking in this way.

Fig. 53.13

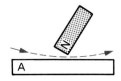

   (b) If such a metal were used, what pole would be produced at A? (*E.A.*)

4. Fig. 53.14 shows an iron rod AB inside a coil of wire. When a current flows through the coil in the direction shown, which end of AB will become a N pole? (*A.L.*)

Fig. 53.14

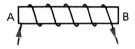

5. Explain why needles hung from either end of a bar magnet held horizontally incline towards one another as shown in Fig. 53.15.

Fig. 53.15

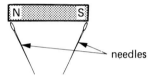

6. Two soft iron rods are placed inside a coil of wire connected to a battery as shown in Fig. 53.16. State the effect on the rods when the switch is closed. (*E.A.*)

Fig. 53.16

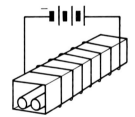

7. The graphs in Fig. 53.17 are for two magnetic materials.
   (a) Does material A or B become the stronger magnet?
   (b) Which material is easier to magnetize?

Fig. 53.17

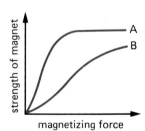

8. **(a)** Describe how electricity may be used to: **(i)** make a temporary magnet, **(ii)** make a permanent magnet, **(iii)** demagnetize a magnet.
   (b) **(i)** Why is there a limit to the strength of the magnet that may be produced? **(ii)** Why does heating reduce the strength of a magnet? (*E.M.*)

check list
p.287

54 Magnetic fields

The space surrounding a magnet where it produces a magnetic force is called a *magnetic field*. The force around a bar magnet can be detected and shown to vary in direction using the apparatus in Fig. 54.1. If the floating magnet is released near the N pole of the bar magnet, it is repelled to the S pole and moves along a curved path known as a *line of force* or a *field line*. It moves in the opposite direction if its south pole is uppermost.

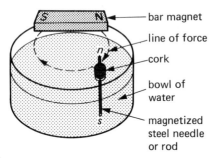

bar magnet

line of force

cork

bowl of water

magnetized steel needle or rod

Fig. 54.1

It is useful to consider that a magnetic field has a direction and to represent the field by lines of force. It has been decided that *the direction of the field at any point should be the direction of the force on a N pole*. To show the direction, arrows are put on the lines of force and point away from a N pole towards a S pole.

〰 Experiment: plotting lines of force

(a) Plotting compass method. A plotting compass is a small pivoted magnet in a glass case with non-magnetic metal walls, Fig. 54.2a.

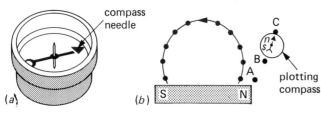

compass needle

C

B

A

plotting compass

(a)   *(b)*   S   N

Fig. 54.2

Lay a bar magnet NS on a sheet of paper, Fig. 54.2b. Place the plotting compass at a point such as A near one pole of the magnet. Mark the position of the poles *n*, *s* of the compass by pencil dots A, B. Move the compass so that pole *s* is exactly over B, mark the new position of *n* by dot C.

Continue this process until the S pole of the bar magnet is reached. Join the dots to give one line of force and show its direction by putting an arrow on

it. Plot other lines by starting at different points round the magnet.

A typical field pattern is shown in Fig. 54.3.

Fig. 54.3

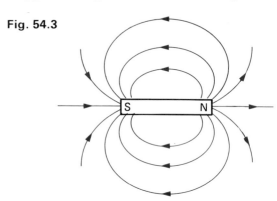

S   N

The combined field due to two neighbouring magnets can also be plotted to give patterns like those in Fig. 54.4a, b. In (a), where two like poles are facing each other, the point X is called a *neutral point*. At X the field due to one magnet cancels out that due to the other and there are no lines of force.

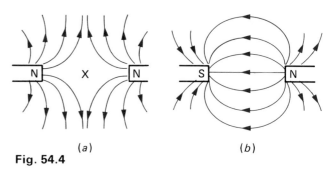

N   X   N   S   N

(a)   *(b)*

Fig. 54.4

(b) Iron filings method. Place a sheet of paper *on top of* a bar magnet and sprinkle iron filings *thinly and evenly* on to the paper from a 'pepper pot'.

Tap the paper gently with a pencil and the filings should form patterns of the lines of force. Each filing is magnetized by induction and turns in the direction of the field when the paper is tapped.

The method is quick but no use for weak fields. Fig. 54.5a,b shows typical patterns with two magnets. Why are they different?

Earth's magnetic field

If lines of force are plotted on a sheet of paper with no magnets near, a set of parallel straight lines is obtained. They run roughly from S to N geographi-

158

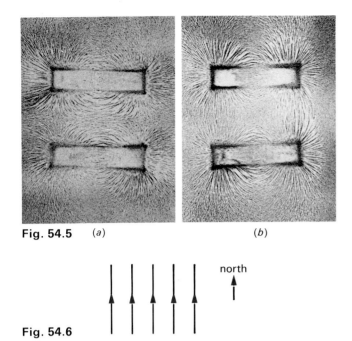

Fig. 54.5 *(a)*   *(b)*

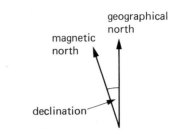

Fig. 54.8

true north is called the *declination*, Fig. 54.8. In London at present (1985) it is 7° W of N and is decreasing. By about the year 2140 it should be 0°.

⊡ Questions

1. Copy Fig. 54.9 which shows a plotting compass and a magnet. Label the N pole of the magnet and draw the field line on which the compass lies.

Fig. 54.9

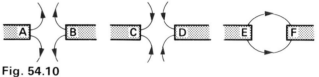

2. The three diagrams in Fig. 54.10 show the lines of force (field lines) between the poles of two magnets. Identify the poles A, B, C, D, E, F.   *(E.M.)*

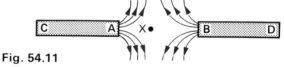

Fig. 54.10

3. The magnetic field between the poles of two bar magnets is shown in Fig. 54.11. The neutral point X is marked.

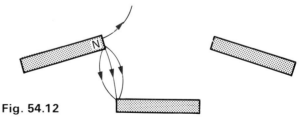

Fig. 54.11

  (a) Explain what is meant by a neutral point.
  (b) C is a S pole. What is **(i)** A, **(ii)** B, **(iii)** D?
  (c) Which is the stronger pole, A or B? Give a reason for your answer.

4. Fig. 54.12 shows three similar magnets and the lines of force in part of the resulting magnetic field.

Fig. 54.12

  (a) Copy and complete the diagram to show the lines of force in the remainder of the field due to the three magnets.
  (b) Mark the polarity of the magnets.
  (c) Mark any neutral points with an X.   *(E.A.)*

cally, Fig. 54.6, and represent a small part of the earth's magnetic field in a horizontal plane.

The *combined* field due to the earth and a bar magnet with its N pole pointing N is shown in Fig. 54.7*a*: it is

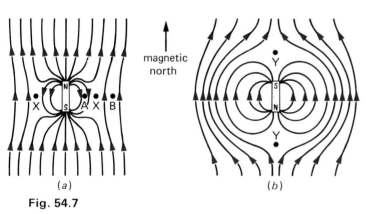

Fig. 54.7

obtained by the plotting compass method. At the points marked X, the fields of the earth and the magnet are equal and opposite and the resultant field is zero. They are *neutral points*. At A the magnet's field is stronger than the earth's field; at B it is weaker.

When the N pole of the magnet points S, the neutral points Y are along the axis of the magnet, Fig. 54.7*b*.

Declination

At most places on the earth's surface a magnetic compass points east or west of true north, i.e. the earth's geographical and magnetic north poles do not coincide. The angle between magnetic north and

159

55 Static electricity

Fig. 55.1

A nylon garment often crackles when it is taken off. We say it has become 'charged with static electricity'; the crackles are caused by tiny electric sparks which can be seen in the dark. Pens and combs made of certain plastics become charged when rubbed on the sleeve and can then attract scraps of paper.

Static electricity is used (i) in coal-burning power stations to stop flue-ash and dust escaping into the atmosphere, (ii) in paint-spraying and (iii) in office-copying machines.

Sparks from static electricity can be dangerous when flammable vapour is present, e.g. in the petrochemical industry. Also, an aircraft in flight may become charged by 'rubbing' the air. Its tyres are of conducting rubber which lets the charge pass harmlessly to earth on landing. Otherwise an explosion could be 'sparked off' when it refuels.

A flash of lightning is nature's most spectacular static electricity effect, Fig. 55.1.

Positive and negative charges

When a strip of polythene (white) is rubbed with a cloth it becomes charged. If it is hung up and another rubbed polythene strip is brought near, repulsion occurs, Fig. 55.2. Attraction occurs when a rubbed cellulose acetate (clear) strip approaches.

Fig. 55.2

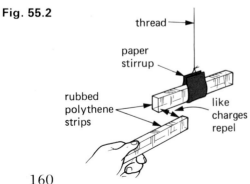

thread
paper stirrup
rubbed polythene strips
like charges repel

This shows there are two kinds of electric charge. That on cellulose acetate is taken as *positive* (+) and that on polythene is *negative* (−). It also shows that

like charges (+ and + or − and −) *repel, unlike charges* (+ and −) *attract.*

The force between electric charges decreases as their separation increases.

Charges, atoms and electrons

There is evidence (Topic 75) that we can picture an atom as made up of a small central nucleus containing positively charged particles called *protons*, surrounded by an equal number of negatively charged *electrons*. The charges on a proton and an electron are equal and so an atom as a whole is normally electrically neutral, i.e. has no net charge.

Hydrogen is the simplest atom with one proton and one electron, Fig. 55.3. A copper atom has 29 protons in the nucleus and 29 surrounding electrons. Every nucleus except hydrogen also contains uncharged particles called *neutrons*.

Fig. 55.3

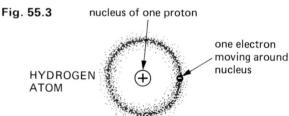

nucleus of one proton

one electron moving around nucleus

HYDROGEN ATOM

The production of charges by rubbing can be explained by supposing that electrons are transferred from one material to the other. For example, when cellulose acetate is rubbed, electrons go from the acetate to the cloth, leaving the acetate short of electrons, i.e. positively charged. The cloth now has more electrons than protons and becomes negatively charged. Note that it is electrons which move, the protons remain in the nucleus.

How does polythene become charged when rubbed?

⎍ Experiment: gold-leaf electroscope

A gold-leaf electroscope consists of a metal cap on a metal rod at the foot of which is a metal plate having a leaf of gold foil attached, Fig. 55.4. The rod is held by an insulating plastic plug in a case with glass sides to protect the leaf from draughts.

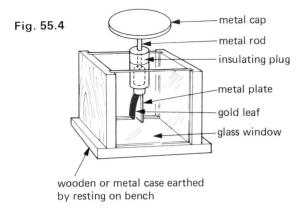

Fig. 55.4

- metal cap
- metal rod
- insulating plug
- metal plate
- gold leaf
- glass window

wooden or metal case earthed
by resting on bench

(a) Detecting a charge. Bring a charged polythene strip towards the cap: the leaf rises away from the plate. On removing the charged strip the leaf falls again. Repeat with a charged acetate strip.

(b) Charging by contact. Draw a charged polythene strip *firmly across the edge* of the cap. The leaf should rise and stay up when the strip is removed. If it does not, repeat the process but press harder. The electroscope has now become negatively charged by contact with the polythene strip.

(c) Finding the sign of a charge. Recharge the polythene strip and bring it near the cap of the negatively charged electroscope. The leaf should rise farther.

Remove the polythene strip and bring up a charged acetate strip. The leaf should fall a bit.

Discharge the electroscope by touching the cap with your hand. Its charge then passes through your body to the earth and the leaf collapses completely.

Charge the electroscope positively by contact with a charged acetate strip. Show that the leaf rises farther when a charged acetate strip approaches the cap but falls when a charged polythene strip approaches.

We can conclude that *if the leaf rises more, the sign of the charge being investigated is the same as the charge on the electroscope.*

If the leaf falls it does not always follow that a charge of the opposite sign has approached the cap. An uncharged body has the same effect, as you can show by holding your hand near the cap. The sure test is when the leaf rises farther, i.e. diverges.

(d) Insulators and conductors. Touch the cap of the charged electroscope with different things, e.g. a piece of paper, a wire, your finger, a comb, a cotton handkerchief, a piece of wood, a glass rod, a plastic pen, rubber tubing.

When the leaf falls, charge is passing to or from the earth through you and the material touching the cap. If the fall is rapid the material is a *good conductor,*

if slow, it is a poor conductor and if the leaf does not alter the material is a *good insulator.* Record your results.

Electrons, insulators and conductors

In an insulator all electrons are bound firmly to their atoms; in a conductor some electrons can move freely from atom to atom. An insulator can be charged by rubbing because the charge produced cannot move from where the rubbing occurs, i.e. the electric charge is static. A conductor will become charged only if it is held in an insulating handle; otherwise the charge passes to earth via our body.

Good insulators are plastics such as polythene, cellulose acetate, Perspex, nylon. All metals and carbon are good conductors. In between are materials that are both poor conductors and (because they conduct to some extent) poor insulators. Examples are wood, paper, cotton, the human body, the earth. Water conducts and if it were not present in materials like wood and on the surface of, for example, glass, these would be good insulators. Dry air insulates well.

Q Questions

1. Two light conducting balls, suspended on nylon threads, come to rest with the threads making equal angles with the vertical, Fig. 55.5. This shows that:

    A the balls are equally and oppositely charged
    B the balls are oppositely charged but not necessarily equally charged
    C one ball is charged and the other is uncharged
    D the balls are equally charged and both carry the same charge
    E one is charged and the other may or may not be charged

Fig. 55.5

2. Explain in terms of electron movement, what happens when a polythene rod becomes charged negatively by being rubbed with a cloth.

3. A metal body, insulated from everything else, is given an electrostatic charge. This body is then touched with a rod of material, one end of which is held in the hand. Which of the following statements is/are true?

    (a) If the body retains its charge, the rod is made from a material which is a good insulator.
    (b) If the body loses its charge slowly, the rod is made from a material which is a poor insulator.
    (c) If the body loses its charge rapidly, the rod is made from a material which is a good conductor.

(J.M.B./A.L./N.W.16+)

check list
P.287

56 More electrostatics

Electrostatics is the study of electric charges at rest.

Electrostatic induction

This is similar to magnetic induction (p. 156) and may be shown by bringing a negatively charged polythene strip near to an insulated metal sphere X which is touching a similar sphere Y, Fig. 56.1a. Electrons in the spheres are repelled to the far side of Y.

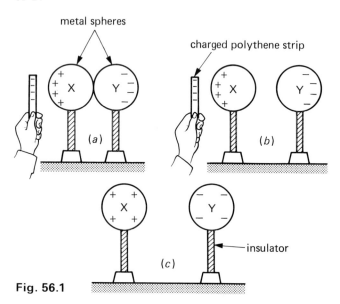

Fig. 56.1

If X and Y are separated, with the charged strip still in position, X is left with a positive charge (deficient of electrons) and Y with a negative charge (excess of electrons), Fig. 56.1b. The signs of the charges can be tested by removing the charged strip and taking X up to the cap of a positively charged electroscope and Y to a negatively charged one, Fig. 56.1c. In both cases the leaf should rise farther.

Attraction of uncharged objects

The attraction of an uncharged object by a charged object near it is due to electrostatic induction.

In Fig. 56.2a a small piece of aluminium foil is attracted to a negatively charged polythene rod held just above it. The charge on the rod pushes free electrons to the bottom of the foil (aluminium is a conductor), leaving the top of the foil short of electrons, i.e. with a net positive charge, and the bottom negatively charged. The top of the foil is nearer the rod than the bottom. Hence the force of

attraction between the negative charge on the rod and the positive charge on the top of the foil is greater than the force of repulsion between the negative charge on the rod and the negative charge on the bottom of the foil. The foil is therefore pulled to the rod.

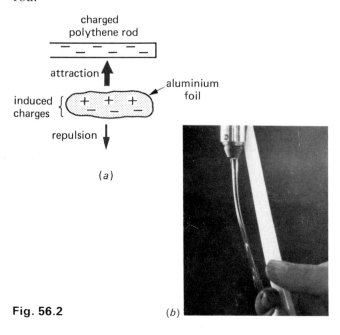

Fig. 56.2   (b)

A small scrap of paper, although an insulator, is also attracted by a charged rod. There are no free electrons in the paper but the charged rod pulls the electrons of the atoms in the paper slightly closer (by electrostatic induction) and so distorts the atoms. In the case of a negatively charged polythene rod, the paper behaves as if it had a positively charged top and a negative charge at the bottom.

In Fig. 56.2b a slow, uncharged stream of water is attracted by a charged polythene strip.

Induction and the electroscope

Electrostatic induction occurs when a charged strip is brought near an electroscope. In Fig. 56.3a electrons in the cap of the *uncharged* electroscope are repelled down to the plate and leaf which both become negatively charged, causing the leaf to rise due to repulsion by the plate.

In Fig. 56.3b electrons are attracted from the plate and leaf of the positively *charged* electroscope up to the cap. The positive charge on the plate and leaf increases and the leaf rises more.

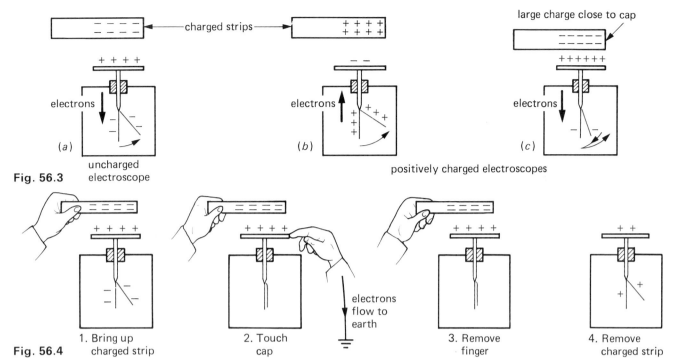

Fig. 56.3

uncharged electroscope

positively charged electroscopes

Fig. 56.4

1. Bring up charged strip
2. Touch cap
3. Remove finger
4. Remove charged strip

electrons flow to earth

If a *large* charge is brought too near the cap of an oppositely charged electroscope, the leaf *rises* after first collapsing. A wrong conclusion may be drawn about the sign of the charge if it approaches too quickly and the initial fall is missed. The effect occurs because, in the case shown in Fig. 56.3c, the number of electrons repelled to the plate and leaf when the large charge is very close, exceeds the original positive charge they had.

Charging by induction

An electroscope can be charged more reliably by induction than by contact. The process is shown in Fig. 56.4. A strip is used with a charge of *opposite* sign to that required on the electroscope.

When the cap is 'earthed' by touching it with the finger, electrons flow from the electroscope to earth through the body. Removal of the negatively charged strip leaves the electroscope with a positive charge.

To charge the electroscope negatively by induction a positively charged acetate strip is used. In this case earthing the electroscope causes electrons to flow to the electroscope from earth.

van de Graaff generator

This produces a continuous supply of charge on a large metal dome when a rubber belt is driven over a Perspex roller, by hand, or by an electric motor. Large versions are used in nuclear research.

Some demonstrations are shown in Fig. 56.5. In (a) sparks jump between the dome and the discharging sphere; in (b) the 'hair' stands on end (why?); in (c) the 'windmill' revolves; in (d) the 'body' on the insulating stool first charges itself by touching the dome and a neon lamp is lit from it.

The dome can be discharged painlessly by bringing your elbow close to it.

Fig. 56.5

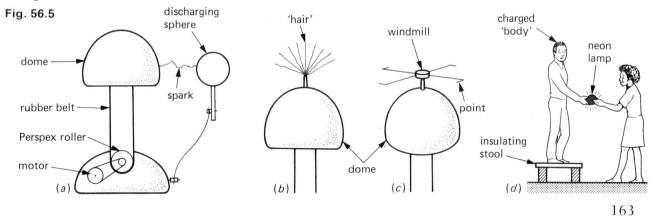

163

Electric fields

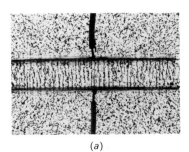
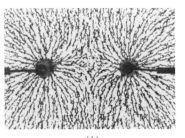
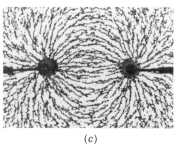

Fig. 56.6

(a)   (b)   (c)

The space around an electric charge where it exerts a force on another charge is an electric field. As with a magnetic field, we represent it by *field lines* or *lines of force*.

The direction of the field at any point is taken as the direction of the force on a positive charge at the point. It is shown by arrows on the lines which we therefore imagine as starting on a positive charge and ending on a negative one.

Electric field patterns similar to the magnetic field patterns given by iron filings can be obtained with semolina powder or grass seed in castor oil. The field is created by charging metal plates or wires dipping in the oil using a van de Graaff generator. Fig. 56.6*a* shows the pattern given by parallel, positive and negative plates. Fig. 56.6*b* is the field of two like charges and Fig. 56.6*c* of two unlike charges.

Lightning conductor

A tall building is protected by a lightning conductor consisting of a thick copper strip on the outside of the building connecting metal spikes at the top to a metal plate in the ground, Fig. 56.7.

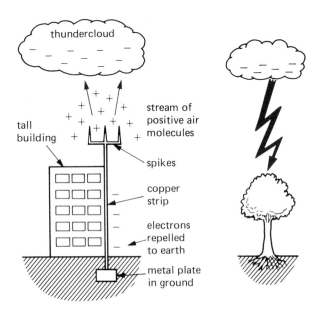

Fig. 56.7

check list p.288

Thunderclouds carry charges and a negatively charged one passing overhead repels electrons from the spikes to earth. The points of the spikes are left with a large positive charge (charge concentrates on sharp points) which removes electrons from nearby air molecules, so charging them positively and causing them to be repelled from the spikes. These form an 'electric wind' that streams upwards to cancel some of the charge on the cloud. If a flash does occur it is less violent and the conductor gives it an easy path to earth.

Ⓠ Questions

1. A negatively charged rod is brought close to an uncharged metal sphere which is held on an insulated stand.

      A   B   C   D

Fig. 56.8

(a) Which of the diagrams in Fig. 56.8 best shows the distribution of charge on the sphere when the rod is near?

(b) What would happen if the sphere was earthed while the negatively charged rod was near?   (*E.A.*)

2. A leaf electroscope may be used as a simple charge detector.

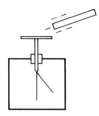

Fig. 56.9

(a) Copy Fig. 56.9 and mark where insulation is required on the electroscope.

(b) Why does the leaf diverge when a negative charge is brought near the cap of an uncharged electroscope?

(c) What happens when the electroscope is weakly positively charged and a strong negative charge is brought near the cap?   (*O.L.E./S.R.16+*)

3. A balloon, if rubbed, will often 'stick' to the wall where it has been rubbed. Why?

57 Capacitors

Quantity of charge

The same amount of charge is carried by every electron, so the more electrons an object loses or gains (e.g. by rubbing), the greater is its positive or negative charge.

Charge is measured in *coulombs* (C), 1 coulomb being the charge on about 6 million million million (6×10^{18}) electrons. For most purposes 1 C is too large and the *microcoulomb* (μC; one millionth of a coulomb) is used.

The negative charge on a polythene rod charged by rubbing is very small, perhaps of the order of a few ten thousandths of a microcoulomb, i.e. about 10^{-4} μC.

Capacitors

A capacitor stores a small quantity of charge. In its simplest form it consists of two parallel metal plates separated by an insulator, called the *dielectric*, Fig 57.1.

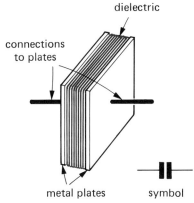

Fig. 57.1
connections to plates   dielectric   metal plates   symbol

The more charge a capacitor can store, the greater is its *capacitance* C. This increases
 (i) as the area of the plates increases,
 (ii) as the separation of the plates decreases,
 (iii) if the dielectric is a solid and not air.

The unit of capacitance is the *farad* (F) but the *microfarad* (μF), which equals one millionth of a farad, is a more convenient unit, i.e. 1 μF = 10^{-6} F.

Practical capacitors (with values ranging from about 0.01 μF to 100 000 μF) often consist of two long strips of metal foil, separated by long strips of dielectric, rolled up like a 'Swiss roll', as in Fig. 57.2. The arrangement allows plates of large area to be close together in a small volume. Plastics (e.g. polyesters) are commonly used as the dielectric, with

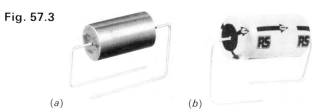

Fig. 57.2
metal foil   dielectric   connections to plates

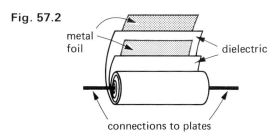

Fig. 57.3

(a)   (b)

films of metal being deposited on the plastic to act as the plates, Fig. 57.3a.

In the *electrolytic* type a very thin dielectric of aluminium oxide is formed chemically between two strips of aluminium foil, giving a high-value (up to 100 000 μF = 0.1 F), compact capacitor, Fig. 57.3b. One plate is marked + and should be charged positively.

A *variable air capacitor* is used to tune a radio receiver. It has two sets of parallel metal plates separated by air, each set consisting of several plates joined together to give, in effect, one large plate, Fig. 57.4.

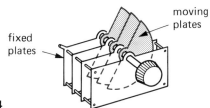

Fig. 57.4
fixed plates   moving plates

One set is fixed, the other can be rotated, thus varying the area of overlap of the plates. The maximum capacitance with the plates fully interleaved may be 0.0005 μF.

The easiest way to charge a capacitor is to connect a suitable battery across it (see Topic 71).

Ⓠ Question

1. (a) What does a capacitor consist of basically and what does it do?
 (b) State two ways of increasing the capacitance of a capacitor.
 (c) What is the unit of (i) charge, (ii) capacitance?

165

check list p.288

58 Electric current

An electric current consists of moving electric charges. In Fig. 58.1 when the van de Graaff is working, the table tennis ball dashes to and fro between the plates and the meter records a small current. As the ball touches each plate it becomes charged and is repelled to the other plate. In this way charge is carried across the gap. This also shows that 'static' charges cause a deflection on a meter just like that produced by a battery in current electricity.

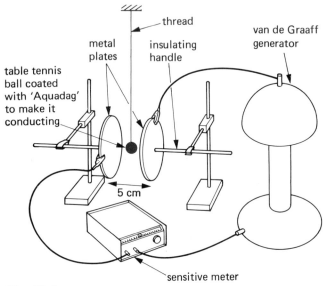

Fig. 58.1

In a metal, each atom has one or more loosely held electrons that are free to move. When a van de Graaff or a battery is connected across the ends of such a conductor, the free electrons drift slowly along it in the direction from the negative to the positive terminal of a battery. There is then a current of negative charge.

Effects of a current

An electric current has three effects that reveal its existence and which can be shown with the circuit of Fig. 58.2.

(a) Heating and lighting. The bulb lights due to a small wire in it (the filament) being made white hot by the current.

(b) Magnetic. The plotting compass is deflected when it is placed in the magnetic field produced round any wire carrying a current.

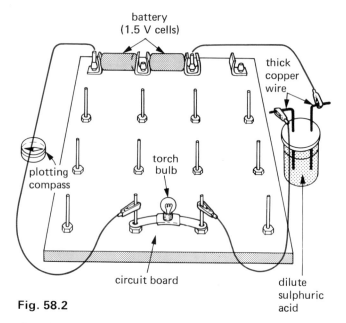

Fig. 58.2

(b) Chemical. Bubbles of gas are given off at the wires in the acid because of the chemical action of the current.

The ampere and the coulomb

The unit of current is the *ampere* (A) which is defined using the magnetic effect. One milliampere (mA) is one-thousandth of an ampere. Current is measured by an *ammeter*.

The unit of charge, the *coulomb* (C), is defined in terms of the ampere.

One coulomb is the charge passing any point in a circuit when a steady current of 1 ampere flows for 1 second.

A charge of 3 C would pass each point in 1 s if the current was 3 A. In 2 s, $3 \times 2 = 6$ C would pass. In general, if a steady current I (amperes) flows for time t (seconds) the charge Q (coulombs) passing any point is given by

$$Q = I \times t$$

This is a useful expression connecting charge and current.

Circuit diagrams

Current must have a complete path (a circuit) of conductors if it is to flow. Wires of copper are used to connect batteries, lamps, etc. in a circuit since copper is a good electrical conductor. If the wires are

166

covered with insulation, e.g. plastic, the ends are bared for connecting up.

The signs or symbols used for various parts of an electrical circuit are shown in Fig. 58.3.

Fig. 58.3

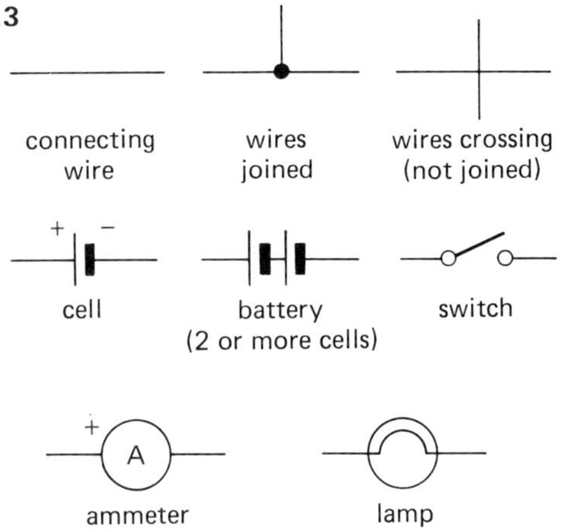

connecting wire   wires joined   wires crossing (not joined)

cell   battery (2 or more cells)   switch

ammeter   lamp

Before the electron was discovered scientists *agreed* to think of current as positive charges moving round a circuit in the direction from positive to negative of a battery. This agreement still stands. Arrows on circuit diagrams show the direction of what we call the *conventional current*, i.e. the direction in which positive charges would flow.

Experiment: measuring current

(a) Connect the circuit of Fig. 58.4a (on a circuit board if possible) ensuring that the + of the cell (the metal stud) goes to the + of the ammeter (marked red). Note the current.

(b) Connect the circuit of Fig. 58.4b. The cells are *in series* (+ of one to − of the other), as are the lamps. Record the current. Measure the current at B, C, and

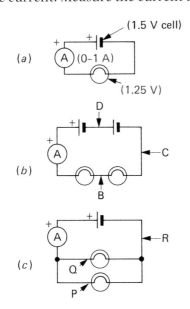

Fig. 58.4

D by disconnecting the circuit at each point in turn and inserting the ammeter. What do you find?

(c) Connect the circuit of Fig. 58.4c. The lamps are *in parallel*. Read the ammeter. Also measure the currents at P, Q and R. What is your conclusion?

Series and parallel circuits

(a) **Series.** In a series circuit, Fig 58.4b, the different parts follow one-after-the-other and there is just one path for the current to follow. You should have found in the previous experiment that the reading on the ammeter when in the position shown (e.g. 0.2 A) is also obtained at B, C and D. That is, current is not used up.

The current is the same at all points in a series circuit.

(b) **Parallel.** In a parallel circuit, Fig. 58.4c, the lamps are side-by-side and alternative paths are provided for the current which splits. Some goes through one lamp and the rest through the other. For example, if the ammeter reading was 0.4 A in the position shown, then if the lamps are identical the reading at P would be 0.2 A, as it would be at Q, giving a total of 0.4 A. Whether the current splits equally or not depends on the lamps (as we will see later); it might divide so that 0.3 A goes one way and 0.1 A by the other branch.

The sum of the currents in the branches of a parallel circuit equals the current entering or leaving the parallel section.

Questions

1. If the current through a floodlamp is 5 A, what charge passes in (a) 1 s, (b) 10 s, (c) 5 minutes?

2. What is the current in a circuit if the charge passing each point is (a) 10 C in 2 s, (b) 20 C in 40 s, (c) 240 C in 2 minutes?

3. (a) Fig. 58.5a shows a cell correctly connected to 2 lamps and a meter. Draw a circuit diagram for the same circuit.

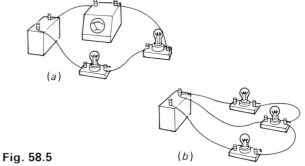

Fig. 58.5   (b)

(b) Fig. 58.5b shows 3 lamps and a cell. Draw a circuit diagram of the same circuit.   (*E.A.*)

4. Study the circuits in Fig. 58.6. The switch S is open (there is a break in the circuit at this point). The circuit in which lamp P would not light but lamps Q and R would is:

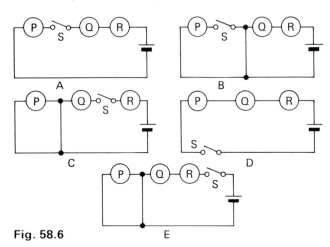

Fig. 58.6

5. Using the circuit in Fig. 58.7, which of the following statements is correct?

(a) When S_1 and S_2 are closed A and B are lit.

(b) With S_1 open and S_2 closed A and B are not lit.

(c) With S_2 open and S_1 closed A lights and B does not light. *(A.L.)*

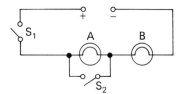

Fig. 58.7

6. In the circuit of Fig. 58.8, A, B, C, D are just positions and the switch is closed. Choose the correct statement.

(a) The current at A will not be the same as the current at B.

(b) The two lamps are wired in parallel.

Fig. 58.8

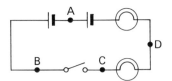

(c) The current reached B before it reached C.

(d) With an extra lamp (at D) the current would be less.

(e) With an extra lamp (at A) the batteries would run down faster. *(S.R.)*

7. If the lamps are all the same in Fig. 58.9 and if A_1 reads 0.50A, what do A_2, A_3, A_4 and A_5 read?

Fig. 58.9

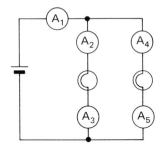

8. Fig. 58.10 represents a junction in an electrical circuit containing four ammeters, the numbers in the circles denoting the reading in amperes. The reading on M would be

A 0   B 2   C 4   D 6   E 12   *(J.M.B.)*

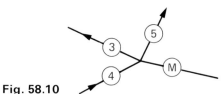

Fig. 58.10

check list p.288

59 Potential difference

A battery changes chemical energy into electrical energy and, because of the chemical action going on inside it, it builds up a surplus of electrons at one of its terminals (the negative) and creates a shortage at the other (the positive). It is then able to maintain a flow of electrons, i.e. an electric current, in any circuit connected across its terminals so long as the chemical action lasts.

The battery is said to have a *potential difference* (*p.d.* for short) at its terminals. Potential difference is measured in *volts* (V) and the term *voltage* is also used instead of p.d. The p.d. of a car battery is 12 V and of the domestic mains supply in the U.K. 240 V.

Energy changes and p.d.

In an electric circuit electrical energy is supplied from a source such as a battery and is changed into other forms of energy by devices in the circuit. A lamp produces heat and light.

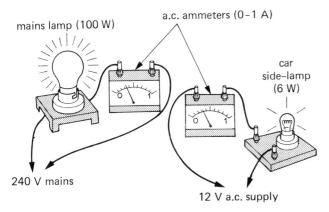

Fig. 59.1

mains lamp (100 W)
a.c. ammeters (0–1 A)
car side-lamp (6 W)
240 V mains
12 V a.c. supply

If the circuits of Fig. 59.1 are connected up, it will be found from the ammeter readings that the current is about the same (0.4 A) in each lamp. However, the mains lamp with a p.d. of 240 V applied to it, gives much more light and heat than the car lamp with 12 V across it. In terms of energy, the mains lamp changes a great deal more electrical energy in a second than does the car lamp.

Evidently the p.d. across a device affects the rate at which it changes electrical energy. This gives us a way of defining the unit of p.d.—the volt.

Model of a circuit

It may help you to understand the definition of the volt, i.e. what a volt is, if you *imagine* that the current in a circuit is formed by 'drops' of electricity, each having a charge of 1 coulomb and carrying equal-sized 'bundles' of electrical energy. In Fig. 59.2, Mr. Coulomb represents one such 'drop'. As a 'drop' moves around the circuit it gives up all its energy which is changed to other forms of energy. Note that electrical energy is 'used up', not charge or current.

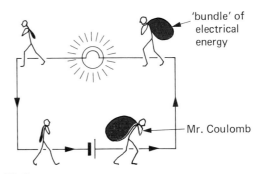

'bundle' of electrical energy

Mr. Coulomb

Fig. 59.2

In our imaginary representation, Mr. Coulomb travels round the circuit and unloads energy as he goes, most of it in the lamp. We think of him receiving a fresh 'bundle' every time he passes through the battery, which suggests he must be travelling very fast. In fact, as we saw earlier (p. 166), the electrons drift along quite slowly. As soon as the circuit is complete, energy is delivered at once to the lamp, not by electrons directly from the battery but from electrons that were in the connecting wires. The model is helpful but not an exact representation.

The volt

The demonstrations of Fig. 59.1 show that the greater the p.d. at the terminals of a supply, the larger is the 'bundle' of electrical energy given to each coulomb and the greater is the rate at which light and heat are produced in a lamp.

The p.d. between two points in a circuit is 1 volt if 1 joule of electrical energy is changed into other forms of energy when 1 coulomb passes from one point to the other.

That is, 1 volt = 1 joule per coulomb (1 V = 1 J/C). If 2 J are given up by each coulomb, the p.d. is 2 V. If 6 J are changed when 2 C pass, the p.d. is 6 J/2 C = 3 V.

In general if W (joules) is the energy changed (i.e. the work done) when charge Q (coulombs) passes between two points, the p.d. V (volts) between the points is given by

$$V = W/Q \quad \text{or} \quad W = Q \times V$$

If Q is in the form of a steady current I (amperes) flowing for time t (seconds) then $Q = I \times t$ (p. 166) and

$$W = I \times t \times V$$

Cells and batteries

Greater p.d.s are obtained when cells are joined in series, i.e. $+$ of one to $-$ of next. In Fig. 59.3a the two 1.5 V cells give a p.d. of 3 V at the terminals A, B. Every coulomb in a circuit connected to this battery will have 3 J of electrical energy.

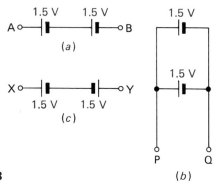

Fig. 59.3 (b)

If two 1.5 V cells are connected in parallel, Fig. 59.3b, the p.d. at terminals P, Q is still 1.5 V but the arrangement behaves like a larger cell and will last longer.

The cells in Fig. 59.3c are in opposition and the p.d. at X, Y is zero.

Experiment: measuring p.d.

A *voltmeter* is an instrument for measuring p.d. It looks like an ammeter but has a scale marked in volts. Whereas an ammeter is inserted *in series* in a circuit to measure the current, a voltmeter is connected across that part of the circuit where the p.d. is required, i.e. *in parallel*. (We will see later that a voltmeter should have a high resistance and an ammeter a low resistance.)

To prevent damage the $+$ terminal (marked red) must be connected to the point nearest the $+$ of the battery.

(a) Connect the circuit of Fig. 59.4a. The voltmeter gives the p.d. across the lamp. Read it.

170

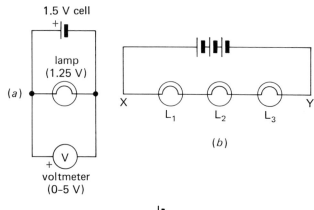

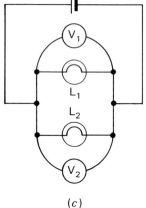

Fig. 59.4 (c)

(b) Connect the circuit of Fig. 59.4b. Measure:
    (i) the p.d. V between X and Y,
    (ii) the p.d. V_1 across lamp L_1,
    (iii) the p.d. V_2 across lamp L_2.
    (vi) the p.d. V_3 across lamp L_3.

How does the value of V compare with $V_1 + V_2 + V_3$?

(c) Connect the circuit of Fig. 59.4c. so that two lamps L_1 and L_2 are in parallel across one 1.5 V cell. Measure the p.d.s, V_1 and V_2, across each lamp in turn.

How do V_1 and V_2 compare?

P.D.s round a circuit

(a) Series. In the previous experiment you should have found in the circuit of Fig. 59.4b that

$$V = V_1 + V_2 + V_3$$

For example, if $V_1 = 1.4$ V, $V_2 = 1.5$ V and $V_3 = 1.6$ V, then V should be $(1.4 + 1.5 + 1.6) = 4.5$ V.

The p.d. at the terminals of a battery equals the sum of the p.d.s across the devices in the external circuit from one battery terminal to the other.

(b) Parallel. In the circuit of Fig. 59.4c $V_1 = V_2$.

The p.d.s across devices in parallel in a circuit are equal.

Q Questions

1. The p.d. across the lamp in Fig. 59.5 is 12 V. How many joules of electrical energy are changed into light and heat when
   (a) a charge of 1 C passes through it,
   (b) a charge of 5 C passes through it,
   (c) a current of 2 A flows through it for 10 s?

Fig. 59.5

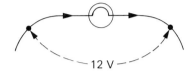

2. Three 2 V accumulators are connected in series and used as the supply for a circuit.
   (a) What is the p.d. at the terminals of the supply?
   (b) How many joules of electrical energy does 1 C gain on passing through (i) one accumulator, (ii) all three accumulators?

3. A set of Christmas tree lights consists of 20 identical lamps connected in series to a 240 V mains supply. What is the voltage (p.d.) across each lamp?

   A 12 V   B 20 V   C 240 V   D 4800 V (E.A.)

4. The symbol for a 1.5 V cell is ⊣⊢. Which of the arrangements in Fig. 59.6 would produce a battery with a p.d. of 6 V?

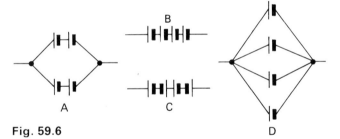

Fig. 59.6

5. The lamps and the cells in all the circuits of Fig. 59.7 are the same. If the lamp in a has its full, normal brightness, what can you say about the brightness of the lamps in b, c, d, e and f?

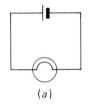 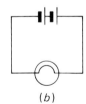 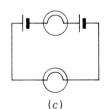 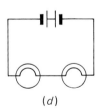 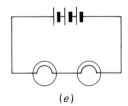 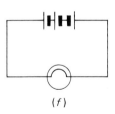

    (a)   (b)   (c)   (d)   (e)   (f)

Fig. 59.7

6. Three voltmeters V, V_1, V_2 are connected as in Fig. 59.8.
   (a) If V reads 18 V and V_1 reads 12 V, what does V_2 read?
   (b) If the ammeter A reads 0.5 A, how much electrical energy is changed to heat and light in L_1 in 1 minute?
   (c) Copy Fig. 59.8 and mark with a +, the positive terminals of the ammeter and voltmeters for correct connection.

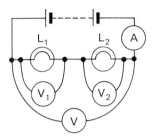

Fig. 59.8

7. Three voltmeters are connected as in Fig. 59.9. What are the voltmeter readings x, y and z in the table below (which were obtained with three different batteries)?

Fig. 59.9

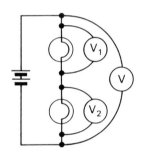

V/V	V_1/V	V_2/V
x	12	6
6	4	y
12	z	4

check list p.288

60 Resistance

Electrons move more easily through some conductors than others when a p.d. is applied. The opposition of a conductor to current is called its *resistance*. A good conductor has a low resistance and a poor conductor has a high resistance. The resistance of a wire of a certain material

(i)   increases as its length increases,
(ii) increases as its cross-section area decreases,
(iii) depends on the material.

A long thin wire has more resistance than a short thick one of the same material.[1] Silver is the best conductor, but copper, the next best, is cheaper and is used for connecting wire and in electric cables.

The ohm

If the current through a conductor is I when the p.d. across it is V, Fig. 60.1a, its resistance R is defined by

$$R = \frac{V}{I}$$

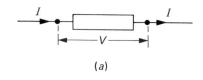

(a)   (b)

Fig. 60.1

This is a reasonable way to measure resistance since the smaller I is for a given V, the greater is R. If V is in volts and I in amperes, R is in *ohms* (Ω: pronounced omega). For example, if $I = 2$ A when $V = 12$ V, then $R = 12/2 = 6\ \Omega$.

The ohm is the resistance of a conductor in which the current is 1 ampere when a p.d. of 1 volt is applied across it.

Alternatively, if R and I are known, V can be found from

$$V = IR$$

Also, knowing V and R, I can be calculated from

$$I = \frac{V}{R}$$

The triangle in Fig. 60.1b is an aid to remembering the three equations. It is used like the 'density triangle' on p. 58.

[1] See Topic 81 for a mathematical treatment.

172

Resistors

Conductors intended to have resistance are called *resistors* (symbol ⌐□⌐) and are made either from wires of special alloys or from carbon. Those in radio and television sets have values from a few ohms up to millions of ohms, Fig. 60.2a.

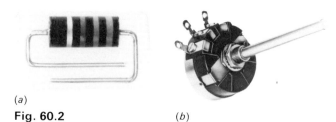

(a)
Fig. 60.2   (b)

Variable resistors are used in electronics (and are then called *potentiometers*) as volume and other controls, Fig. 60.2b. Larger current versions are useful in laboratory experiments and consist of a coil of constantan wire (an alloy of 60% copper, 40% nickel) wound on a tube with a sliding contact on a metal bar above the tube, Fig. 60.3.

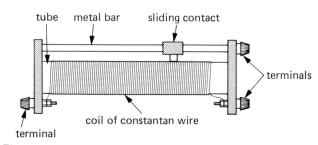

Fig. 60.3

There are two ways of using a variable resistor. It may be used as a *rheostat* for changing the current in a circuit; only one end connection and the sliding contact are then required. In Fig. 60.4a moving the sliding contact to the left reduces the resistance and increases the current. It can also act as a *potential divider* for changing the p.d. applied to a device, all

Fig. 60.4

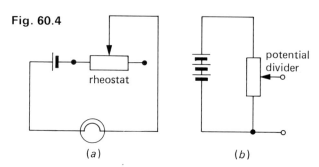

(a)   (b)

three connections being used. In Fig. 60.4*b* any fraction from the total p.d. of the battery to zero can be tapped off by moving the sliding contact down.

Experiment: measuring resistance

The resistance R of a conductor can be found by measuring the current I through it when a p.d. V is applied across it and then using $R = V/I$. This is called the *ammeter-voltmeter method*.

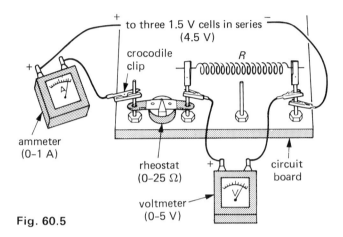

Fig. 60.5

Set up the circuit of Fig. 60.5 in which the unknown resistance R is 1 metre of S.W.G. 34 constantan wire. Altering the rheostat changes both the p.d. V and the current I. Record in a table, with three columns, five values of I (e.g. 0.10, 0.15, 0.20, 0.25 and 0.30 A) and the corresponding values of V.

Work out R for each pair of readings.

Repeat the experiment but instead of the wire use (**i**) a torch bulb (e.g. 2.5 V, 0.3 A), (**ii**) a semiconductor diode (e.g. 1 N4001) connected first one way then the other way round, (**iii**) a thermistor (e.g. TH 7).

I–V graphs: Ohm's law

The results of the previous experiment allow graphs of I against V to be plotted for different conductors.

(**a**) **Metallic conductors.** Metals and some alloys give I–V graphs which are a straight line through the origin, Fig. 60.6*a*, so long as their temperature is constant. I is directly proportional to V, i.e. $I \propto V$. Doubling V doubles I, etc. Such conductors obey *Ohm's law*, stated as follows.

The current through a metallic conductor is directly proportional to the p.d. across its ends if the temperature and other conditions are constant.

They are called *ohmic* or *linear* conductors and since $I \propto V$, it follows that $V/I = $ a constant (obtained from the slope of the I–V graph). The resistance of an

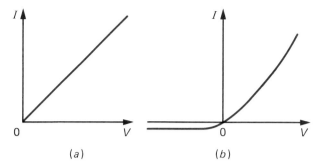

Fig. 60.6

ohmic conductor therefore does not change when the p.d. does.

(**b**) **Semiconductor diodes.** The typical I–V graph in Fig. 60.6*b* shows that current passes when the p.d. is applied in one direction but is almost zero when it acts in the opposite direction. A diode has a small resistance when connected one way round but a very large resistance when the p.d. is reversed. It conducts in one direction only and is a *non-ohmic* conductor. This makes it useful as a *rectifier* for changing alternating current (a.c.) to direct current (d.c.).

(**c**) **Filament lamp.** For a filament lamp, e.g. a torch bulb, the I–V graph bends over as V and I increase, Fig. 60.7*a*. That is, the resistance (V/I) increases as I

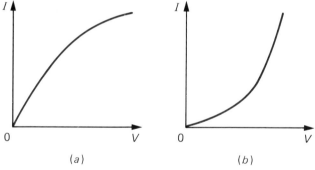

Fig. 60.7

increases and makes the filament hotter.

In general, an increase of temperature increases the resistance of metals but decreases the resistance of semiconductors. The resistance of most thermistors decreases if their temperature rises, i.e. their I–V graph bends up, Fig. 60.7*b*.

Resistors in series

The resistors in Fig. 60.8 are in series. The *same* current I flows through each and the total p.d. V across all three equals the separate p.d.s across them, i.e.

$$V = V_1 + V_2 + V_3$$

173

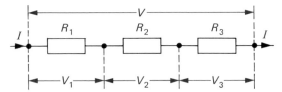

Fig. 60.8

But $V_1 = IR_1$, $V_2 = IR_2$ and $V_3 = IR_3$. Also if R is the combined resistance, $V = IR$ and so

$$IR = IR_1 + IR_2 + IR_3$$

Dividing both sides by I

$$R = R_1 + R_2 + R_3$$

Resistors in parallel

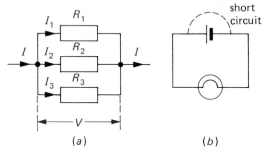

Fig. 60.9

The resistors in Fig. 60.9*a* are in parallel. The *p.d. V between the ends of each is the same* and the total current I equals the sum of the currents in the separate branches, i.e.

$$I = I_1 + I_2 + I_3$$

But $I_1 = V/R_1$, $I_2 = V/R_2$ and $I_3 = V/R_3$. Also if R is the combined resistance, $V = IR$ and so

$$\frac{V}{R} = \frac{V}{R_1} + \frac{V}{R_2} + \frac{V}{R_3}$$

Dividing both sides by V

$$\frac{1}{R} = \frac{1}{R_1} + \frac{1}{R_2} + \frac{1}{R_3}$$

For the simpler case of *two* resistors in parallel

$$\frac{1}{R} = \frac{1}{R_1} + \frac{1}{R_2} = \frac{R_2}{R_1 R_2} + \frac{R_1}{R_1 R_2}$$

$$\therefore \frac{1}{R} = \frac{R_2 + R_1}{R_1 R_2}$$

Inverting both sides

$$R = \frac{R_1 R_2}{R_1 + R_2} = \frac{\text{product of resistances}}{\text{sum of resistances}}$$

174

Short circuit

In Fig. 60.9*b* if a short length of copper wire is connected across the battery (the dotted line), the lamp goes out. A large current flows in the copper wire which gets very hot. It acts as a 'short circuit' and provides an easy path for the current because of its low resistance. Short circuits are to be avoided (see p. 183).

✎ Worked example

A p.d. of 24 V from a battery is applied to the network of resistors in Fig. 60.10a.

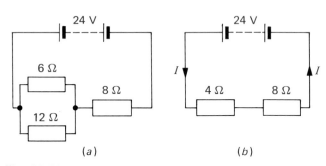

Fig. 60.10

(a) What is the combined resistance of the 6 Ω and 12 Ω resistors in parallel?
(b) What is the current in the 8 Ω resistor?
(c) What is the p.d. across the parallel network?
(d) What is the current in the 6 Ω resistor?

(a) Let R_1 = resistance of 6 Ω and 12 Ω in parallel

$$\therefore \frac{1}{R_1} = \frac{1}{6} + \frac{1}{12} = \frac{2}{12} + \frac{1}{12} = \frac{3}{12}$$

$$\therefore R_1 = \frac{12}{3} = 4 \ \Omega$$

(b) Let R = *total* resistance of circuit = $4 + 8 = 12 \ \Omega$. The equivalent circuit is shown in Fig. 60.10*b* and if I is the current in it then since $V = 24$ V

$$I = \frac{V}{R} = \frac{24}{12} = 2 \text{ A}$$

$\therefore$ Current in 8 Ω resistor = 2 A.

(c) Let V_1 = p.d. across parallel network

$$\therefore V_1 = I \times R_1 = 2 \times 4 = 8 \text{ V}$$

(d) Let I_1 = current in 6 Ω resistor, then since $V_1 = 8$ V

$$I_1 = \frac{V_1}{6} = \frac{8}{6} = \frac{4}{3} \text{ A}$$

Ⓠ Questions

1. What is the resistance of a lamp when a p.d. of 12 V across it causes a current of 4 A?

2. Calculate the p.d. across a 10 Ω resistor carrying a current of 2 A.

3. The p.d. across a 2 Ω resistor is 4 V. What is the current flowing (in ampere)?

    A $\frac{1}{2}$ B 1   C 2   D 6   E 8
                          (*J.M.B./A.L./N.W.16+*)

4. Three 5 Ω resistors are connected as shown in Fig. 60.11. Their effective total resistance is

    A $\frac{3}{5}$Ω   B $1\frac{2}{3}$Ω   C 5 Ω   D 15 Ω   E 125 Ω
                          (*O.L.E./S.R.16+*)

Fig. 60.11   5 Ω   5 Ω   5 Ω

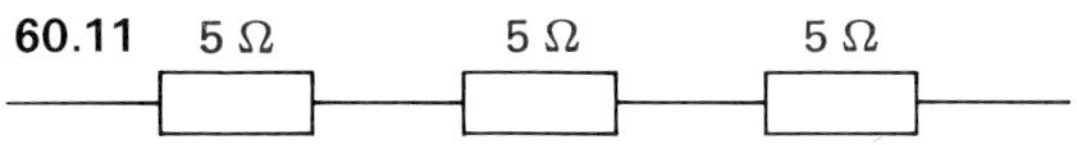

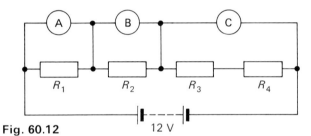

5. The resistors R_1, R_2, R_3 and R_4 in Fig. 60.12 are all equal in value. What would you expect the voltmeters A, B and C to read, assuming that the connecting wires in the circuit have negligible resistance?

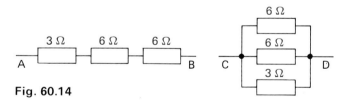

Fig. 60.12

6. Calculate the effective resistance between A and B in Fig. 60.13. *(E.M.)*

Fig. 60.13

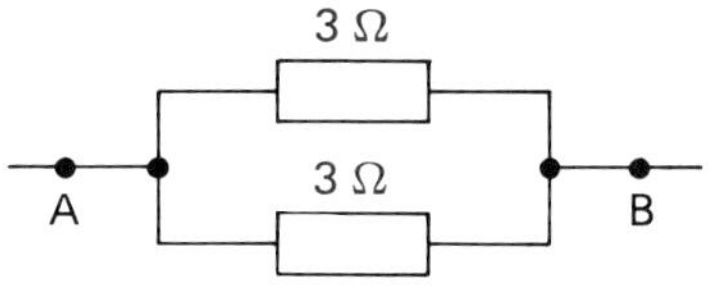

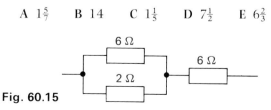

7. What is the effective resistance in Fig. 60.14 between **(a)** A and B, **(b)** C and D?

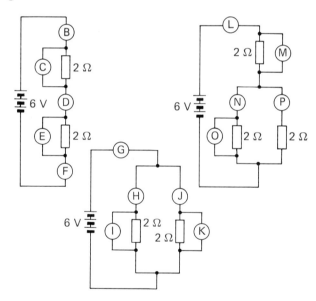

Fig. 60.14

8. Fig. 60.15 shows three resistors. Their combined resistance in ohm is

    A $1\frac{5}{7}$ B 14   C $1\frac{1}{5}$ D $7\frac{1}{2}$ E $6\frac{2}{3}$

Fig. 60.15

9. Fig. 60.16 shows a 2 V cell connected to an arrangement of resistors part in series and part in parallel.

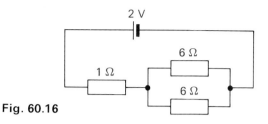

Fig. 60.16

    (a) What is the total resistance of the two 6 Ω resistors in parallel?
    (b) What is the current flowing in the 1 Ω resistor?
    (c) What is the current in one of the 6 Ω resistors?
                           (*N.W.*)

10. A 4 Ω coil and a 2 Ω coil are connected in parallel. What is their combined resistance? A total current of 3 A passes through the coils. What current flows through the 2 Ω coil?

11. In Fig 60.17 the circles are either ammeters or voltmeters. State what each is and the reading it would show. *(E.A)*

Fig. 60.17

12. A boy is given a 12 V lamp and decides to measure the resistance of the lamp filament using the voltmeter-ammeter method. He decides to apply various voltages to the lamp and to measure the current in each case.

    (a) Draw a circuit diagram, showing clearly where the voltmeter and ammeter are placed in the circuit.
    (b) Two of the boy's results are given below:
        voltmeter reading 2.0 V   12 V
        ammeter reading 1.0 A   2.0 A
    Calculate the resistance of the lamp filament in each case.
    (c) Explain why the resistance of the lamp filament is different in the two cases.

175

check list p.289

61 Electromotive force

E.M.F. and terminal p.d.

A voltmeter connected across the terminals of a cell measures the *terminal p.d.* of the cell. The reading depends on the current being supplied; it is greatest when the cell is on 'open circuit', i.e. not supplying current, Fig. 61.1a.

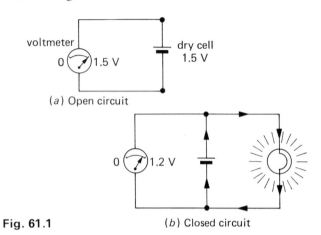

Fig. 61.1

(a) Open circuit

(b) Closed circuit

When a circuit is connected to the cell, it is on 'closed circuit' and the reading is less, Fig. 61.1b. It decreases as the current supplied increases. The terminal p.d. of a cell on open circuit is thus greater than its terminal p.d. on closed circuit.

If we look upon a cell as a device which *supplies* electrical energy, we can define its terminal p.d. on open circuit as the number of joules of electrical energy it gives to each coulomb. If a voltmeter across the cell reads 1.5 V then each coulomb is supplied with 1.5 J of electrical energy and the cell is said to have an *electromotive force* (e.m.f., denoted by E) of 1.5 V.

The e.m.f. of a source of electrical energy is its terminal p.d. on open circuit.

A high resistance voltmeter must be used to measure the e.m.f. of a cell. Otherwise appreciable current flows through the voltmeter and the terminal p.d. measured is not that on open circuit.

Like p.d., e.m.f. is stated in volts and it too is often called a 'voltage': it has the same value for all cells made of the same materials.

The term e.m.f. is misleading to some extent; it measures energy per unit charge and not force but it does cause charges to move in the circuit. Electromotive energy would be better.

176

Internal resistance

The terminal p.d. of a cell on closed circuit is also the p.d. applied to the external circuit. For example, in Fig. 61.1b, the voltmeter reading is the same (i.e. 1.2 V) if it is connected across the lamp terminals instead of the cell terminals. (The resistance of the connecting wires from the cell to the lamp is negligible, as is the p.d. across them.)

In an external circuit electrical energy is *changed* into other forms of energy and we regard the terminal p.d. of a cell on closed circuit as being the number of joules of electrical energy changed by each coulomb in the external circuit. If the terminal p.d. on closed circuit is 1.2 V then each coulomb changes 1.2 J of electrical energy.

Not all the electrical energy supplied by a cell to each coulomb is changed in the external circuit. The 'lost' energy per coulomb is due to the cell itself having resistance. Each coulomb has to 'waste' some energy—0.3 joule in the above example—to get through the cell itself and so less is available for the external circuit. The resistance of a cell is called its *internal resistance* (r) and depends among other things on its size.

The circuit equation

Taking stock of the energy changes in a complete circuit, including the cell, we can say from the principle of conservation of energy:

energy supplied per coulomb by cell	=	energy changed per coulomb in external circuit	+	energy *wasted* per coulomb on cell resistance

Or, from the definitions of e.m.f. and p.d.,

e.m.f. = useful p.d. + 'lost' p.d.   (1)

Hence the e.m.f. in a circuit equals the p.d. across the external resistance + the p.d. across the internal resistance.

If E is the e.m.f. of a cell of internal resistance r, Fig. 61.2a, and V is its terminal p.d. when sending a current I through a resistor R, Fig. 61.2b, then from equation (1),

$$E = V + v \qquad (2)$$

where v is the 'lost' p.d. It is a quantity which cannot be measured directly by a voltmeter but is only found by subtracting V from E. In equation (2), called the *circuit equation*, we can write $V = IR$ and $v = Ir$ since the current all round the circuit is I.

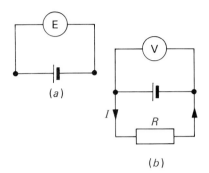

Fig. 61.2

Suppose $E = 1.5$ V, $V = 1.2$ V and $I = 0.30$ A, then replacing in (2),

$$1.5 = 1.2 + v \quad \therefore \quad v = 1.5 - 1.2 = 0.30 \text{ V}$$

But

$$v = Ir$$
$$\therefore \ 0.30 = 0.30 \times r$$
$$\therefore \quad r = 1.0 \ \Omega$$

Also

$$V = IR$$
$$\therefore \ 1.2 = 0.30 \times R$$
$$\therefore \quad R = 1.2/0.30 = 4.0 \ \Omega$$

The effect of internal resistance can be seen when a car starts with its lights on. Suppose the electric motor which starts the engine needs a current $I = 100$ A and is connected to a battery of e.m.f. $E = 12$ V and internal resistance $r = 0.040 \ \Omega$. The 'lost' p.d. $v = Ir = 100 \times 0.040 = 4.0$ V. The terminal p.d. of the battery when it sends 100 A through the starter motor $= V = E - v = 12 - 4.0 = 8.0$ V. If the lights are meant to be fully bright when there is a p.d. of 12 V across them, they will be dim on 8.0 V, as is observed.

Q Questions

1. (a) What current will flow in the circuit of Fig. 61.3?
(b) What is the p.d. across the 3 Ω resistor in this circuit?

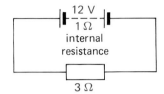

Fig. 61.3

2. Fig. 61.4 shows a battery of e.m.f. 24 V and internal resistance 3 Ω in series with a resistor of resistance 9 Ω. The current will be

A $\frac{3}{8}$ A   B $\frac{1}{2}$ A   C 2 A   D $3\frac{2}{3}$ A   E 8 A
(O.L.E./S.R.16+)

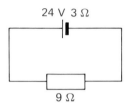

Fig. 61.4

3. In Fig. 61.5, the cell has an e.m.f. of 1.5 V. The voltmeter reading is 1.35 V. The ammeter records the value of the current as 0.30 A. Calculate (a) the internal resistance of the cell, (b) the value of the resistance R.
(J.M.B./A.L./N.W.16+)

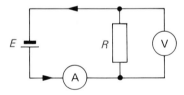

Fig. 61.5

4. A high resistance voltmeter reads 3.0 V when connected across the terminals of a battery on open circuit and 2.6 V when the battery sends a current of 0.20 A through a lamp.

What is (a) the e.m.f. of the battery, (b) the terminal p.d. of the battery when supplying 0.20 A, (c) the p.d. across the lamp, (d) the 'lost' p.d., (e) the internal resistance of the battery and (f) the resistance of the lamp?

check list
p.289

62 Additional questions

Core level (all students)

Permanent magnets: magnetic fields

1. A bar magnet with north and south poles marked N and S respectively and a bar of iron PQ are fixed a short distance apart as shown in Fig. 62.1. A compass needle is brought in turn to ends P, Q, N and S. Its N pole will point towards

    A both ends of the iron but only to the N pole of the magnet
    B both ends of the iron but only to the S pole of the magnet
    C end P only of the iron and only to the S pole of the magnet
    D end Q only of the iron and only to the N pole of the magnet
    E end Q only of the iron and only to the S pole of the magnet.   *(O.L.E./S.R.16+)*

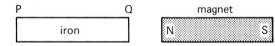

Fig. 62.1

2. How could you use a solenoid to **(i)** magnetize a bar of steel so that its polarity is known, **(ii)** demagnetize a magnet?

Explain in terms of the simple domain theory of ferromagnetism, the process of magnetization to saturation.   *(J.M.B.)*

Static electricity: more electrostatics

3. In the process of induction in electrostatics

    A a conductor is rubbed with an insulator
    B a charge is produced by friction
    C negative and positive charges are separated
    D a positive charge induces a positive charge
    E electrons are 'sprayed' into an object.   *(J.M.B.)*

Electric current: potential difference: resistance

4. If two resistors of 5 Ω and 15 Ω are joined together in series and then placed in parallel with a 20 Ω resistor, the effective resistance in ohms of the combination is

    A 0.1   B 10   C 20   D 40   E 400   *(N.I.)*

5. V_1, V_2, V_3 are the p.d.s across the 2 Ω, 3 Ω and 4 Ω resistors respectively in Fig. 62.2 and the current is 5 A. Which one of the columns A to E shows the correct values of V_1, V_2 and V_3 measured in volts?

	A	B	C	D	E
V_1	2.0	10	0.4	2.5	3.0
V_2	3.0	15	0.6	1.6	2.0
V_3	4.0	20	0.8	1.25	1.0

   (J.M.B.)

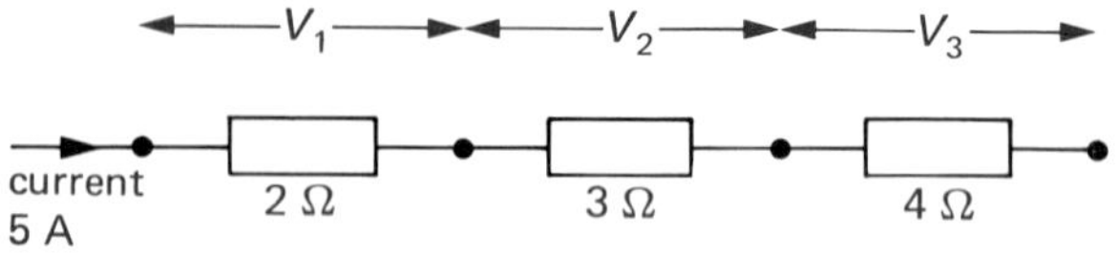

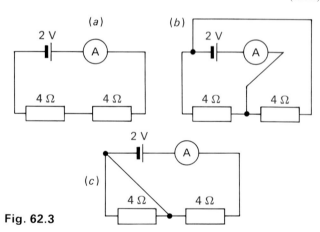

Fig. 62.2

6. Two 4 Ω resistors are connected to a 2 V cell and an ammeter as shown in Fig. 62.3*a,b,c*. Find the reading on the ammeter in each case. The resistance of the connecting wires, the meter and the cell may be neglected.   *(E.A.)*

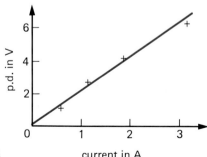

Fig. 62.3

7. (a) The graph in Fig. 62.4 illustrates how the p.d. across the ends of a conductor is related to the current flowing through it. **(i)** What law may be deduced from the graph? **(ii)** What is the resistance of the conductor? **(b)** Draw diagrams to show how six 2 V lamps could be lit to normal brightness when using a **(i)** 2 V supply, **(ii)** 6 V supply and **(iii)** 12 V supply.

Fig. 62.4

8. When a 6 Ω resistor is connected across the terminals of a 12 V battery, the number of coulombs passing through the resistor per second is

A 0.5   B 2   C 6   D 12   E 72   (N.I.)

9. In Fig. 62.5, what is the voltmeter reading if the ammeter reads 2 A and the resistance of the voltmeter is 900 Ω?

A 50   B 180   C 200   D 450   E 1800
(N.I.)

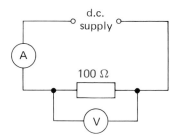

Fig. 62.5

Further level (higher grade students)

Permanent magnets: magnetic fields

10. A powerful bar magnet is placed horizontally with its poles in line with the magnetic meridian, and the two small regions are found on the axis of the magnet where a small plotting compass sets in a random manner. With the aid of a diagram on which the poles of the magnet are marked, account for this effect. On another diagram show how similar regions could be produced on a horizontal line through the mid-point of the magnet and perpendicular to its axis.   (L.)

Static electricity: more electrostatics

11. Fig. 62.6 shows two similar small conducting spheres suspended by light insulating threads from a point P. The spheres carry equal negative charges.

(a) Copy the figure and show by arrows drawn on it, the directions of the forces acting on sphere A.
(b) Against each arrow, write the name of the force it represents.
(c) If a large vertical earthed metal plate were placed between the spheres in the position MN, state and explain what would happen.   (O.L.E.)

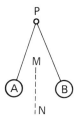

Fig. 62.6

Electric current: potential difference: resistance

12. Two electric light bulbs, both marked 0.3 A, 4.5 V are connected (a) in parallel, (b) in series across a 4.5 V battery of negligible internal resistance. Assume that the resistance of the filament does not change. In each case (a) and (b), (i) state what might be seen, (ii) calculate the currents through each bulb, and (iii) calculate the current supplied by the battery.   (J.M.B.)

13. Resistors are connected as in Fig. 62.7 and a p.d. of 12 V is applied across them. S1 and S2 are switches. Calculate
(a) the total resistance when both switches S1 and S2 are closed,
(b) the total current supplied when both switches S1 and S2 are closed,
(c) the current in the 3 Ω resistor when switch S1 is closed and switch S2 open,
(d) the current supplied when both switches are open.   (W.)

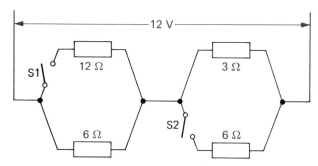

Fig. 62.7

14. In the circuit of Fig. 62.8, C is a 2 V cell of negligible internal resistance. S_1 and S_2 are switches. The current taken by the voltmeter and the resistance of the ammeter can both be neglected.
(a) Calculate the readings of the ammeter and the voltmeter when the switches are as shown with S_1 closed and S_2 open. Explain your calculation.
(b) What will the meters read when S_1 and S_2 are both closed? Explain.
(c) What will the meters read when S_1 and S_2 are both open? Explain.   (O.L.E.)

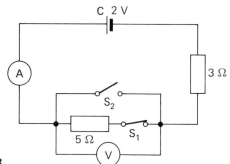

Fig. 62.8

179

15. Fig. 62.9 shows two ammeters X and Y arranged in parallel and connected to a battery and a resistor in order to determine the current supplied by the battery. Both give a full-scale deflection; Y reads 3.0 A, X reads 2.0 A.

(a) What is the current supplied by the battery?
(b) Explain why neither ammeter, used alone, is suitable for finding the current supplied.
Given that the resistance of Y is 0.04 Ω, calculate
(c) the p.d. between the points labelled P and Q,
(d) the resistance of X. 					(C.)

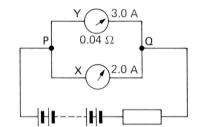

Fig. 62.9

16. Describe with the aid of a circuit diagram how you would measure the resistance of a coil of wire using a voltmeter/ammeter method.

Calculate the combined resistance of 3 resistors, each 6 Ω, when connected (a) in series, (b) in parallel. Draw a sketch showing one other way in which these 3 resistors could be connected and calculate the combined resistance.

(S. part qn.)

Electromotive force

17. A battery consists of 3 cells, each of *electromotive force* $E = 1.5$ V and *internal resistance* $r = 0.10$ Ω, connected in series. Explain the meaning of the terms in italics.

Is either of these affected by altering the dimensions of the cell?

A resistor of 1.48 Ω is put in series with the battery, and an ammeter of resistance 0.02 Ω is connected so as to measure the current through it. What is the ammeter reading?

18. Describe, giving a circuit diagram, an experiment which would show whether a particular cell has an appreciable internal resistance. Indicate briefly what readings would be needed in order to estimate its value. (You are not required to derive the equation for this.)

Explain why a check against a short-circuit should be made when using secondary cells such as lead plate accumulators, whereas this is not so important when dry cells are used.

Resistances of 2 Ω and 3 Ω are connected in series with a cell. A high resistance voltmeter connected across the 3 Ω resistor reads 1.0 V but this increases to 1.2 V when an extra 2 Ω resistor is connected in parallel with the first 2 Ω resistor. Calculate the *electromotive force* and *internal resistance* of the cell. 					(L.)

19. A cell in series with a 2 Ω resistor and a switch has a high-resistance voltmeter across it, Fig. 62.10. The voltmeter reads 1.5 V with the switch open, and 1.2 V with it closed.

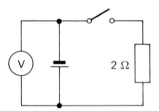

Fig. 62.10

(a) What is the electromotive force of the cell? Explain.
(b) What is the current through the 2 Ω resistor with the switch closed?
(c) A further resistor is added in series with the 2 Ω resistor making the current 0.25 A with the switch closed. Calculate its value.
(d) What current would flow if the cell were short-circuited?
(e) What would the voltmeter read if the cell were short-circuited? 					(O.L.E./S.R.16+)

180

63 Electric power

Power in electric circuits

In many circuits it is important to know the rate at which electrical energy is being changed into other forms of energy. Earlier (p. 77) we said that *energy changes were measured by the work done* and power was defined by the equation

$$power = \frac{work\ done}{time\ taken} = \frac{energy\ change}{time\ taken}$$

In symbols

$$P = \frac{W}{t} \qquad (1)$$

where P is in watts (W) if W is in joules (J) and t in seconds (s).

From the definition of p.d. (pp. 169–70) we saw that if W is the electrical energy changed when a steady current I (in amperes) passes for time t (in seconds) through a device (e.g. a lamp) with a p.d. V (in volts) across it, Fig. 63.1, then

$$W = ItV \qquad (2)$$

Substituting for W in (1) we get

$$P = \frac{W}{t} = \frac{ItV}{t} = IV$$

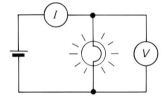

Fig. 63.1

Therefore to calculate the power P of an electrical appliance we multiply the current I through it by the p.d. V across it. For example if a lamp on a 240 V supply has a current of 0.25 A through it, its power is $240 \times 0.25 = 60$ W. The lamp is changing 60 J of electrical energy into heat and light each second. Larger units of power are the *kilowatt* (kW) and the *megawatt* (MW) where

$$1\ kW = 1000\ W \quad and \quad 1\ MW = 1\,000\,000\ W$$

In units

$$WATTS = AMPERES \times VOLTS \qquad (3)$$

It follows from (3) that since

$$VOLTS = \frac{WATTS}{AMPERES} \qquad (4)$$

the volt can be defined as a *watt per ampere* and p.d. calculated from (4).

If all the energy is changed to heat in a resistor of resistance R, then $V = IR$ and the rate of production of heat is given by

$$P = V \times I = IR \times I = I^2R$$

That is, if the current is doubled, four times as much heat is produced per second. Also, $P = V^2/R$.

Experiment: measuring electric power

(a) Lamp. Connect the circuit of Fig. 63.2. Note the ammeter and voltmeter readings and work out the electric power supplied to the bulb in watts.

Fig. 63.2

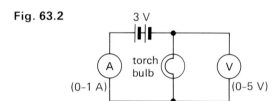

(b) Motor. Replace the bulb in Fig. 63.2 by a small electric motor. Attach a known mass m (in kg) to the axle of the motor with a length of thin string and find the time t (in s) required to raise the mass through a known height h (in m) at a steady speed. Then the power output P_o (in W) of the motor is given by

$$P_o = \frac{work\ done\ in\ raising\ mass}{time\ taken} = \frac{mgh}{t}$$

If the ammeter and voltmeter readings I and V are noted while the mass is being raised, the power input P_i (in W) can be found from

$$P_i = IV$$

The efficiency of the motor is given by

$$efficiency = \frac{P_o}{P_i} \times 100\%$$

Also investigate the effect of a *greater mass* on **(i)** the speed, **(ii)** the power output and **(iii)** the efficiency, of the motor at its rated p.d.

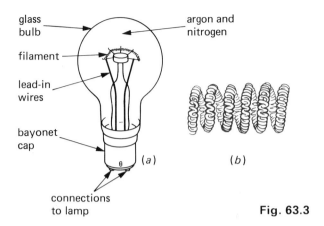

Fig. 63.3

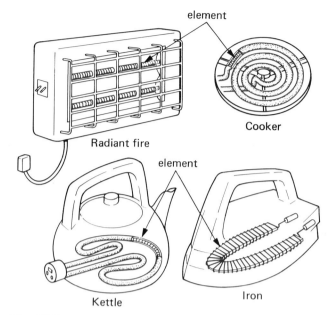

Radiant fire

Cooker

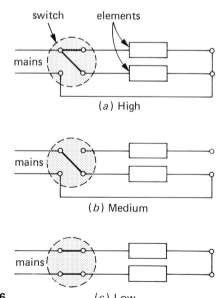

Kettle

Iron

Fig. 63.5

Electric lighting

(a) Filament lamps, Fig. 63.3a. The filament is a small coiled coil of tungsten wire, Fig. 63.3b, which becomes white hot when current flows through it. The higher the temperature of the filament the greater is the proportion of electric energy changed to light and for this reason it is made of tungsten, a metal with a high melting point (3400 °C).

Most lamps are gas-filled and contain nitrogen and argon, not air. This reduces evaporation of the tungsten which would otherwise condense on the bulb and blacken it. The coiled coil, being compact, is cooled less by convection currents in the gas.

(b) Fluorescent lamps. A filament lamp changes only 10% of the electrical energy supplied into light; the other 90% becomes heat. Fluorescent lamps are five times as efficient and may last 3000 hours compared with the 1000-hour life of filament lamps. They cost more to install but running costs are less.

Fig. 63.4

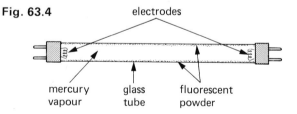

A simplified diagram of a fluorescent lamp is shown in Fig. 63.4. When the lamp is switched on, the mercury vapour emits ultraviolet radiation (invisible) which makes the powder on the inside of the tube fluoresce (glow), i.e. light (visible) is emitted. Different powders give different colours.

Electric heating

(a) Heating elements. In domestic appliances such as electric fires, cookers, kettles and irons the 'elements', Fig. 63.5, are made from Nichrome wire.

This is an alloy of nickel and chromium which does not oxidize (and become brittle) when the current makes it red hot.

The elements in *radiant* electric fires are at red heat (about 900 °C) and the radiation they emit is directed into the room by polished reflectors. In *convector* types the element is below red heat (about 450 °C) and is designed to warm air which is drawn through the heater by natural or forced convection. In *storage* heaters the elements heat fire-clay bricks during the night using 'off-peak' electricity. On the following day these cool down, giving off the stored heat to warm the room.

Fig. 63.6

(a) High

(b) Medium

(c) Low

(b) Three-heat switch. This is sometimes used to control heating appliances. It has three settings and uses two identical elements. On 'high', the elements are in parallel across the supply voltage, Fig. 63.6a; on 'medium', current only passes through one, Fig. 63.6b; on 'low', they are in series, Fig. 63.6c.

(c) Fuses. A fuse is a short length of wire of material with a low melting point (often tinned copper), which melts and breaks the circuit when the current through it exceeds a certain value. Two reasons for excessive currents are 'short circuits' due to worn insulation on connecting wires, and overloaded

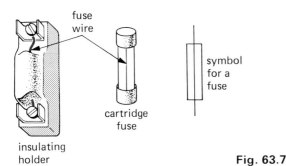

insulating holder

Fig. 63.7

circuits. Without a fuse the wiring would become hot in these cases and could cause a fire. A fuse should ensure that the current-carrying capacity of the wiring is not exceeded. In general the thicker a cable is, the more current it can carry, but each size has a limit.

Two types of fuse are shown in Fig. 63.7. *Always switch off before replacing a fuse.*

Ⓠ Questions

1. How much electrical energy in *joules* does a 100 watt lamp change in **(a)** 1 second, **(b)** 5 seconds, **(c)** 1 minute?

2. **(a)** What is the power of a lamp rated at 12 V 2 A? **(b)** How many joules of electrical energy are changed per second by a 6 V 0.5 A lamp?

3. The largest number of 100 W bulbs which can safely be run from a 240 V supply with a 5 A fuse is
    A 2   B 5   C 10   D 12   E 20

4. What is the maximum power in kilowatts of the appliance(s) that can be connected safely to a 13 A 240 V mains socket?

5. A current of 2 A passes through a resistance of 4 Ω. Calculate **(a)** the p.d. between the ends of the resistance, **(b)** the power used by the resistance.   *(E.M.)*

6. A bulb is labelled 12 V 36 W. When used on a 12 V supply, **(a)** what current will it take, **(b)** what is its resistance?   *(A.L.)*

7. A 3 kW electric fire is designed to be run on a 250 V supply. Assuming it to be operating at the correct voltage, **(a)** what current will it draw from the supply, **(b)** what is the resistance of the element of the fire?   *(N.W.)*

check list P.289

64 Electricity in the home

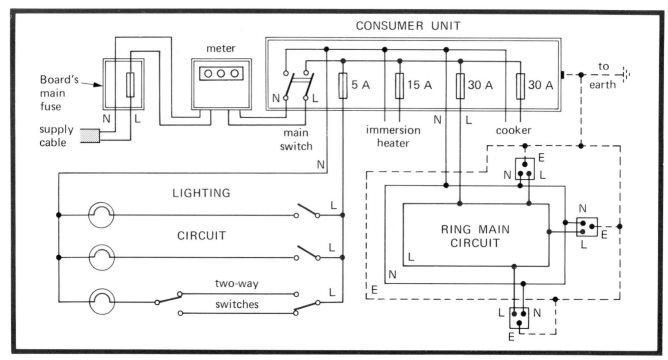

Fig. 64.1

House circuits

Electricity usually comes to our homes by an underground cable containing two wires, the *live* (L) and the *neutral* (N). The neutral is earthed at the local sub-station and so there is no p.d. between it and earth. The supply is a.c. (p. 203) and the live wire is alternately positive and negative. Study the modern house circuit shown in Fig. 64.1.

(a) Circuits in parallel. Every circuit is connected in parallel with the supply, i.e. across the live and neutral, and receives the full mains p.d. of 240 V.

(b) Switches and fuses. These are always in the live wire. If they were in the neutral, lamp and power sockets would be 'live' when switches were 'off' or fuses 'blown'. A shock (fatal) could then be obtained by, for example, touching the element of an electric fire when it was switched off.

(c) Staircase circuit. The lamp is controlled from two places by the two two-way switches.

(d) Ring main circuit. The live and neutral wires each run in two complete rings round the house, and the power sockets, each rated at 13 A, are tapped off from them. Thinner wires can be used since the current to each socket flows by two paths, i.e. in the

whole ring. The ring has a 30 A fuse and if it has ten sockets all can be used so long as the total current does not exceed 30 A.

(e) Fused plug. Only one type of plug is used in a ring main circuit. It is wired as in Fig. 64.2 and has its own cartridge fuse, 3 A (red) for appliances with powers up to 720 W and 13 A (brown) for those between 720 W and 3 kW.

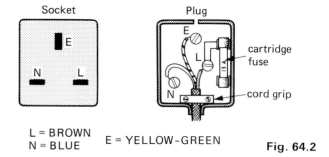

L = BROWN
N = BLUE   E = YELLOW–GREEN

Fig. 64.2

(f) Earthing and safety. A ring main has a third wire which goes to the top sockets on all power points, Fig. 64.1, and is earthed by being connected either to a *metal* water pipe in the house or to an earth connection on the supply cable. This third wire is a safety precaution to prevent electric shock should an appliance develop a fault.

The earth pin on a three-pin plug is connected to the metal case of the appliance which is thus joined to earth by a path of almost zero resistance. If then, for example, the element of an electric fire breaks or sags and touches the case, a large current flows to earth and 'blows' the fuse. Otherwise the case becomes 'live' and anyone touching it receives a shock *which might be fatal*, especially if they were 'earthed' by, say, standing in a damp environment, e.g. on a wet concrete floor.

(g) Circuit breakers. Fig. 64.3. These are used in consumer units instead of fuses. They contain an electromagnet (Topic 66) which, when the current exceeds the rated value of the circuit breaker, becomes strong enough to separate a pair of contacts and breaks the circuit. They operate much faster than fuses and are reset by pressing a button.

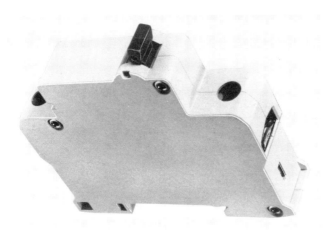

Fig. 64.3

The *residual current circuit breaker* (R.C.C.B.), also called a *residual current device* (R.C.D.), is an adapted circuit breaker which is used when the resistance of the earth path between the consumer and the sub-station is not small enough for a fault-current to blow the fuse (or circuit breaker). It works by detecting any difference between the currents in the live and neutral wires; when these become unequal due to an earth fault (i.e. some of the current returns to the sub-station via the case of the appliance and earth) it breaks the circuit before there is any danger.

An R.C.D. should be plugged into a socket supplying power to a portable appliance such as an electric lawnmower or hedge trimmer. In these cases the risk of electrocution is greater because the user is generally making a good earth connection through his feet.

(h) Double insulation. Appliances such as vacuum cleaners, hair dryers and food mixers are usually double-insulated. Connection to the supply is by a 2-core insulated cable, with no earth wire, and the appliance is enclosed in an insulating plastic case. Any metal attachments which the user might touch are fitted into this case so that they do not make a direct connection with the internal electrical parts, e.g. a motor. There is then no risk of a shock should a fault develop.

(i) Radial circuit. Some appliances, e.g. electric cookers, are fed *directly* from the consumer unit by a *radial circuit*. This is a cable which, because it may have to carry 30 A or more, is thicker to prevent overheating and the risk of fire.

Experiment: house circuits

On a circuit board connect up and investigate **(a)** the staircase circuit of Fig. 64.4*a* and **(b)** the ring main circuit of Fig. 64.4 *b*.

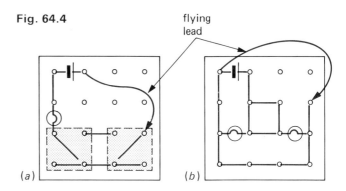

Fig. 64.4

Paying for electricity

Electricity Boards charge for the *electrical energy* they supply. A joule is a very small amount of energy and a larger unit, the *kilowatt-hour* (kWh) is used.

A kilowatt-hour is the electrical energy used by a 1 kW appliance in 1 hour.

A 3 kW electric fire working for 2 hours uses 6 kWh of electrical energy—usually called 6 'units'. Electricity meters are marked in kWh: the latest have digital readouts like the one in Fig. 64.5. At present a 'unit' costs about 5p.

Typical powers of some appliances are:

Lamps	60, 100 W	Fire	1, 2, 3 kW
Fridge	150 W	Kettle	2–3 kW
TV set	200 W	Immersion heater	3 kW
Iron	750 W	Cooker	8 kW

Note that

$$1 \text{ kWh} = 1000 \text{ J/s} \times 3600 \text{ s}$$
$$= 3\,600\,000 \text{ J} = 3.6 \text{ MJ}$$

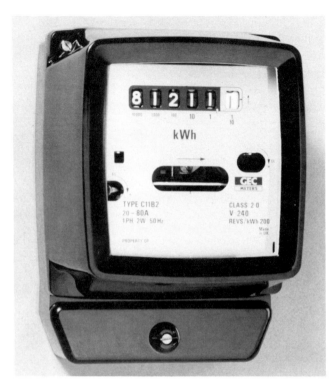

Fig. 64.5

Ⓠ Questions

1. The circuits of Fig. 64.6a,b, show 'short circuits' between the live (L) and neutral (N) wires. In both, the fuse has blown but whereas (a) is now safe, (b) is still dangerous even though the lamp is out and suggests the circuit is safe. Explain.

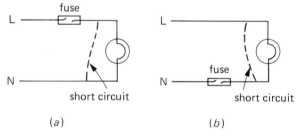

Fig. 64.6

2. What steps should be taken before replacing a blown fuse in a plug?

3. What size fuse (3 A or 13 A) should be used in a plug connected to (a) a 150 W TV set, (b) a 750 W electric iron, (c) a 2 kW electric kettle, if the supply is 240 V?

4. An electric cooker has an oven rated at 3 kW, a grill rated at 2 kW and two rings each rated at 500 W. The cooker operates from 240 V mains.
   (a) Are the different heating elements connected in series or in parallel? Give a reason for your answer.
   (b) Would a 30 A fuse be suitable for the cooker, assuming that all parts are switched on? Show clearly how you obtain your answer.
   (c) What is the cost of operating all the parts for 30 minutes if electricity costs 5p per unit?   (E.A.)

5. Copy the diagrams of the electrical fitting of a domestic installation shown in Fig. 64.7. Connect up these fittings so that (i) the three lamps will be controlled separately, (ii) a supply is taken to the 13 A socket.

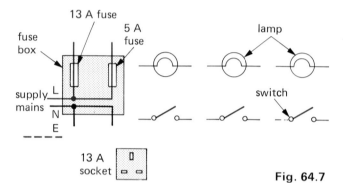

Fig. 64.7

6. If a 2 kW electric heater is used for 10 hours and electricity costs 5p per unit, (a) how many units (kWh) are used, (b) what is the cost?   (A.L.)

7. What is the cost of heating a tank of water with a 3000 W immersion heater for 80 minutes if electricity costs 5p per kWh?

8. (a) A lamp is marked 240 V 60 W. What exactly does this tell you about the bulb?
   (b) A boy has a large number of 240 V 60 W coloured bulbs he wishes to use for decorations so that the bulbs operate normally.
      (i) How many can he connect to a 240 V supply through a 5 A fuse?
      (ii) What would be the total power of the circuit?
      (iii) Draw a diagram of the circuit he would employ.
   (c) If electrical energy costs 5p per unit, what will be the cost of running the above circuit for 5 hours a night for 14 nights?   (S.E.)

check list p.289

65 Electric cells

In an electric cell an *electrolyte* reacts chemically with two *electrodes*, making one have a positive electric charge and the other a negative charge. Chemical energy becomes electrical energy.

Primary cells[1]

A primary cell is one which is discarded when the chemicals are used up.

(a) Simple cell, Fig. 65.1. It has an e.m.f. (voltage) of about 1.0 V but stops working after a short time due to 'polarization', i.e. the collection of hydrogen gas bubbles on the copper plate. The cell is depolarized by

Fig. 65.1

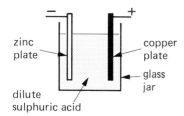

zinc plate

copper plate

glass jar

dilute sulphuric acid

adding potassium dichromate which oxidizes the hydrogen to water. A second defect is 'local action'. This is due to impurities in the zinc and results in the zinc being used up even when current is not supplied. The simple cell is no longer used.

(b) Zinc-carbon (Leclanché or dry) cell, Fig. 65.2. It has a zinc negative electrode, a manganese(IV) oxide positive electrode (which also acts as a slow depolarizer) and the electrolyte is a solution of ammonium

Fig. 65.2

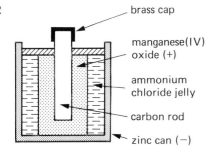

brass cap

manganese(IV) oxide (+)

ammonium chloride jelly

carbon rod

zinc can (−)

chloride. The carbon rod is in contact with the positive electrode (but is not involved in the chemical reaction) and is called the 'current collector'. The e.m.f. is 1.5 V and the internal resistance about 0.5 Ω. This is the most popular cell for low current (e.g. 0.3 A) or occasional use, e.g. in torches.

[1] Topic 81 describes some new types of cell.

Secondary cells

Secondary cells or accumulators can be recharged[2] by passing a current through them in the opposite direction to that in which they supply one. They are much more expensive than primary cells.

In the *lead-acid cell*, the positive plate is lead(IV) oxide (brown) and the negative one lead (grey). The electrolyte is dilute sulphuric acid, Fig. 65.3. During discharge, both electrodes change to lead sulphate and the acid becomes more dilute. When fully charged the density of the acid (measured by a hydrometer, p. 92) is 1.25 g/cm^3, which falls to 1.18 at full discharge. A 'flat' cell should not be left discharged or the lead sulphate hardens and cannot be changed back to lead and lead oxide.

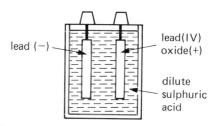

lead (−)

lead(IV) oxide(+)

dilute sulphuric acid

Fig. 65.3

The e.m.f. is steady at 2.0 V and the low internal resistance of 0.01 Ω allows quite large continuous currents to be supplied. A 12 V car battery consists of six lead-acid cells in series and is recharged by an alternator or dynamo (Topic 69) driven by the engine.

Maintenance-free sealed types are now available.

Q Questions

1. **(a)** What is the voltage of **(i)** a simple cell, **(ii)** a zinc-carbon cell, **(iii)** a lead-acid cell?
   (b) What materials are used for the positive and negative electrodes and the electrolyte of each of the cells in **(a)**?

2. **(a)** What is the difference between a primary and a secondary cell?
   (b) State advantages and disadvantages of each type.

[2] Topic 81 deals with battery charging.

check list
p.289

66 Electromagnets

Oersted's discovery

In 1819 Oersted accidentally discovered the magnetic effect of an electric current. His experiment can be repeated by holding a wire over and parallel to a compass needle which is pointing N and S, Fig. 66.1. The needle moves when the current is switched on. Reversing the current causes the needle to point in the opposite direction.

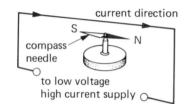

Fig. 66.1

Evidently around a wire carrying a current there is a *magnetic field*. As with the field due to a permanent magnet, we represent the field due to a current by *field lines* or *lines of force*. Arrows on the lines show the direction of the field, i.e. the direction in which a N pole points.

Different field patterns are given by differently shaped conductors.

Field due to a straight wire

If a straight vertical wire passes through the centre of a piece of card held horizontally and a current is passed through the wire, iron filings sprinkled on the card set in concentric circles when the card is tapped, Fig. 66.2.

Fig. 66.2

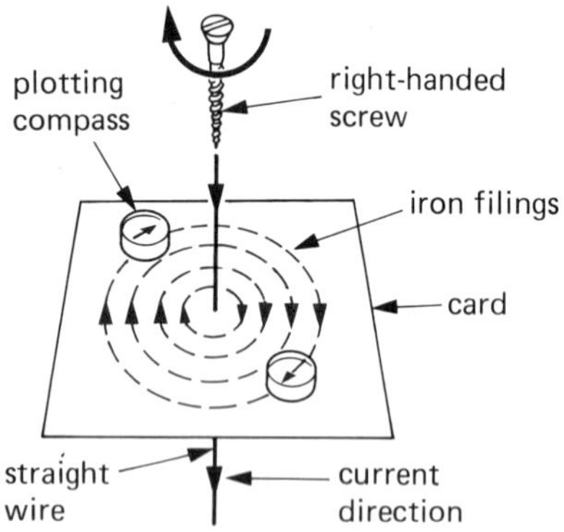

Plotting compasses placed on the card set along the field lines and show the direction of the field at different points. When the current direction is

reversed, the compasses point in the opposite direction showing that the direction of the field reverses when the current reverses.

If the current direction is known, the direction of the field can be predicted by the *Right-hand screw rule*.

If a right-handed screw moves forward in the direction of the current (conventional), the direction of rotation of the screw gives the direction of the field.

Field due to a circular coil

The field pattern is shown in Fig. 66.3. At the centre of the coil the field lines are straight and at right angles to the plane of the coil. The Right-hand screw rule again gives the direction of the field at any point.

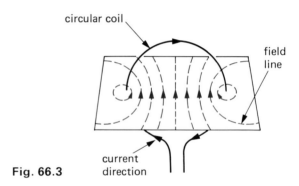

Fig. 66.3

Field due to a solenoid

A solenoid is a long cylindrical coil. It produces a field similar to that of a bar magnet; in Fig. 66.4, end A behaves like a N pole and end B like a S pole. The polarity is found as before by applying the Right-hand screw rule to a short length of one turn of the solenoid. Alternatively the *Right-hand grip rule* can be used (p. 155).

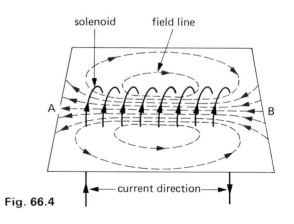

Fig. 66.4

The field inside a solenoid can be made very strong if it has large numbers of turns and a large current flows. Previously we used it to magnetize materials (p. 155): permanent magnets are now made by allowing the molten metal to solidify in such fields.

⎍ Experiment: simple electromagnet

An electromagnet is a coil of wire wound on a soft iron core. A 5 cm iron nail and 3 m of rayon-covered copper wire (SWG 26) are needed.

(a) Leave about 25 cm at one end of the wire (for connecting to the circuit) and then wind about 50 cm as a single layer on the nail. *Keep the turns close together and always wind in the same direction.* Connect the circuit of Fig. 66.5, setting the rheostat at its *maximum* resistance.

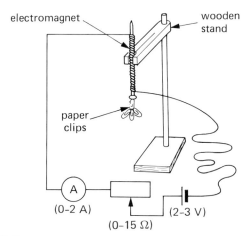

Fig. 66.5

Find the number of paper clips the electromagnet can support when known currents between 0.25 and 2.0 A pass through it. Record the results in a table. How does the 'strength' of the electromagnet depend on the current?

(b) Wind on another two layers of wire on the nail in the *same direction* as the first layer. Repeat the experiment. What can you say about the 'strength' of an electromagnet and the number of turns of wire?

(c) Place the electromagnet on the bench and under a sheet of paper. Sprinkle iron filings on the paper, tap it gently and observe the field pattern. How does it compare with that given by a bar magnet?

(d) Use the Right-hand screw (or grip) rule to predict which end of the electromagnet is a N pole. Check with a plotting compass.

Electromagnets

The magnetism of an electromagnet is *temporary* and can be switched on and off, unlike that of a permanent magnet. It has a core of *soft iron* which is magnetized only when current flows in the surrounding coil.

The strength of an electromagnet increases if

(i) the current in the coil increases,
(ii) the number of turns on the coil increases,
(iii) the poles are closer together.

In C-core (or horseshoe) electromagnets condition **(iii)** is achieved, Fig. 66.6. Note that the coil is wound in *opposite* directions on each limb of the core.

Fig. 66.6

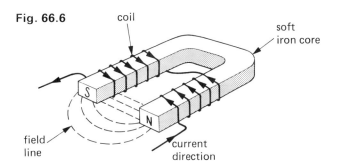

Apart from being used as a crane to lift iron objects, Fig. 66.7, scrap iron, etc., an electromagnet is an essential part of many electrical devices.

Fig. 66.7

189

Electric bell

When the circuit in Fig. 66.8 is completed, current flows in the coils of the electromagnet which becomes magnetized and attracts the soft iron bar (the armature). The hammer hits the gong but the circuit is now broken at the point C of the contact screw.

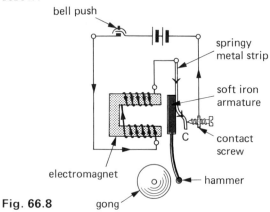

Fig. 66.8

The electromagnet loses its magnetism and no longer attracts the armature. The springy metal strip is then able to pull the armature back, remaking contact at C and so completing the circuit again. This cycle is repeated so long as the bell push is depressed and continuous ringing occurs.

Relay: reed switch

(a) Relay. This is a switch worked by an electromagnet. It is useful if we want one circuit to control another, especially if the current and power are larger in the second circuit (see Qn. 7, p. 206). Fig. 66.9 shows a typical relay. When current flows in the coil from the circuit connected to AB, the soft iron core is magnetized and attracts the L-shaped iron armature. This rocks on its pivot and closes the contacts at C in the circuit connected to DE. The relay is then 'energized' or 'on'.

Fig. 66.9

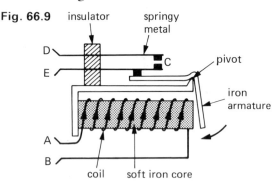

The current needed to operate a relay is called the *pull-on* current and the *drop-off* current is the smaller current in the coil when the relay just stops working. If the coil resistance R of a relay is 185 Ω and its

operating p.d. V is 12 V, the pull-on current $I = V/R = 12/185 = 0.065\,A = 65\,mA$. The symbols for relays with normally open and normally closed contacts are given in Fig. 66.10a,b.

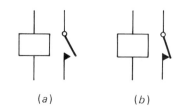

Fig. 66.10 (a) (b)

(b) Reed switch. One is shown in Fig. 66.11a. When current flows in the coil, the magnetic field produced magnetizes the strips (called 'reeds') of magnetic material. The ends become opposite poles and one reed is attracted to the other so completing the circuit connected to AB. The reeds separate when the current in the coil is switched off.

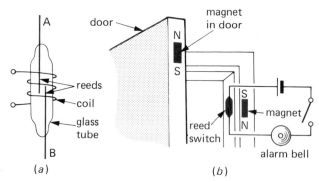

Fig. 66.11

Reed switches are also operated by permanent magnets. Fig. 66.11b shows the use of a normally open reed switch as a burglar alarm. How does it work?

Telephone

A telephone contains a microphone at the speaking end and a receiver at the listening end.

(a) Carbon microphone, Fig. 66.12. When someone speaks, sound waves cause the diaphragm to move backwards and forwards. This varies the pressure on the carbon granules between the movable carbon dome which is attached to the diaphragm and the

Fig. 66.12

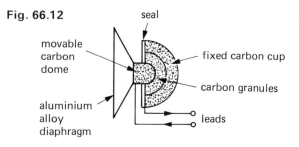

fixed carbon cup at the back. When the pressure increases, the granules are squeezed closer together and their electrical resistance decreases. A decrease of pressure has the opposite effect. If there is a current passing through the microphone (from a battery), this too will vary in a similar way to the sound wave variations.

(b) Receiver. The coils are wound in opposite directions on the two S poles of the magnet, Fig. 66.13.

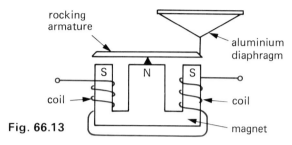
Fig. 66.13

Therefore if current goes round one in a clockwise direction, it goes round the other anticlockwise so making one S pole stronger and the other weaker. This causes the iron armature to rock on its pivot towards the stronger S pole. When the current reverses, the armature rocks the other way due to the S pole which was the stronger before becoming the weaker. These armature movements are passed on to the diaphragm, making it vibrate and produce sound of the same frequency as the a.c. in the coil (received from the microphone).

Q Questions

1. The vertical wire in Fig. 66.14 is at right angles to the card. In what direction will a plotting compass at A point when **(a)** there is no current in the wire, **(b)** current flows upwards?

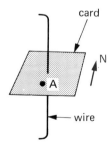
Fig. 66.14

2. Fig. 66.15 shows a solenoid wound round a core of soft iron. Will the end A be a N pole or S pole when the current (conventional) flows in the direction shown?

(J.M.B./A.L./N.W.16+)

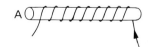
Fig. 66.15

3. A small electromagnet, used for lifting and then releasing a small steel ball, is made in the laboratory, Fig. 66.16.

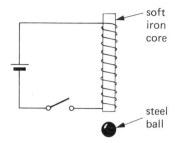
Fig. 66.16

(a) Explain why soft iron is a better material than steel to use for the core.
(b) In order to lift a slightly larger ball it is necessary to make a stronger electromagnet. State TWO ways in which an electromagnet could be made more powerful.

4. You are asked to carry out an experiment to find how the lifting power of a horseshoe electromagnet depends upon the current flowing through the coils.
(a) Draw a clear diagram of the circuit you would use showing clearly the direction of the current, the direction of the winding of the coils of the electromagnet and the polarity of one end of the magnet.
(b) Eventually when the current has reached a certain value any further increase in the current does not result in a further increase in the lifting power. Why is this? (E.M.)

5. Fig. 66.17 shows an arrangement for lighting three lamps, A, B and C, only one of which is controlled directly by the switch.
(a) Which of the lamps is directly controlled by the switch?
(b) What is the name given to this use of an electromagnet?
(c) Which lamps can be on at once?
(d) Explain how lamp C comes on. (S.R.)

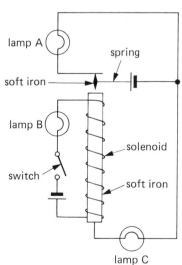
Fig. 66.17

191

check list
p.289

67 Electric motors

Electric motors form the heart of a whole host of electrical devices ranging from domestic appliances such as vacuum cleaners and washing machines to electric locomotives and lifts. In a car the windscreen wipers are usually driven by one and the engine is started by another.

The motor effect

A wire carrying a current in a magnetic field experiences a force. If the wire can move it does.

(a) Demonstration. In Fig. 67.1 the flexible wire is loosely supported in the strong magnetic field of a C-shaped magnet (permanent or electro). When the switch is pressed, current flows in the wire which jumps upwards as shown. If either the direction of the current or the direction of the field is reversed, the wire moves downwards. *The force increases if the strengths of the field and the current increase.*

Fig. 67.1

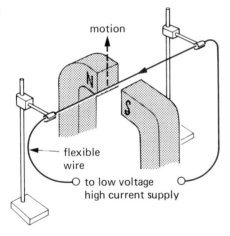

(b) Explanation. Fig. 67.2*a* is a side-view of the magnetic field lines due to the wire and the magnet. Those due to the wire are circles and we will suppose their directions are as shown. The dotted lines represent the field lines of the magnet and their direction is to the right.

Fig. 67.2

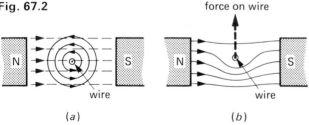

(a)   (b)

The resultant field obtained by combining both fields is shown in Fig. 67.2*b*. There are more lines below than above the wire since both fields act in the same direction below but in opposition above. If we *suppose* the lines are like stretched elastic, those below will try to straighten out and in so doing will exert an upwards force on the wire.

Fleming's left-hand rule

The direction of the force or thrust on the wire can be predicted by this rule which is also called the 'Motor rule', Fig. 67.3.

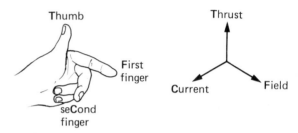

Fig. 67.3

Hold the thumb and first two fingers of the left hand at right angles to each other with the First finger pointing in the direction of the Field and the seCond finger in the direction of the Current, then the Thumb points in the direction of the Thrust.

If the wire is not at right angles to the field, the force is smaller and is zero if it is parallel to the field.

Simple d.c. electric motor

A simple motor to work from direct current (d.c.) consists of a rectangular coil of wire mounted on an axle which can rotate between the poles of a C-shaped magnet, Fig. 67.4. Each end of the coil is connected to half of a split ring of copper, called the

Fig. 67.4

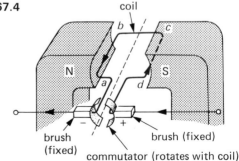

commutator, which rotates with the coil. Two carbon blocks, the *brushes*, are pressed lightly against the commutator by the springs. The brushes are connected to an electrical supply.

If Fleming's left-hand rule is applied to the coil in the position shown, we find that side *ab* experiences an upward force and side *cd* a downward force. (No forces act on *ad* and *bc* since they are parallel to the field.) These two forces form a *couple* which rotates the coil in a clockwise direction until it is vertical.

The brushes are then in line with the gaps in the commutator and the current stops. However, because of its inertia, the coil overshoots the vertical and the commutator halves change contact from one brush to the other. This reverses the current through the coil and so also the directions of the forces on its sides. Side *ab* is on the right now, acted on by a downward force, whilst *cd* is on the left with an upward force. The coil thus carries on rotating clockwise.

⌇⌇ Experiment: a model motor

The motor is shown in Fig. 67.5 and is made from a kit of parts.

1. Wrap Sellotape round one end of the metal tube which passes through the wooden block.

2. Cut two rings off a piece of narrow rubber tubing; slip them onto the Sellotaped end of the metal tube.

3. Remove the insulation from one end of a 1½ metre length of SWG 26 PVC-covered copper wire and fix it under both rubber rings so that it is held tight against the Sellotape. This forms one end of the coil.

4. Wind 10 turns of the wire in the slot in the wooden block and finish off the second end of the coil by removing the PVC and fixing this too under the rings but on the *opposite* side of the tube from the first end. The bare ends act as the *commutator*.

5. Push the axle through the metal tube of the wooden base so that the block spins freely.

6. Arrange two ½ metre lengths of wire to act as *brushes* and leads to the supply, as shown. Adjust the brushes so that they are vertical and each touches one bare end of the coil when the plane of the coil is horizontal. *The motor will not work if this is not so.*

7. Slide the base into the magnet with *opposite poles facing.* Connect to a 3 V battery (or other low voltage d.c. supply) and a slight push of the coil should set it spinning at high speed.

Practical motors

Practical motors have:

(a) *a coil of many turns wound on a soft-iron cylinder or core* which rotates with the coil. This makes it more powerful. The coil and core together are called the *armature.*

(b) *several coils* each in a slot in the core and each having a pair of commutator segments. This gives increased power and smoother running. The motor of an electric drill is shown in Fig. 67.6.

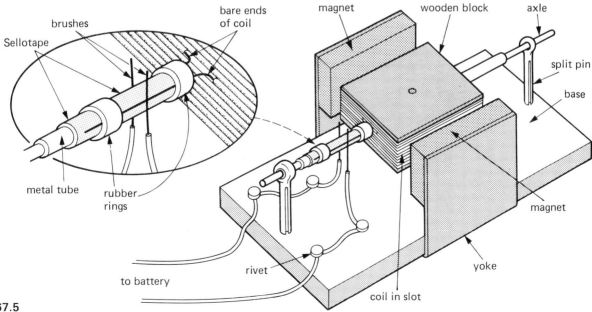

Fig. 67.5

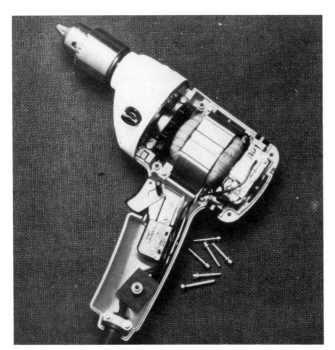

Fig. 67.6

(c) *an electromagnet* (usually) to produce the field in which the armature rotates.

Most electric motors used in industry are *induction motors*. They work off a.c. (alternating current) on a different principle to the d.c. motor.

Moving coil loudspeakers

Varying currents from a radio, record-player etc. pass through a short cylindrical coil whose turns are at right angles to the magnetic field of a magnet with a central pole and a surrounding ring pole, Fig. 67.7a.

Fig. 67.7

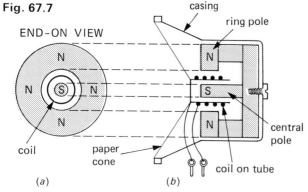

(a)   (b)

A force acts on the coil which, according to Fleming's left-hand rule, makes it move in and out. A paper cone attached to the coil moves with it and sets up sound waves in the surrounding air, Fig. 67.7b.

check list
p.289

Q Questions

1. A wire carries a current horizontally between the magnetic poles N and S which face each other on a table, Fig. 67.8. The direction of the force on the wire due to the magnets is

   A from N to S
   B from S to N
   C opposite to the current direction
   D in the direction of the current
   E vertically upwards   (*O.L.E./S.R.16+*)

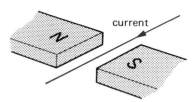

Fig. 67.8

2. In Fig. 67.9 AB is a copper wire hanging from a pivot at A and dipping in mercury in a copper dish at B so that it hangs between the poles of a powerful magnet.
   (a) Copy the diagram and mark in the lines of force representing the field between the poles.
   (b) Mark in the direction of the conventional current flow in AB when the switch is closed.
   (c) What else will happen when the switch is closed?
                          (*S.E.*)

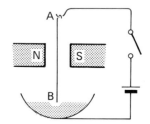

Fig. 67.9

3. In the simple electric motor of Fig. 67.10, will the coil rotate clockwise or anticlockwise as seen by the eye from the position X? The arrows on the diagram indicate the direction of the conventional current flow.
                         (*J.M.B./A.L./N.W.16+*)

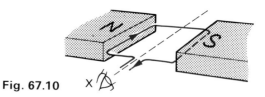

Fig. 67.10

4. An electric motor is a device which converts

   A mechanical energy into electrical energy
   B heat energy into electrical energy
   C electrical energy into heat only
   D heat energy into mechanical energy
   E electrical energy into mechanical energy and heat.

68 Electric meters

Moving coil galvanometer

A galvanometer detects small currents or small p.d.s, often of the order of milliamperes (mA) or millivolts (mV).

In the moving coil *pointer-type* meter, a coil is pivoted on jewelled bearings between the poles of a permanent magnet, Fig 68.1a. Current enters and leaves the coil by hair springs above and below it. When current flows, a couple acts on the coil (as in an electric motor), causing it to rotate until stopped by the springs. The greater the current the greater the deflection which is shown by a pointer attached to the coil.

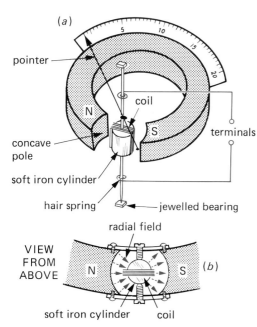

Fig. 68.1

The soft iron cylinder at the centre of the coil is *fixed* and along with the concave poles of the magnet it produces a *radial* field, Fig. 68.1b, i.e. the field lines are directed to the centre of the cylinder. The scale on the meter is then even or linear, i.e. all divisions are the same size.

The sensitivity of a galvanometer is increased by having

(i) more turns on the coil,
(ii) a stronger magnet,
(iii) weaker hair springs or a wire suspension,
(iv) as a pointer, a long beam of light reflected from a mirror on the coil.

The last two are used in *light-beam* meters which have a full-scale deflection of a few microamperes (μA). (1 μA $= 10^{-6}$ A.)

Ammeters and shunts

An ammeter is a galvanometer having a known low resistance (a *shunt*) in parallel with it to take most of the current, Fig. 68.2. An ammeter is placed in *series* in a circuit and must have a *low resistance* otherwise it changes the current to be measured.

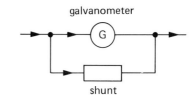

Fig. 68.2

Voltmeters and multipliers

A voltmeter is a galvanometer having a known high resistance (a *multiplier*) in series with it, Fig. 68.3. A voltmeter is placed in *parallel* with the part of the circuit across which the p.d. is to be measured and must have a *high resistance*—otherwise the total resistance of the whole circuit is reduced so changing the current and the p.d. required.

Fig. 68.3

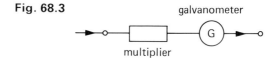

Ⓠ Questions

1. What does a galvanometer do?

2. Why should the resistance of **(i)** an ammeter be very small, **(ii)** a voltmeter be very large?

check list
p.290

69 Generators

An electric current creates a magnetic field. The reverse effect of producing electricity from magnetism was discovered in 1831 by Faraday and is called *electromagnetic induction*. It led to the construction of generators for producing electrical energy in power stations.

Electromagnetic induction

Two ways of investigating the effect follow.

(a) Straight wire and U-shaped magnet, Fig. 69.1. First the wire is held at rest between the poles of the magnet and the galvanometer observed. It is then moved in each of the six directions shown. Only when *it is moving* upwards (direction 1) or downwards (direction 2) is there a deflection on the galvanometer, indicating an induced current in the wire. The deflection is in opposite directions in each case and only lasts while the wire is in motion.

Fig. 69.1

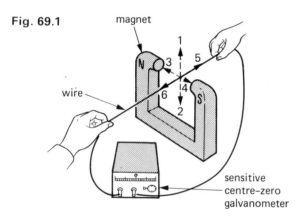

magnet

wire

sensitive centre–zero galvanometer

(b) Bar magnet and coil, Fig. 69.2. The magnet is pushed into the coil, one pole first, then held still inside it. It is next withdrawn. The galvanometer shows that current is induced in the coil in one direction as the magnet *moves in* and in the opposite direction as it is *removed*. There is no deflection when

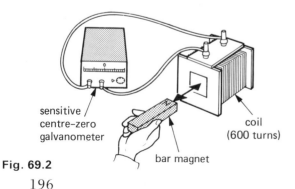

sensitive centre–zero galvanometer

coil (600 turns)

bar magnet

Fig. 69.2
196

the magnet is at rest. The results are the same if the coil is moved instead of the magnet, i.e. only *relative motion* is needed.

Faraday's law

To 'explain' electromagnetic induction Faraday suggested that an e.m.f. is induced in a conductor whenever it 'cuts' magnetic field lines, i.e. moves *across* them, but not when it moves along them or is at rest. If the conductor forms part of a complete circuit, an induced current is also produced.

Faraday found, and it can be shown with apparatus like that in Fig. 69.2, that the induced e.m.f. increases with increases of

(i) the speed of motion of the magnet or coil,
(ii) the number of turns on the coil,
(iii) the strength of the magnet.

These facts led him to state a law.

The size of the induced e.m.f. is directly proportional to the rate at which the conductor cuts magnetic field lines.

Lenz's law

The direction of the induced current can be predicted by a law due to the Russian scientist, Lenz.

The direction of the induced current is such as to oppose the change causing it.

In Fig. 69.3*a* the magnet approaches the coil, north pole first. According to Lenz's law the induced current should flow in a direction which makes the coil behave like a magnet with its top a north pole. The downward motion of the magnet will then be opposed.

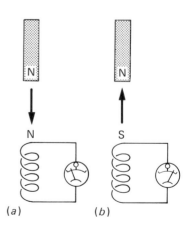

Fig. 69.3   (*a*)   (*b*)

When the magnet is withdrawn, the top of the coil should become a south pole, Fig. 69.3*b*, and attract the north pole of the magnet, so hindering its removal. The induced current is thus in the opposite direction to that when the magnet approaches.

Lenz's law is an example of the principle of conservation of energy. If the currents caused opposite poles to those that they do, electrical energy would be created from nothing. As it is, mechanical energy is provided by whoever moves the magnet, to overcome the forces that arise.

For a straight wire moving at right angles to a magnetic field a more useful form of Lenz's law is *Fleming's right-hand rule* (the 'Dynamo rule'), Fig. 69.4.

Fig. 69.4

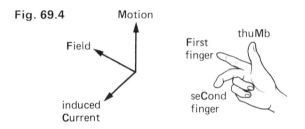

Hold the thumb and first two fingers of the right hand at right angles to each other with the First finger pointing in the direction of the Field and the thuMb in the direction of Motion of the wire, then the seCond finger points in the direction of the induced Current.

Simple a.c. generator (alternator)

The simplest alternating current (a.c.) generator consists of a rectangular coil between the poles of a C-shaped magnet, Fig. 69.5*a*. The ends of the coil are joined to two *slip rings* on the axle and against which carbon *brushes* press.

When the coil is rotated it cuts the field lines and an e.m.f. is induced in it. Fig. 69.5*b* shows how the e.m.f. varies over one complete rotation.

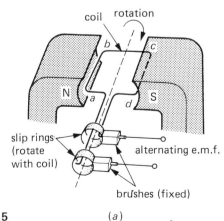

Fig. 69.5   (*a*)

As the coil moves through the vertical position with *ab* uppermost, *ab* and *cd* are moving along the lines (*bc* and *da* do so always) and no cutting occurs. The induced e.m.f. is zero.

During the first quarter rotation the e.m.f. increases to a maximum when the coil is horizontal. Sides *ab* and *dc* are then cutting the lines at the greatest rate.

In the second quarter rotation the e.m.f. decreases again and is zero when the coil is vertical with *dc* uppermost. After this, the direction of the e.m.f. reverses because, during the next half rotation, the motion of *ab* is directed upwards and *dc* downwards.

An alternating e.m.f. is generated which acts first in one direction and then the other; it would cause a.c. to flow in a circuit connected to the brushes. The *frequency* of an a.c. is the number of complete cycles it makes each second (c/s) and is measured in *hertz* (Hz), i.e. 1 c/s = 1 Hz. If the coil rotates twice per second, the a.c. has frequency 2 Hz. The mains supply is a.c. of frequency 50 Hz.

Simple d.c. generator (dynamo)

An a.c. generator becomes a direct current (d.c.) one if the slip rings are replaced by a *commutator* like that in a d.c. motor, Fig. 69.6*a*.

The brushes are arranged so that as the coil goes through the vertical, changeover of contact occurs from one half of the split ring of the commutator to the other. In this position the e.m.f. induced in the coil reverses and so one brush is always positive and the other negative.

The e.m.f. at the brushes is shown in Fig. 69.6*b*; although varying in value, it never changes direction and would produce a direct current (d.c.) in an external circuit.

In construction the simple d.c. dynamo is the same as the simple d.c. motor and one can be used as the other. When an electric motor is working it acts as a

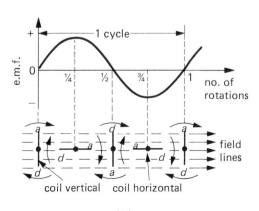

(*b*)

Fig. 69.6

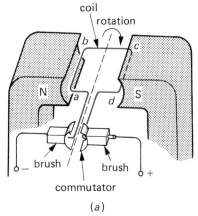

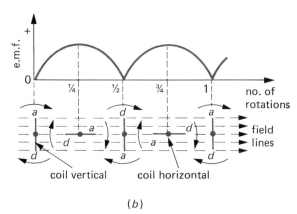

(a)

(b)

dynamo and creates an e.m.f., called a *back e.m.f.*, opposing and nearly equal to the applied p.d. The current in the coil is therefore much less once the motor is running.

Practical generators

In actual generators several coils are wound in evenly-spaced slots in a soft iron cylinder and electromagnets usually replace permanent magnets.

(a) Power stations. In power station alternators the electromagnets rotate (the *rotor*) while the coils and their iron core are at rest (the *stator*). The large p.d.s and currents (e.g. 25 kV at several thousand amperes) induced in the stator are led away through stationary cables, otherwise they would quickly destroy the slip rings by sparking. Instead the relatively small d.c. required by the rotor is fed via the slip rings from a small dynamo (the *exciter*) which is driven by the same steam turbine as the rotor.

The turbine is rotated by high pressure steam obtained by heating water in a coal or oil-fired boiler or in a nuclear reactor. Block and energy flow diagrams of a power station are shown in Fig. 69.7.

(b) Cars. Most are now fitted with alternators because they give a greater output than dynamos at low engine speeds.

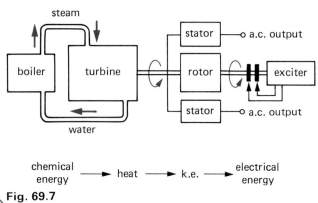

Fig. 69.7

198

(c) Bicycles. The rotor is a permanent magnet and the e.m.f. is induced in the coil which is at rest, Fig. 69.8.

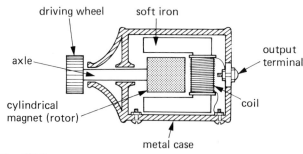

Fig. 69.8

Q Questions

1. **(a)** A coil of copper wire is connected to a centre-zero galvanometer. Explain what would be observed if a bar magnet was pushed into the coil so that the north pole of the magnet entered first.
   (b) What would be observed if the magnet was
   (i) held at rest inside the coil?
   (ii) pulled out again?
   (iii) pushed in faster than before?
   (iv) pushed in so that the south pole entered first?
   (E.A.)

2. A simple generator is shown in Fig. 69.9.
   (a) What are A and B called and what is their purpose?
   (b) What changes can be made to increase the e.m.f. generated?

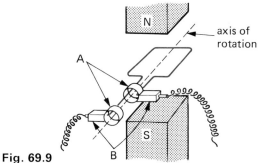

Fig. 69.9

check list p.290

70 Transformers

Mutual induction

When the current in a coil is switched on or off or changed, an e.m.f. and current are induced in a neighbouring coil. The effect, called *mutual induction*, is an example of electromagnetic induction not using a permanent magnet and can be shown with the arrangement of Fig. 70.1. Coil A is the *primary* and coil B the *secondary*.

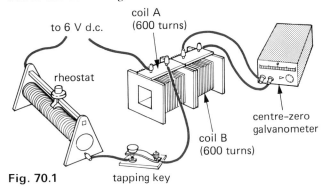

Fig. 70.1

Switching on the current in the primary sets up a magnetic field and as its field lines 'grow' outwards from the primary they 'cut' the secondary. An e.m.f. is induced in the secondary until the current in the primary reaches its steady value. When the current is switched off in the primary, the magnetic field dies away and we can imagine the field lines cutting the secondary as they collapse, again inducing an e.m.f. in it. Changing the primary current by *quickly* altering the rheostat has the same effect.

The induced e.m.f. is increased by having a soft iron rod in the coils, or better still, by using coils wound on a complete iron ring. More field lines then cut the secondary due to the magnetization of the iron.

Experiment: mutual induction with a.c.

An alternating current is changing all the time and if it flows in a primary coil, an alternating e.m.f. and current are induced in a secondary coil.

Connect the circuit of Fig. 70.2. The 1 V high current power unit supplies a.c. to the primary and the lamp detects the secondary current.

Find the effect on the brightness of the lamp of

(i) pulling the C-cores apart slightly,
(ii) increasing the secondary turns to 15,
(iii) decreasing the secondary turns to 5.

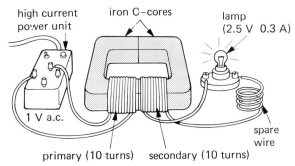

Fig. 70.2

Transformer equation

A transformer transforms (changes) an *alternating p.d.* (voltage) from one value to another of greater or smaller value. It has primary and secondary coils wound on a complete soft iron core, either one on top of the other, Fig. 70.3a, or on separate limbs of the core, Fig. 70.3b.

Fig. 70.3

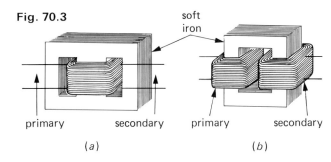

(a)   (b)

An alternating voltage applied to the primary induces an alternating voltage in the secondary whose value can be shown for an ideal transformer, to be given by

$$\frac{\text{secondary voltage}}{\text{primary voltage}} = \frac{\text{secondary turns}}{\text{primary turns}}$$

In symbols

$$\frac{V_s}{V_p} = \frac{N_s}{N_p}$$

A 'step-up' transformer has more turns on the secondary than the primary and V_s is greater than V_p, Fig. 70.4a. For example, if the secondary has

Fig. 70.4

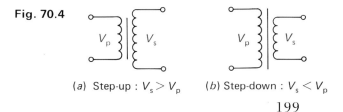

(a) Step-up : $V_s > V_p$ (b) Step-down : $V_s < V_p$

199

twice as many turns as the primary, V_s is about twice V_p. In a 'step-down' transformer there are fewer turns on the secondary than the primary and V_s is less than V_p, Fig. 70.4*b*.

Energy losses in a transformer

If the p.d. is stepped-up in a transformer the current is stepped-down in proportion. This must be so if we assume that all the electrical energy given to the primary appears in the secondary, i.e. that energy is conserved and the transformer is 100% efficient (many approach this). Then

power in primary = power in secondary
$$V_p \times I_p = V_s \times I_s$$

where I_p and I_s are the primary and secondary currents respectively.

$$\therefore \frac{I_s}{I_p} = \frac{V_p}{V_s}$$

So, for the ideal transformer, if the p.d. is doubled the current is halved. In practice, it is more than halved due to small energy losses in the transformer arising from three causes.

(a) Resistance of windings. The windings of copper wire do have some resistance and heat is produced by the current in them.

(b) Eddy currents. The iron core is in the changing magnetic field of the primary, and currents, called eddy currents, are induced in it which cause heating. These are reduced by using a *laminated* core made of sheets, insulated from each other to have a high resistance.

(c) Leakage of field lines. All the lines produced by the primary may not 'cut' the secondary, especially if the core has an air-gap or is badly designed.

A large transformer like that in Fig. 70.5 has to be oil-cooled to prevent overheating.

Fig. 70.5

200

The stepping-up of current can be demonstrated with the 100:1 *step-down* transformer of Fig. 70.6 in which the nail melts spectacularly due to the very large secondary current.

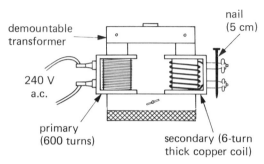

Fig. 70.6

✎ Worked example

A transformer steps down the mains supply from 240 V to 12 V to operate a 12 V lamp.

(a) What is the turns ratio of the transformer windings?
(b) How many turns are on the primary if the secondary has 100 turns?
(c) What is the current in the primary if the transformer is 100% efficient and the current in the lamp is 2 A?

(a) Primary voltage $= V_p = 240$ V
Secondary voltage $= V_s = 12$ V
Turns ratio $= N_s/N_p = V_s/V_p = 12/240$
$= 1/20$

(b) Secondary turns $= N_s = 100$

From (a), $\dfrac{N_s}{N_p} = \dfrac{1}{20}$

$\therefore N_p = 20\,N_s = 20 \times 100$
$= 2000$ turns

(c) Efficiency = 100%
$\therefore$ power in primary = power in secondary
$V_p \times I_p = V_s \times I_s$

$$\therefore I_p = \frac{V_s \times I_s}{V_p} = \frac{12\text{ V} \times 2\text{ A}}{240\text{ V}}$$

$$= 1/10 = 0.1 \text{ A}$$

Note. In this ideal transformer the current is stepped up in the same ratio as the voltage is stepped down.

Transmission of electrical power

(a) Grid system. This is a network of cables, most supported on pylons, which connect about 160 power stations throughout the country to con-

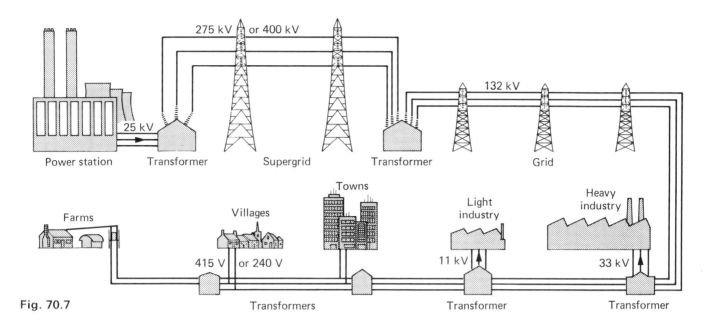

275 kV or 400 kV

25 kV

132 kV

Power station · Transformer · Supergrid · Transformer · Grid

Towns

Farms · Villages · Light industry · Heavy industry

415 V or 240 V · 11 kV · 33 kV

Fig. 70.7 · Transformers · Transformer · Transformer

sumers. In the largest modern stations, electricity is generated at 25 000 V (25 kilovolts = 25 kV) and stepped up at once in a transformer to 275 or 400 kV to be sent over long distances on the Supergrid. Later, the p.d. is reduced by sub-station transformers for distribution to local users, Fig. 70.7.

At the Area Control Centres engineers direct the flow and re-route it when breakdown occurs.

This makes the supply more reliable, and cuts costs by enabling smaller, less efficient stations to be shut down at off-peak periods.

(b) Use of high alternating p.d.s. If 400 000 W of electrical power has to be sent through cables it can be done as 400 000 V at 1 A or 400 V at 1000 A (since watts = amperes × volts). But the amount of electrical energy changed to unwanted heat (due to resistance of the cables) is proportional to the *square of the current* and so the power loss (I^2R) is less if transmission occurs at high voltage (high tension) and low current. On the other hand high p.d.s need good insulation. The efficiency with which transformers step alternating p.d.s up and down accounts for the use of a.c. rather than d.c.

The advantages of 'high' alternating voltage power transmission may be shown using the apparatus in Fig. 70.8, *so long as the power line is well insulated* (using insulated eureka wire). The eureka resistance wires represent long transmission cables. Without the transformer at each end the lamp at the 'village' end glows dimly due to the power loss in the wires.

When the transformers are connected as shown, the 12 V a.c. supply voltage is stepped up to about 240 V (since $N_p/N_s = 1/20$) at the 'power station' end,

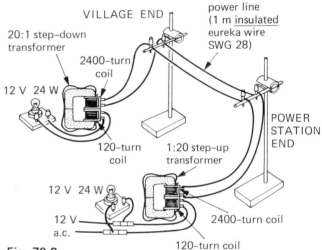

VILLAGE END

power line (1 m insulated eureka wire SWG 28)

20:1 step-down transformer

2400-turn coil

12 V 24 W

120-turn coil

1:20 step-up transformer

POWER STATION END

12 V 24 W

12 V a.c.

2400-turn coil

120-turn coil

Fig. 70.8

then stepped down at the 'village' end to 12 V (since $N_s/N_p = 20/1$). The current in the wires is then much less than without the transformers and so the lamp at the 'village' end is fully lit due to the smaller power loss during transmission.

Car ignition system

A transformer does not work off d.c. but the *induction coil*, which uses the same principle, does. It is used in a car to produce the high voltage that causes the sparking plug to spark and ignite the mixture of petrol vapour and air.

The primary circuit is broken when the rotating cam, Fig. 70.9, separates the 'points'. A high p.d. is induced in the secondary which is applied at the right time to each plug in turn by the rotor arm of the distributor. When the 'points' open, the capacitor acts as a reservoir and stops sparking at them which

201

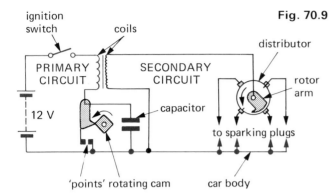

ignition switch

coils

PRIMARY CIRCUIT

SECONDARY CIRCUIT

distributor

rotor arm

12 V

capacitor

to sparking plugs

'points' rotating cam

car body

Fig. 70.9

would cause 'pitting'. It also ensures that the primary current falls rapidly, so giving a larger induced e.m.f. in the secondary.

Eddy current applications

Eddy currents are the currents induced in a piece of metal when it cuts magnetic field lines. They can be quite large due to the low resistance of the metal. They have their uses as well as their disadvantages.

(a) Car speedometer. The action depends on the eddy currents induced in a thick aluminium disc when a permanent magnet, near it but *not touching it*, is rotated by a cable driven from the gearbox of the car, Fig. 70.10. The eddy currents in the disc make it rotate in an attempt to reduce the *relative motion* between it and the magnet (see Topic 69). The extent to which the disc can turn however is controlled by a spring. The faster the magnet rotates the more does the disc turn before it is stopped by the spring. A pointer fixed to the disc moves over a scale marked in mph (or km/h) and gives the speed of the car.

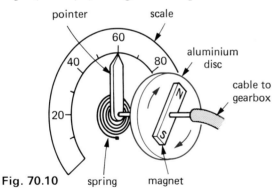

pointer   scale

60

40   80

aluminium disc

cable to gearbox

20

N

S

Fig. 70.10   spring   magnet

(b) Electromagnetic damping. This is used in moving coil meters to make the coil take up its deflected position quickly without overshooting and oscillating about its final reading. The movement is said to be 'dead-beat'. In most pointer instruments the coil is wound on a metal frame in which large eddy currents are induced and cause opposition to the motion of the coil as it cuts across the radial magnetic field of the permanent magnet.

202

check list p.290

Q Questions

1. Two coils of wire, A and B, are placed near one another, Fig. 70.11. Coil A is connected to a switch and battery. Coil B is connected to a centre-reading moving coil galvanometer.

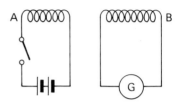

A   B

G

Fig. 70.11

(a) If the switch connected to coil A were closed for a few seconds and then opened, the galvanometer connected to coil B would be affected. Explain and describe, step by step, what would actually happen.
(b) What changes would you expect if a bundle of soft iron wires were placed through the centre of the coils? Give a reason for your answer.
(c) What would happen if more turns of wire were wound on the coil B?

2. The main function of a step-up transformer is to
A increase current
B increase voltage
C change a.c. to d.c.
D change d.c. to a.c.
E increase the resistance of a circuit   (S.E.)

3. (a) Calculate the number of turns on the secondary of the step-down transformer, which would enable a 12 V bulb to be used with a 240 V a.c. mains power, if there are 480 turns on the primary.
(b) What current will flow in the secondary when the primary current is 0.50 A? Assume there are no energy losses.   (N.W.)

4. The block diagram in Fig. 70.12 represents the generation and distribution of electricity.

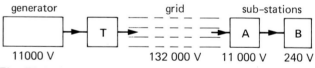

generator   grid   sub-stations

T   A   B

11000 V   132 000 V   11 000 V   240 V

Fig. 70.12

(a) T is a 'step-up' transformer which has a core made up of 'laminations'.
   (i) Explain what is meant by a 'step-up' transformer.
   (ii) What is meant by 'laminations'?
   (iii) What material is used for the laminations?
   (iv) Why is the core built in this way?
(b) If the transformer T has 1000 turns on its primary coil, how many turns must it have on its secondary coil? (Assume that there are no losses in the transformer.)
(c) Why is it an advantage to transmit electric power at high voltage?
(d) What will be the main item of equipment in the sub-stations A and B?

71 Alternating current

Difference between d.c. and a.c.

In a direct current (d.c.) the electrons flow in one direction only. Graphs for steady and varying direct currents are shown in Fig. 71.1. In an alternating current (a.c.) the direction of flow reverses regularly, Fig. 71.2. The sign for a.c. is ∼.

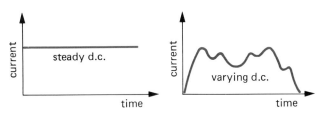

Fig. 71.1

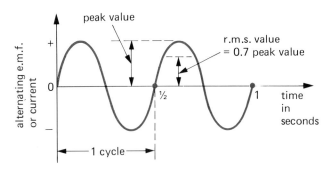

Fig. 71.2

The pointer of a moving coil meter is deflected one way by d.c.; a.c. makes it move to and fro about the zero if the changes are slow enough, otherwise there is no deflection. Batteries give d.c.; generators can produce either d.c. or a.c. The mains supply is a.c.

For heating and lighting a.c. and d.c. are equally satisfactory but radio and television sets need d.c., as do processes such as battery charging and electroplating. a.c. can be *rectified* to give d.c.

Frequency of a.c.

The number of complete alternations or cycles in 1 second is the *frequency* of the a.c. The unit of frequency is the *hertz* (Hz) (formerly the cycle per second, c/s). The frequency of the a.c. in Fig. 71.2 is 2 Hz.

The mains supply in many countries including Britain has frequency 50 Hz. There are therefore 50 cycles in a second and one cycle lasts $1/50 = 0.02$ s.

Root mean square values

An alternating voltage or current is measured by its *effective* or *root mean square* (r.m.s.) value. This is *the steady direct voltage or current which would give the same heating effect.*

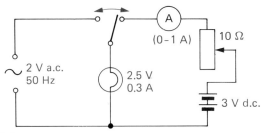

Fig. 71.3

For example, if the lamp in Fig. 71.3 is lit first by a.c. and its brightness noted, then if 0.3 A d.c. produces the same brightness, the r.m.s. value of the a.c. is 0.3 A. It can be shown that

r.m.s. value = 0.7 × peak value

The r.m.s. voltage of the mains supply is 240 V; the peak value is therefore $240/0.7 = 340$ V.

Meters for a.c.

Most a.c. meters are marked in r.m.s. values. In a rectifier type a rectifier (e.g. a germanium diode) allows current to pass in one direction only. When

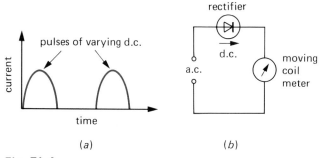

Fig. 71.4

connected to an a.c. supply it produces pulses of varying d.c., Fig. 71.4*a*, and in a rectifier meter these cause a deflection in a moving coil meter, Fig. 71.4*b*.

Capacitors in d.c. and a.c. circuits

Capacitors were introduced in Topic 57 as devices that store electric charge. They have other properties as well.

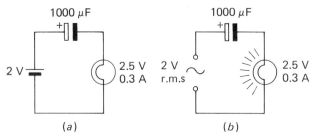

Fig. 71.5

(a) In d.c. circuits. In the circuit of Fig. 71.5a the supply is d.c. and the lamp does not light, i.e. *a capacitor blocks d.c.* Initially there is a brief flow of charge, i.e. a momentary current, which charges the capacitor until the p.d. across it is equal but opposite to that of the supply (here 2 V). Charge flow then stops.

If a charge Q (in coulombs) is stored when a p.d. V (in volts) is applied across a capacitor, its *capacitance* C (in farads) is defined by

$$C = \frac{Q}{V} \quad \text{or} \quad Q = VC$$

In Fig. 71.5a, we have $V = 2$ V, $C = 1000\,\mu$F, therefore $Q = VC = 2 \times 1000 = 2000\,\mu$C $= 0.002$ C.

Every capacitor has a *working voltage*, and any larger p.d. should not be applied to it or the dielectric is damaged.

(b) In a.c. circuits. In the circuit of Fig. 71.5b the supply is a.c. and the lamp lights, i.e. *a capacitor seems to let a.c pass.* The a.c. causes the capacitor to be charged, discharged, charged in the opposite direction (i.e. other plate positive) and discharged again, 50 times a second.

No current actually passes through the capacitor (its plates are separated by an insulator) but electrons flow to and fro so rapidly in the wires joining the plates (as the polarity of the a.c. supply reverses) that it *seems* to do so. The result is just as if it did and an a.c. ammeter would give a reading.

The current increases, i.e. more electrons flow on and off the plates per second, if

(i) the capacitance C increases,
(ii) the frequency f of the a.c. increases.

In other words, the *opposition* of the capacitor *decreases* if C or f increase.

204

Capacitor networks

(a) Parallel. Capacitors connected in parallel behave like one capacitor with larger plates. In Fig. 71.6a the combined capacitance C can be shown to be given by

$$C = C_1 + C_2 + C_3$$

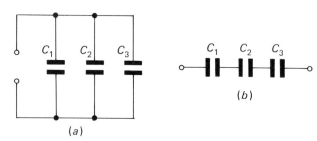

Fig. 71.6

(b) Series. In Fig. 71.6b the combined capacitance C is less than that of the smallest capacitor and is found from

$$\frac{1}{C} = \frac{1}{C_1} + \frac{1}{C_2} + \frac{1}{C_3}$$

Capacitor charge and discharge through a resistor

These processes do not happen instantaneously but take a time which depends on the values of the capacitor and resistor. Using the circuit of Fig. 71.7,

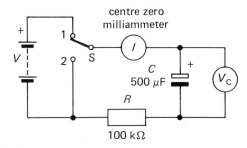

Fig. 71.7

current (I) and p.d. (V_c) readings can be obtained for charging (with S in position 1) and discharging (with S in position 2). If graphs of these quantities are plotted as in Fig. 71.8a,b, they show that during:

(a) charging
(i) I has its maximum value at the start and falls more and more slowly to zero as C charges up,
(ii) V_c rises rapidly from zero and slowly approaches the charging p.d. V which it equals when C is fully charged.

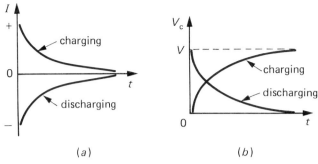

Fig. 71.8

(b) discharging

(i) I has its maximum value at the start but is in the opposite direction to the charging current,

(ii) V_c falls to zero.

In both processes the larger the values of C and R, the more slowly do they occur.

Inductors in a.c. circuits

An *inductor* or *choke* is a coil of wire with an air or an iron core. It opposes or 'chokes' a changing current and is said to have *inductance*.

(a) In d.c. circuits. In Fig. 71.9 R is adjusted to have the same resistance as the iron-cored inductor L. When the current is switched on, the lamp in series with L lights up a second or two *after* that in series with R. L delays the rise of the d.c. to its steady value.

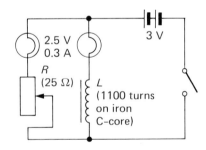

Fig. 71.9

(b) In a.c. circuits. If the 3 V d.c. supply is replaced by a 3 V a.c. supply, the lamp in series with L never lights. The current is changing all the time and the resulting opposition of L makes the current too small to light the lamp.

The current through an inductor increases if

(i) the inductance L decreases, i.e. fewer turns of wire or removal of any iron core,

(ii) the frequency f of the a.c. decreases.

In other words, the *opposition* of an inductor *decreases* if L or f decrease.

Q Questions

1. (a) How does a.c. differ from d.c.?
   (b) Name a source of (i) d.c., (ii) a.c.
   (c) Name a device which (i) only works on a.c., (ii) works equally well on a.c. or d.c.

2. Sketch a graph of current value against time for two complete cycles of an alternating current of peak value 0.5 A and frequency 50 Hz. Label your axes and make a correct scale of values on these axes (you must calculate the time for one cycle in order to mark the time axis).

Indicate with a horizontal line on the graph the approximate value of the r.m.s. current.

(*J.M.B./A.L./N.W.16+*)

3. An a.c. supply lights a lamp with the same brightness as does a 12 V battery. What is (a) the r.m.s. voltage, (b) the peak voltage, of the a.c. supply?

4. Fig. 71.10 shows a circuit containing a capacitor, an inductor and two lamps. Explain what happens when the circuit is connected to (i) a.c., (ii) d.c.   (*E.A.*)

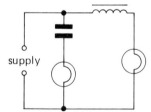

Fig. 71.10

205

check list p.290

72 Additional questions

Core level (all students)

Electric power

1. (a) Below is a list of wattages of various appliances. State which is most likely to be the correct one for each of the appliances named.

   60 W   250 W   750 W   1.5 kW   3 kW

   (i) Immersion heater (for a bath)
   (ii) Table lamp
   (iii) Iron.

   (b) What current will be taken by a 1 kW appliance if the supply voltage is 250?   (*J.M.B./A.L./N.W.16+*)

2. A heating coil has a resistance of 4 Ω and is connected to a 12 V supply. It is placed in a container holding 100 g of water at 20 °C. Calculate (a) the current, (b) the power of the circuit, (c) the energy produced in 5 minutes, (d) the rise in the temperature of the water, assuming no heat is lost. (Specific heat capacity of water is 4200 J/(kg °C) or 4.2 J/(g °C).)   (*E.A.*)

3. A 2 kW electric fire is used for 10 hours each week and a 100 W lamp is used for 10 hours each day. Find the total energy consumed each week and the total cost for each week if 1 kWh of electricity costs 5p.   (*J.M.B.*)

4. (a) Fig. 72.1*a* represents a simple lighting circuit. (i) What is wrong with this circuit? (ii) Explain why this is dangerous.

   (b) Fig. 72.1*b* represents an incomplete two-way lighting circuit. Copy and complete the circuit in such a way that the lamp would be lit with the circuit working correctly.   (*J.M.B./A.L./N.W.16+*)

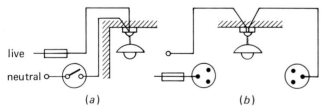

Fig. 72.1

5. Two precautions usually taken when using electrical equipment connected to the mains are to (a) earth the case (if metal), (b) put a fuse in the circuit.

Explain clearly why each of these precautions is desirable.   (*L.*)

206

Electromagnets

6. Fig. 72.2*a* to *d* show identical vertical insulated coils with many turns of wire and the same large direct current passing through each. The pieces of iron and steel are attached as shown. When the current is switched off what will happen to each piece of iron and steel?   (*W.M.*)

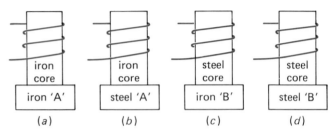

Fig. 72.2

7. Part of the electrical system of a car is shown in Fig. 72.3.

   (a) Why are connections made to the car body?
   (b) There are *two* circuits in parallel with the battery. What are they?
   (c) Why is wire A thicker than wire B?
   (d) Why is a relay used?

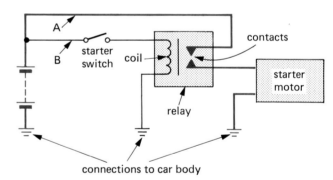

Fig. 72.3

8. Fig. 72.4 represents a coil of copper wire XY connected to a small battery, fixed to a cork and floating in a bowl of water.

Sketch the form of the magnetic field produced by the coil. Indicate the direction of the field.

How would you use this arrangement to find the direction of magnetic north?   (*L.*)

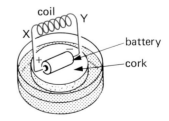

Fig. 72.4

Electric meters

9. A moving coil galvanometer is shown in Fig. 72.5.

(a) Why is it an advantage for the galvanometer to have many turns?

(b) What accounts for the scale being linear?

(c) If the galvanometer were a centre-zero type, how would it respond to slow and rapid a.c.?

(d) Explain why a good voltmeter needs to have a high resistance and state what design changes would need to be made for this.

Fig. 72.5

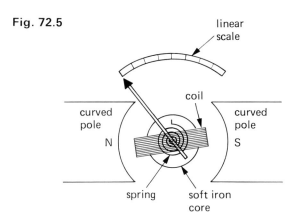

10. Describe the deflections observed on the sensitive, centre-zero galvanometer G when the copper rod XY in Fig. 72.6 is connected to its terminals and is made to vibrate up and down (as shown by the arrows), between the poles of a U-shaped magnet so that it is at right angles to the magnetic field.

Explain what is happening

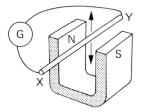

Fig. 72.6

Transformers

11. A transformer has 2000 turns on the primary coil. The voltage applied to the primary coil is 240 V a.c. How many turns are on the secondary coil if the output voltage is 48 V a.c.?

    A 40   **B** 400   **C** 4000   **D** 5000
    E 10 000   (*O.L.E./S.R.16+*)

Alternating current

12. The two circuits in Fig. 72.7 show a parallel plate capacitor in an a.c. circuit and in a d.c. circuit. What would happen to the bulb and ammeter in each circuit when the switch is closed?   (*A.L.*)

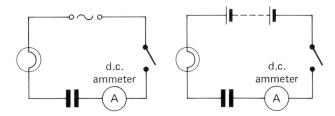

Fig. 72.7

Further level (higher grade students)

Electric motors

13. (a) Draw a labelled diagram of the essential components of a simple direct current electric motor. Explain clearly how continuous rotation is produced and show how the direction of rotation is related to the direction of the current.

(b) State what would happen to the direction of rotation of the motor you have described if **(i)** the current were reversed, **(ii)** the magnetic field were reversed, **(iii)** both current and field were reversed simultaneously.

(c) Why is a lead-acid accumulator preferred to a Leclanché (dry) battery in motor cars?   (*J.M.B.*)

14. (a) In Fig. 72.8, AB is a copper rod resting on two horizontal copper bars between the poles of a horseshoe magnet. Describe and account for what will happen to AB when the key K is closed.

(b) A working electric motor takes a current of 1.5 A when the p.d. across its terminals is 250 V. Determine the efficiency of the motor if the power output is 200 W.

If the resistance of the armature of the motor is about 1 Ω why is the current as little as 1.5 A when the motor is running?   (*L.*)

(*Note.* Efficiency = power output/power input. For last part, see p. 198.)

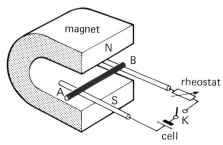

Fig. 72.8

207

Generators

15. In a simple a.c. generator a rectangular coil is rotated in a uniform magnetic field about an axis which is perpendicular to the field. The diagrams in Fig. 72.9 show end-on views of the coil which rotates about an axis through O in a uniform magnetic field produced between the poles of the permanent magnet N–S.

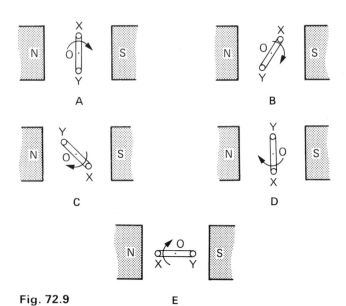

Fig. 72.9   E

In each of the questions following select the coil position which corresponds to the situation described in the question. Each position may be used once, more than once or not at all.

(a) The induced e.m.f. is at its peak value.

(b) The induced e.m.f. is between zero and its peak value and is *increasing*.

(c) The induced e.m.f. is between its peak value and zero and is decreasing.

(d) When the external circuit is completed, the direction of the e.m.f. induced in the coil is such as to give a current in YX *from* Y *to* X.   (*N.I.*)

16. The power output of a generator which supplies power to a consumer along cables of total resistance $16\,\Omega$ is 30 kW at 6 kV. What is (a) the current in the cables, (b) the power loss in the cables and (c) the drop of voltage between the ends of the cables?

17. Describe, with the aid of a labelled diagram, a simple form of generator. Explain how it may be used to generate (a) an alternating current, (b) a direct current.

In the case of (a) sketch a graph of the variation with time of the p.d. between the ends of the coil. Use the same axes to show clearly the effects of:

(i) doubling the number of turns of the coil and keeping the rotation speed and the magnetic field constant,

(ii) doubling the speed of rotation and keeping the number of turns and the magnetic field constant.
                              (*W.*)

Transformers

18. (a) (i) Draw a labelled diagram showing the essential features of a voltage step-down transformer. Explain how the transformer works.

(ii) State one source of energy loss in the transformer and say how it may be reduced.

(b) A transformer designed to operate a 12 V lamp from a 240 V supply has 1200 turns on the primary coil. Assuming that the transformer is 100% efficient, calculate (i) the number of turns on the secondary coil, (ii) the current passing through the primary coil when the 12 V lamp has a current of 2 A flowing through it.   (*J.M.B.*)

19. A transformer is used on the 240 V a.c. supply to deliver 9.0 A at 80 V to a heating coil. If 10% of the energy taken from the supply is dissipated in the transformer itself, what is the current in the primary winding?
                              (*O. and C.*)

Alternating current

20. (a) One capacitor has a value of 10 μF, a second has a value of 15 μF. Calculate the total capacitance of the two in parallel.

(b) Calculate the value of X in Fig. 72.10 if the total value of the three components is 15 μF.
                    (*J.M.B./A.L./N.W.16+*)

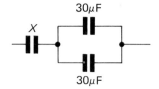

Fig. 72.10

Electrons and atoms

73 Electrons

The discovery of the electron was a landmark in physics and led to great technological advances. Evidence for its existence will now be considered.

Thermionic emission

The evacuated bulb in Fig. 73.1 contains a small coil of wire, the *filament*, and a metal plate called the *anode* because it is connected to the positive of the 400 V d.c. power supply. The negative of the supply is joined to the filament which is also called the *cathode*. The filament is heated by current from a 6 V supply (a.c. or d.c.).

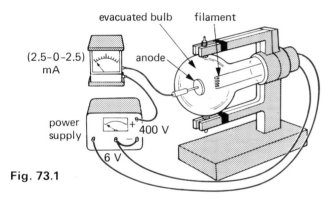

Fig. 73.1

With the circuit as shown, the meter deflects, indicating current flow in the circuit containing the gap between anode and cathode. The current stops if *either* the 400 V supply is reversed to make the anode negative, *or* the filament is not heated.

This demonstration supports the view that negative charges, in the form of electrons, escape from the filament when it is hot because they have enough energy to get free from the metal surface. The process is known as *thermionic emission*. The electrons are attracted to the anode if it is positive and are able to reach it because there is a vacuum in the bulb.

Cathode rays

Streams of electrons moving at high speed are called *cathode rays*. Their properties can be studied using two special cathode ray tubes.

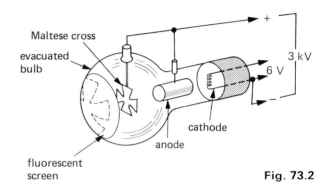

Fig. 73.2

(a) Maltese cross tube, Fig. 73.2. Electrons emitted by the hot cathode are accelerated towards the anode but most pass through the hole in it and travel on along the tube. Those that miss the cross, cause the screen to fluoresce with a green or blue light and cast a shadow of the cross on it. The cathode rays evidently travel in straight lines.

If the N pole of a magnet is brought up to the neck of the tube, the rays (and the fluorescent shadow) move upwards. The rays are deflected by a magnetic field and, using Fleming's left-hand rule (p. 192), we see that they behave like conventional current (positive charge flow) travelling from anode to cathode, i.e. like negative charge moving from cathode to anode.

The optical shadow of the cross due to the light also emitted by the cathode, is unaffected by the magnet.

(b) Deflection tube, Fig. 73.3. This can be used to show the deflection of cathode rays in electric and magnetic fields. Electrons from a hot cathode strike a fluorescent screen S set at an angle. A p.d. applied

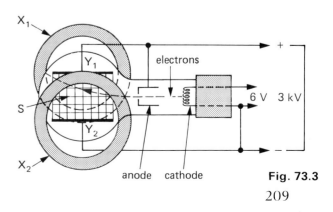

Fig. 73.3

across two horizontal metal plates Y_1Y_2 creates a *vertical* electric field which deflects the rays upwards if Y_1 is positive (as shown) and downwards if it is negative.

When current flows in the two coils X_1X_2 (in series) outside the tube, a *horizontal* magnetic field is produced across the tube. It can be used instead of a magnet to deflect the rays, or to cancel the deflection due to an electric field.

Specific charge e/m

The ratio of the charge e of an electron to its mass m is called its *specific charge e/m* and can be found from experiments in which cathode rays are deflected by electric and magnetic fields. This was first done by J. J. Thomson in 1897 using a deflection-type tube. His work is regarded as proving the existence of the electron as a negatively charged particle of very small mass (about 1/2000 of that of the hydrogen atom) and not, as some scientists thought, a form of electromagnetic radiation like light.

The modern value of e/m for an electron is 1.76×10^{11} C/kg.

Electronic charge e

The actual value of e was first measured accurately by Millikan in 1911. The apparatus he used is shown simplified in Fig. 73.4. Tiny oil drops, which became charged when they were sprayed, were observed as they fell through the air between two horizontal metal plates connected to a high p.d. Their time of fall in the electric field due to the p.d. depended on the charge they carried.

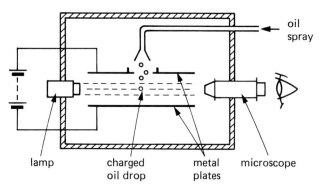

lamp   charged   metal   microscope
         oil drop   plates

Fig. 73.4

Millikan found that, for a large number of different drops, the charge was always a multiple of 1.6×10^{-19} C. Since the drops gain or lose a certain number of electrons when they become charged, he

210

concluded that the charge on *one* electron, i.e. the electronic charge, was 1.6×10^{-19} C.

Knowing e/m and e, m can be calculated. Its value is 9.1×10^{-31} kg.

Cathode ray oscilloscope (C.R.O.)

The C.R.O. is one of the most important scientific instruments ever to be developed. It contains, like a television set, a cathode ray tube which has three main parts, Fig. 73.5.

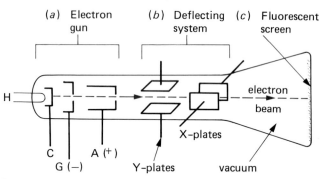

Fig. 73.5

(a) Electron gun. This consists of a *heater* H, a *cathode* C, another electrode called the *grid* G and two or three *anodes* A. G is at a negative voltage with respect to C and controls the number of electrons passing through its central hole from C to A; it is the *brilliance* or *brightness* control. The anodes are at high positive voltages relative to C; they accelerate the electrons along the highly evacuated tube and also *focus* them into a narrow beam.

(b) Fluorescent screen. A bright spot of light is produced on the screen where the beam hits it.

(c) Deflecting system. Beyond A are two pairs of deflecting plates to which p.d.s can be applied. The *Y-plates* are horizontal but create a vertical electric field which deflects the beam vertically. The *X-plates* are vertical and deflect the beam horizontally.

The p.d. to create the electric field between the Y-plates is applied to the *Y-input* terminals (often marked 'high' and 'low') on the front of the C.R.O. The input is usually amplified by an amount which depends on the setting of the *Y-amp gain* control, before it is applied to the Y-plates. It is then large enough to give a suitable vertical deflection of the beam.

In Fig. 73.6*a* the p.d. between the Y-plates is zero as is the deflection. In Fig. 73.6*b* the d.c. input p.d.

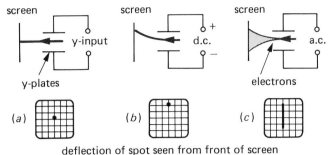

Fig. 73.6

deflection of spot seen from front of screen

makes the upper plate positive and attracts the beam of negatively charged electrons upwards. In Fig. 73.6c the 50 Hz a.c. input makes the beam move up and down so rapidly that it produces a continuous vertical line (whose length increases if the *Y-amp gain* is turned up).

The p.d. applied to the X-plates is also via an amplifier, the X-amplifier, and can either be from an external source connected to the *X-input terminal* or, what is commoner, from the time base circuit in the C.R.O.

The time base deflects the beam horizontally in the X-direction and makes the spot sweep across the screen from left to right at a steady speed determined by the setting of the *time base* controls (usually coarse and fine). It must then make the spot 'fly' back very rapidly to its starting point, ready for the next sweep. The p.d. from the time base should therefore have a sawtooth waveform like that in Fig. 73.7a. Since AB is a straight line, the distance moved by the spot is directly proportional to time and the horizontal deflection becomes a measure of time, i.e. a time axis or base.

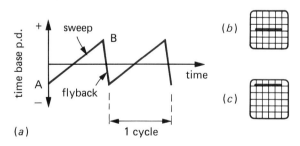

Fig. 73.7

In Fig. 73.7b, c, the time base is on, in (b) the Y-input p.d. is zero, in (c) the Y-input is d.c. which makes the upper Y-plate positive. In both cases the spot traces out a horizontal line which appears to be continuous if the time base frequency is high enough.

Fig. 73.8

Uses of the C.R.O

A small C.R.O is shown in Fig. 73.8.

(a) Practical points. The *brilliance* control, which is usually the *on/off* switch as well, should be as low as possible when there is just a spot on the screen. Otherwise screen 'burn' occurs which damages the fluorescent material. If possible it is best to defocus the spot or draw it into a line by running the time base.

When preparing the C.R.O. for use, set the *brilliance*, *focus*, X- and Y-shift controls (which allow the spot to be moved 'manually' over the screen in the X and Y directions respectively) to their mid-positions. The *time base* and *Y-amp gain* controls can then be adjusted to suit the input.

When the *a.c./d.c. selector* switch is in the 'd.c.' (or 'direct') position, both d.c. and a.c. can pass to the Y-input. In the 'a.c.' (or 'via C') position, a capacitor blocks d.c. in the input but allows a.c. to pass.

(b) Measuring p.d.s. A C.R.O. can be used as a d.c./a.c. voltmeter if the p.d. to be measured is connected across the Y-input terminals; *the deflection of the spot is proportional to the p.d.*

For example, if the *Y-amp gain* control is on, say, 1 V/div, a deflection of 1 vertical division on the screen graticule (like graph paper with squares for measuring deflections) would be given by a 1 V d.c. input. A line 1 division long would be produced by an a.c. input of 1 V peak-to-peak, i.e. peak p.d. = 0.5 V and r.m.s. value = $0.7 \times 0.5 = 0.35$ V.

(c) Displaying waveforms. In this widely used role, the time base is on and the C.R.O. acts as a 'graph-plotter' to show the waveform, i.e. the variation with time, of the p.d. applied to its Y-input. The displays in Fig. 73.9*a,b* are of alternating p.d.s with sine waveforms. In (*a*), the time base frequency *equals* that of the input and one complete wave is obtained. In (*b*) it is *half* that of the input and two waves are formed.

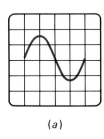 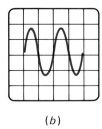

(*a*)   (*b*)

Fig. 73.9

Sound waveforms can be displayed if a microphone is connected to the Y-input terminals (see Topic 17).

(d) Measuring time intervals and frequency. These can be measured if the C.R.O. has a calibrated time base. For example, when the time base is set on 10 ms/div, the spot takes 10 milliseconds to move 1 division horizontally across the screen graticule. If this is the time base setting for the waveform in Fig. 73.9*b* then since 1 complete wave occupies 2 horizontal divisions, we can say

time for 1 complete wave = 2 divs × 10 ms/div
$$= 20 \text{ ms}$$
$$= 20/1000 = 1/50 \text{ s}$$

∴   number of complete waves per second = 50

∴   frequency of a.c. applied to Y-input = 50 Hz

Mass spectrometer

This is an instrument for finding the mass of individual atoms. A stream of positively charged ions (called *positive rays*) of the atoms involved is obtained by knocking electrons out of the atoms in an evacuated tube. The ions are then deflected by electric and magnetic fields and their mass found from measurements on the deflection.

1. Cathode rays consist of

   A beams of fast moving electrons
   B fluorescent particles
   C light rays from a screen
   D light rays from a hot filament
   E infrared rays from a hot filament
   (O.L.E./S.R.16+)

2. **(a)** In Fig. 73.10*a* to which terminals on the power supply must plates A and B be connected to deflect the cathode rays downwards?
   (b) In Fig. 73.10*b* in which direction will the cathode rays be deflected?

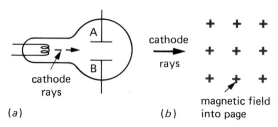

(*a*)   (*b*)   magnetic field into page

Fig. 73.10

3. **(a)** Fig. 73.11*a* shows a cathode ray tube fitted with one pair of plates. **(i)** What name is given to part P? **(ii)** What does part P produce?

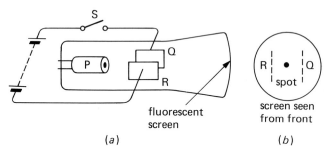

fluorescent screen   screen seen from front

(*a*)   (*b*)

Fig. 73.11

   (b) With switch S off a spot is seen in the centre of the screen as shown in Fig. 73.11*b*. **(i)** What happens to this spot when switch S is closed? **(ii)** What would happen to the spot if the e.m.f. of the battery were increased? **(iii)** What would you *see* on the screen if a low frequency alternating e.m.f. of about 1 Hz were connected to the plates Q and R? **(iv)** What would you *see* on the screen if a higher frequency alternating e.m.f. of about 50 Hz were connected to the plates Q and R?
   (c) Describe another way in which the spot can be made to behave exactly as it does when the switch S is closed as in **(b)(i)**. Illustrate your answer with a diagram. *(S.E.)*

74 Radioactivity

The discovery of radioactivity in 1896 by the French scientist Becquerel was accidental. He found that uranium compounds emitted radiation which **(i)** affected a photographic plate even when wrapped in black paper and **(ii)** ionized a gas. Soon afterwards Madame Curie discovered radium. Today radioactivity is used widely in industry, medicine and research.

We are all exposed to *background radiation* caused partly by radioactive materials in rocks, the air and our bodies, and partly by cosmic rays from outer space.

Ionizing effect of radiation

A charged electroscope discharges when a lighted match or a radium source (*held in forceps*) is brought near the cap, Fig. 74.1a,b.

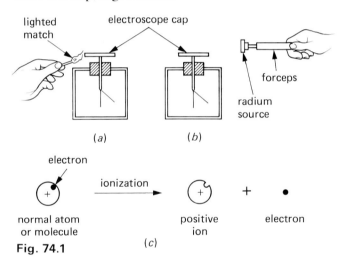

Fig. 74.1

In the first case the flame knocks electrons out of surrounding air molecules leaving them as positively charged air *ions*, i.e air molecules which have lost one or more electrons, Fig. 74.1c; in the second case radiation causes the same effect, called *ionization*. The positive air ions are attracted to the cap if it is negatively charged; if it is positively charged the electrons are attracted. As a result, the charge on the electroscope is neutralized, i.e. it loses its charge.

Geiger–Müller (G–M) tube

The ionizing effect is used to detect radiation.

When radiation enters a G–M tube, Fig. 74.2, either through a thin end-window made of mica, or, if it is

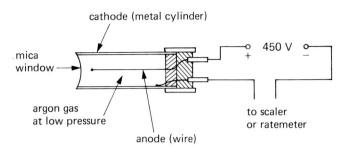

Fig. 74.2

very penetrating, through the wall, it creates argon ions and electrons. These are accelerated towards the electrodes and cause more ionization by colliding with other argon atoms.

On reaching the electrodes, the ions produce a current pulse which is amplified and fed either to a *scaler* or a *ratemeter*. A scaler counts the pulses and shows the total received in a certain time. A ratemeter has a meter marked in 'counts per second (or minute)' from which the average pulse-rate can be read. It usually has a loudspeaker which gives a 'click' for each pulse.

Alpha, beta, gamma rays

Experiments to study the penetrating power, ionizing ability and behaviour in magnetic and electric fields, show that a radioactive substance emits one or more of three types of radiation—called alpha (α), beta (β) and gamma (γ) rays.

Penetrating power can be investigated as in Fig. 74.3 by observing the effect on the count-rate of placing in turn between the G–M tube and the lead, a sheet of **(i)** thick paper (the radium source, lead and tube must be *close together* for this part), **(ii)** aluminium 2 mm thick and **(iii)** lead 2 cm thick. Other sources can be tried, e.g. americium, strontium and cobalt.

Fig. 74.3

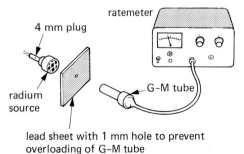

lead sheet with 1 mm hole to prevent overloading of G-M tube

(a) Alpha rays are stopped by a thick sheet of paper and have a range in air of a few centimetres since they cause intense ionization in a gas due to frequent collisions with gas molecules, They are deflected by electric and *strong* magnetic fields in a direction and by an amount which suggests they are helium atoms minus two electrons, i.e. *helium ions with a double positive charge.* From a particular substance, they are all emitted with the same speed (about 1/20th of that of light).

Americium (Am 241) may be used as a pure α source.

(b) Beta rays are stopped by a few millimetres of aluminium and some have a range in air of several metres. Their ionizing power is much less than that of α particles. As well as being deflected by electric fields, they are more easily deflected by magnetic fields and measurements show they are streams of *high-energy electrons,* like cathode rays, emitted with a range of speeds, up to that of light.

Strontium (Sr 90) emits β rays only.

The magnetic deflection of β particles can be shown as in Fig. 74.4. With the G–M tube at A and without the magnet, the count-rate is noted. Inserting the magnet reduces the count-rate but it increases again when the G–M tube is at B.

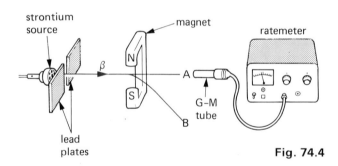

Fig. 74.4

(c) Gamma rays are the most penetrating and are stopped only by many centimetres of lead. They ionize a gas even less than β particles and are not deflected by electric and magnetic fields. They give interference and diffraction effects and are *electromagnetic radiation* travelling at the speed of light. Their wavelengths are those of very short X-rays from which they differ only because they arise *in* atomic nuclei whereas X-rays come from energy changes in the electrons *outside* the nucleus.

Cobalt (Co 60) is a pure γ source. Radium (Ra 226) emits α, β and γ rays.

A G–M tube detects β and γ rays and energetic α particles; a charged electroscope detects α only. All three types of rays cause fluorescence.

214

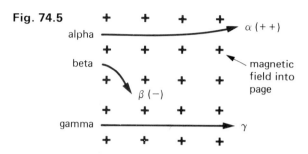

Fig. 74.5

The behaviour of the three kinds of radiation in a magnetic field is summarized in Fig. 74.5. The deflections are not to scale and are found from Fleming's left-hand rule (taking negative charge moving to the right as equivalent to positive (conventional) current to the left).

Cloud chambers

When air containing vapour, e.g. alcohol, is cooled enough, saturation occurs. If ionizing radiation passes through the air, further cooling causes the saturated vapour to condense on the air ions created. The resulting white line of tiny liquid drops shows up as a track when illuminated.

In a *diffusion cloud chamber*, Fig. 74.6, vapour from alcohol in the felt ring diffuses downwards, is cooled by the 'dry ice' (solid carbon dioxide at −78 °C) in the lower section and condenses near the floor on air ions formed by radiation from the source in the

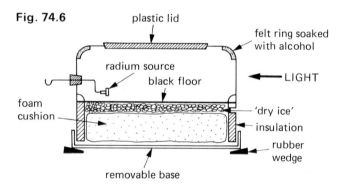

Fig. 74.6

chamber. Tracks are produced continuously which are sharp if an electric field is created by frequently rubbing the plastic lid of the chamber with a cloth.

α particles give straight, thick tracks, Fig. 74.7a. Very fast β particles produce thin, straight tracks; slower ones give short, twisted, thicker tracks, Fig. 74.7b. γ rays eject electrons from air molecules; the electrons behave like β particles and produce their own tracks spreading out from the γ rays.

The *bubble chamber*, in which the radiation leaves a trail of bubbles in liquid hydrogen, has now replaced the cloud chamber in research work.

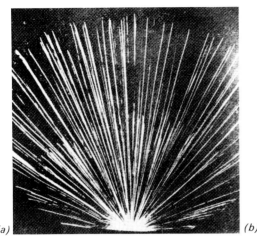

Fig. 74.7

(a)　　　(b)

Radioactive decay: half-life

Radioactive atoms change or *decay* into atoms of different elements when they emit α or β particles (p. 218). These changes are spontaneous and cannot be controlled; also, it does not matter whether the material is pure or combined chemically with something else.

(a) Half-life. The *rate of decay* is unaffected by temperature but every radioactive element has its own definite decay rate, expressed by its *half-life*. This is *the time for half the atoms in a given sample to decay*. It is difficult to know when a substance has lost all its radioactivity, but the time for its activity to fall to half its value can be found more easily.

(b) Decay curve. The average number of disintegrations (i.e. decaying atoms) per second of a sample is its *activity*. If it is measured at different times (e.g. by finding the count-rate using a G–M tube and ratemeter), an *activity* against *time* decay curve can be plotted. The one in Fig. 74.8 shows that the activity decreased by the *same* fraction in successive equal time intervals. It falls from 80 to 40 disintegrations per second in 10 minutes, from 40 to 20 in the next 10 minutes, from 20 to 10 in the third 10 minutes and so on. The half-life is 10 minutes.

Half-lives vary from millionths of a second to millions of years. For radium it is 1600 years.

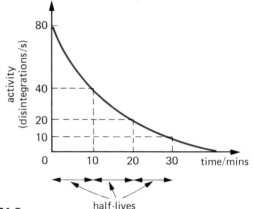

Fig. 74.8

(c) Experiment. The half-life of the α-emitting gas *thoron* can be found as in Fig. 74.9a by squeezing the thoron bottle 3 or 4 times to transfer some thoron to the flask. The clips are then closed, the bottle removed and the stopper replaced by a G–M tube so that it seals the top, Fig. 74.9b.

When the ratemeter reading has reached its maximum and started to fall, the count-rate is noted every 15 s for 2 minutes and then every 60 s for the next few minutes. (The G–M tube is left in the flask for at least 1 hour until the radioactivity has decayed.)

A graph of *count-rate* (from which the background count-rate, found separately, has been subtracted)

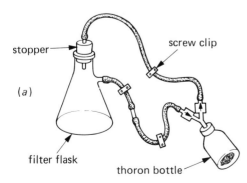

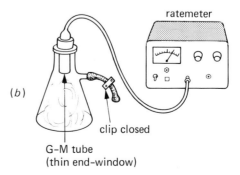

Fig. 74.9

215

against *time* is plotted and the half-life (52 s) estimated from it.

(d) Random nature. During the previous experiment it becomes evident that the count-rate varies irregularly: the loudspeaker of the ratemeter 'clicks' erratically, not at a steady rate. This is because radioactive decay is a *random* process, in that it is a matter of pure chance whether or not a particular nucleus will decay during a certain period of time. All we can say is that about half the nuclei in a sample will decay during the half-life. We cannot say which nuclei these will be, nor can we influence the process in any way.

Uses of radioactivity

Radioactive substances, called *radioisotopes* (p. 218), are now made in nuclear reactors (p. 219) and have many uses.

(a) Thickness gauge. If a radioisotope is placed on one side of a moving sheet of material and a G–M tube on the other, the count-rate decreases if the thickness increases. This technique is used to control automatically the thickness of paper, plastic and metal sheets during manufacture, Fig. 74.10.

(b) Tracers. The progress of a small amount of a weak radioisotope injected into a system can be 'traced' by a G–M tube or other detector. The method is used in medicine to detect brain tumours, in agriculture to study the uptake of fertilizers by plants and in industry to measure fluid flow in pipes.

(c) Radiotherapy. Gamma rays from strong cobalt radioisotopes are replacing X-rays in the treatment of cancer.

(d) Archaeology. A radioisotope of carbon, present in the air, is taken in by plants and trees and is used to date archaeological remains of wood and linen.

Fig. 74.10

Dangers and safety

The danger from α particles (due, like all radiation, to their ionizing effect) is small unless the source enters the body. β and γ rays can cause radiation burns (i.e. redness and sores on the skin) and delayed effects such as cancer and eye cataracts. Fall-out from atomic explosions contains highly active elements (e.g. strontium), with long half-lives, that are absorbed by the bones.

The weak sources used at school *should always be lifted with forceps, never held near the eyes and should be kept in their boxes when not in use.* In industry sources are handled by long tongs and transported in thick lead containers. Workers are protected by lead and concrete walls.

✎ Worked example

A radioactive source has a half-life of 20 minutes. What fraction is left after 1 hour?

After 20 minutes, fraction left $= \frac{1}{2}$
After 40 minutes, fraction left $= \frac{1}{2} \times \frac{1}{2} = \frac{1}{4}$
After 60 minutes, fraction left $= \frac{1}{2} \times \frac{1}{4} = \frac{1}{8}$

Q Questions

1. Which type of radiation from radioactive materials
   (a) has a positive charge?
   (b) is the most penetrating?
   (c) is easily deflected by a magnetic field?
   (d) consists of waves?
   (e) causes the most intense ionization?
   (f) has the shortest range in air?
   (g) has a negative charge?
   (h) is not deflected by an electric field?

2. What is the effect of radiation on
   (a) a charged electroscope?
   (b) a photographic film?
   (c) a fluorescent screen?

3. **(a)** What do you understand by 'background radiation'? State two sources of this radiation.
   (b) State three safety precautions to be observed when using radioactive sources.
   (c) Describe briefly two uses of radioactive sources.
   (d) How would you test to distinguish between two radioactive sources, one of which emits only alpha particles and the other which emits only beta particles?
   (A.L.)

4. In an experiment to find the half-life of radioactive iodine, the count-rate falls from 200 counts per second to 25 counts per second in 75 minutes. What is its half-life?

5. If the half-life of a radioactive gas is 2 minutes, then after 8 minutes the activity will have fallen to a fraction of its initial value. This fraction is

   A $\frac{1}{4}$ **B** $\frac{1}{6}$ **C** $\frac{1}{8}$ **D** $\frac{1}{16}$ **E** $\frac{1}{32}$

check list
p.291

75 Atomic structure

The discovery of the electron and of radioactivity seemed to indicate that atoms contained negatively and positively charged particles and were not indivisible as was previously thought. The questions then were 'how are the particles arranged inside an atom', and 'how many are there in the atom of each element?'

An early theory, called the 'plum-pudding' model, regarded the atom as a positively charged sphere in which the negative electrons were distributed all over it (like currants in a pudding) and in sufficient numbers to make the atom electrically neutral. Doubts arose about this model.

Nuclear model

Whilst investigating radioactivity Rutherford noticed that not only could α particles pass straight through very thin metal foil as if it wasn't there but also that some were deflected from their initial direction. With the help of Geiger (of tube fame) and Marsden, Rutherford investigated this in detail at Manchester University using the arrangement in Fig. 75.1. The fate of the α particles after striking the gold foil was detected by the scintillations (flashes of light) they produced on a glass screen coated with zinc sulphide and fixed to a rotatable microscope.

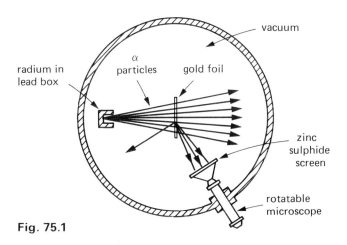

Fig. 75.1

They found that most of the α particles were undeflected, some were scattered by appreciable angles and a few (about 1 in 8000) surprisingly 'bounced' back. To explain these results Rutherford proposed in 1911 a nuclear model of the atom in which *all the positive charge and most of the mass of an atom* formed a dense core or *nucleus*, of very small size

compared with the whole atom. The electrons surrounded the nucleus some distance away.

He derived a formula for the number of α particles deflected at various angles, assuming that the electrostatic force of repulsion between the positive charge on an α particle and the positive charge on the nucleus of a gold atom obeyed an inverse square law (i.e. the force increases four times if the separation is halved). Geiger and Marsden's results completely confirmed Rutherford's formula and supported the view that an atom is mostly empty space. In fact the nucleus and electrons occupy about one million millionth of the volume of an atom. Putting it another way, the nucleus is like a sugar lump in a very large hall and the electrons a swarm of flies.

The paths of three α particles are shown in Fig. 75.2. ① is clear of all nuclei and passes straight *through* the gold atoms. ② suffers some deflection. ③ approaches a gold nucleus so closely as to be violently repelled by it and 'rebounds', appearing to have had a head-on 'collision'.

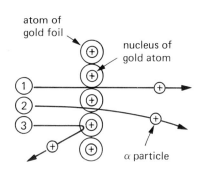

Fig. 75.2

Protons and neutrons

We now believe as a result of other experiments, in some of which α and other high-speed particles were used as atomic probes, that atoms contain three basic particles—protons, neutrons and electrons.

A *proton* is a hydrogen atom minus an electron, i.e. a positive hydrogen ion. Its charge is equal in size but opposite in sign to that of an electron but its mass is about 2000 times greater.

A *neutron* is uncharged with almost the same mass as a proton.

Protons and neutrons are in the nucleus and are called *nucleons*. Together they account for the mass of the nucleus (and most of that of the atom); the protons account for its positive charge. These facts are summarized in Table 1.

Table 1

Particle	Relative mass	Charge	Location
Proton	1836	$+e$	In nucleus
Neutron	1839	0	In nucleus
Electron	1	$-e$	Outside nucleus

In a neutral atom the number of protons equals the number of electrons surrounding the nucleus. Table 2 shows the particles in some atoms. Hydrogen is simplest with 1 proton and 1 electron. Next is the inert gas helium with 2 protons, 2 neutrons and 2 electrons. The soft white metal lithium has 3 protons and 4 neutrons.

Table 2

	Hydrogen	Helium	Lithium	Oxygen	Copper
Protons	1	2	3	8	29
Neutrons	0	2	4	8	34
Electrons	1	2	3	8	29

The atomic or proton number Z of an atom is the number of protons in the nucleus.

It is also the number of electrons in the atom. The electrons determine the chemical properties of an atom and when the elements are arranged in atomic number order in the Periodic table, they fall into chemical families.

The mass or nucleon number A of an atom is the number of nucleons in the nucleus.

In general

$$A = Z + N$$

where N is the *neutron number* of the element.

Atomic *nuclei* are represented by symbols. Hydrogen is written ^1_1H, helium ^4_2He, lithium ^7_3Li and in general atom X is written ^A_ZX where A is the nucleon number and Z the proton number.

Isotopes and nuclides

Isotopes of an element are atoms which have the same number of protons but different numbers of neutrons. That is, their proton numbers are the same but not their nucleon numbers.

Isotopes have identical chemical properties since they have the same number of electrons and occupy the same place in the Periodic table. (In Greek, *isos* means same and *topos* means place.)

Few elements consist of identical atoms; most are mixtures of isotopes. Chlorine has two isotopes; one has 17 protons and 18 neutrons (i.e. $Z = 17$, $A = 35$) and is written $^{35}_{17}Cl$, the other has 17 protons and 20 neutrons (i.e. $Z = 17$, $A = 37$) and is written $^{37}_{17}Cl$. They are present in ordinary chlorine in the ratio of three atoms of $^{35}_{17}Cl$ to one atom of $^{37}_{17}Cl$, giving chlorine an average atomic mass of 35.5.

Hydrogen has three isotopes: ^1_1H with 1 proton, *deuterium* ^2_1D with 1 proton and 1 neutron and *tritium* ^3_1T with 1 proton and 2 neutrons. Ordinary hydrogen contains 99.99 per cent of ^1_1H atoms. Water made from deuterium is called heavy water (D_2O); it has a density of 1.108 g/cm^3, it freezes at $3.8\,°C$ and boils at $101.4\,°C$.

Each form of an element is called a *nuclide*. Nuclides with the same Z but different A are isotopes.

Radioactive decay

The emission of an α or β particle from a nucleus produces an atom of a different element, which may itself be unstable. After a series of changes a stable end-element is formed.

(a) **Alpha decay.** An α particle is a helium nucleus having 2 protons and 2 neutrons and when an atom decays by α emission, its nucleon number decreases by 4 and its proton number by 2. For example, when radium of nucleon number 226 and proton number 88 emits an α particle, it decays to radon of nucleon number 222 and proton number 86. We can write:

$$^{226}_{88}Ra \rightarrow {}^{222}_{86}Rn + {}^4_2He$$

The values of A and Z must balance on both sides of the equation since nucleons and charge are conserved.

(b) **Beta decay.** Here a neutron changes to a proton and an electron. The proton remains in the nucleus and the electron is emitted as a β particle. The new nucleus has the same nucleon number, but its proton number increases by one since it has one more proton. Radioactive carbon, called carbon 14, decays by β emission to nitrogen.

$$^{14}_{6}C \rightarrow {}^{14}_{7}N + {}^{0}_{-1}e$$

(c) Gamma emission. After emitting an α or β particle some nuclei are left in an 'excited' state. Rearrangement of the protons and neutrons occurs and a burst of γ rays is released.

Models of the atom

(a) Rutherford–Bohr model. Shortly after Rutherford proposed his nuclear model of the atom, Bohr, a Danish physicist, developed it to explain how an atom emits light. He suggested that the electrons circled the nucleus at high speed being kept in *certain orbits* by the electrostatic attraction of the nucleus for them. He pictured atoms as miniature solar systems. Fig. 75.3 shows the models for three elements.

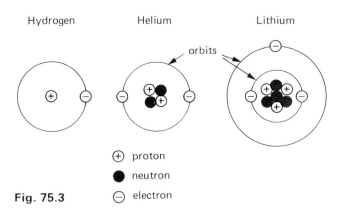

Hydrogen   Helium   Lithium

orbits

⊕ proton
● neutron
⊖ electron

Fig. 75.3

Normally the electrons remain in their orbits but if the atom is given energy, e.g. by being heated, electrons may jump to an outer orbit. The atom is then said to be *excited*. Very soon afterwards the electrons return to an inner orbit and as they do, they emit energy in the form of bursts of electromagnetic radiation (called *photons*), e.g. as light, infrared, ultraviolet or X-rays, Fig. 75.4. The wavelength of the radiation emitted depends on the two orbits between which the electrons jump.

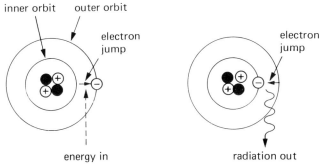

inner orbit   outer orbit

electron jump

electron jump

energy in   radiation out

Fig. 75.4

If an atom gains enough energy for an electron to escape altogether, the atom becomes an ion and the energy needed to achieve this is called the *ionization energy* of the atom.

(b) Modern model. Although it is still useful for some purposes, the Rutherford–Bohr model has now been replaced by a *mathematical model* which is not easy to picture. The best we can do, without using advanced mathematics, is to say that the atom consists of a nucleus surrounded by a hazy cloud of electrons. Regions of the atom where the mathematics predicts electrons are likely to be found, are represented by dense shading, Fig. 75.5.

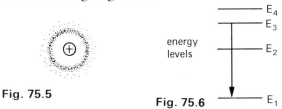

energy levels

E_4
E_3
E_2
E_1

Fig. 75.5

Fig. 75.6

The modern theory does away with the idea of electrons moving in definite orbits and replaces them by *energy levels*. When an electron jumps from one level, say E_3 in Fig. 75.6, to a lower one E_1, a photon of electromagnetic radiation is emitted with energy equal to the difference in energy of the two levels.

Nuclear energy

(a) $E = mc^2$. Einstein predicted that if the energy of a body changes by an amount E, its mass changes by an amount m given by the equation

$$E = mc^2$$

where c is the speed of light (3×10^8 m/s). The implication is that any reaction in which there is a decrease of mass, called a *mass defect*, is a source of energy. The energy and mass changes in physical and chemical changes are very small; those in some nuclear reactions, e.g. radioactive decay, are millions of times greater. It appears that mass (matter) is a very concentrated form of energy.

(b) Fission. The heavy metal uranium is a mixture of isotopes of which $^{235}_{92}U$, called uranium 235, is the most important. Some atoms of this isotope decay quite naturally, emitting high-speed neutrons. If one of these hits the nucleus of a neighbouring uranium 235 atom (being uncharged the neutron is not repelled by the nucleus), this may break (*fission*) into two nearly equal radioactive nuclei, often of barium and krypton, with the production of two or three more neutrons.

$$^{235}_{92}U + {}^{1}_{0}n \rightarrow {}^{144}_{56}Ba + {}^{90}_{36}Kr + 2{}^{1}_{0}n$$

neutron   fission fragments   neutrons

219

The mass defect is large and appears mostly as k.e. of the fission fragments. These fly apart at great speed, colliding with surrounding atoms and raising their average k.e., i.e. their temperature, so producing heat.

If the fission neutrons split other uranium 235 nuclei, a *chain reaction* is set up, Fig. 75.7. In practice some fission neutrons are lost by escaping from the surface of the uranium before this happens. The ratio of those escaping to those causing fission decreases as the mass of uranium 235 increases. This must exceed a certain *critical* value to start the reaction.

Fig. 75.7

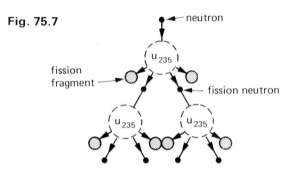

(c) Nuclear reactor. In a nuclear power station a nuclear reactor produces the steam for the turbines instead of a coal- or oil-burning furnace. Fig. 75.8 is a simplified diagram of a reactor.

The chain reaction occurs at a steady rate which is controlled by inserting or withdrawing neutron-absorbing rods of boron among the uranium rods. The graphite core is called the *moderator* and slows

In an *atomic bomb* an increasing uncontrolled chain reaction occurs when two pieces of uranium 235 come together and exceed the critical mass.

(d) Fusion. The union of light nuclei into heavier ones can also lead to a loss of mass and, as a result, the release of energy. At present, research is being done on the controlled fusion of isotopes of hydrogen (deuterium and tritium) to give helium. Temperatures of about 100 million °C are required. Fusion is believed to be the source of the sun's energy.

Ⓠ Questions

1. What are the particles you would expect to find in an atom? Give some idea of their relative masses and state what electrical charge, if any, each kind has. (*S.E.*)

2. **(a)** An atom of cobalt has an atomic number of 27 and a mass number of 59. Describe simply the structure of the cobalt atom.
   (b) What are isotopes?
   (c) Why are isotopes difficult to separate by chemical methods? (*W.M.*)

3. Uranium 238 and uranium 235 are 'isotopes' of uranium and have the same atomic number, 92.
   (a) What do the numbers 238 and 235 represent?
   (b) **(i)** What does the number 92 tell you about the nucleus of either of these two atoms?
   (ii) What else does the number 92 tell you about the atom as a whole?
   (c) In what way does the nucleus of uranium 238 differ from the nucleus of uranium 235?
   (*J.M.B./A.L./N.W.16+*)

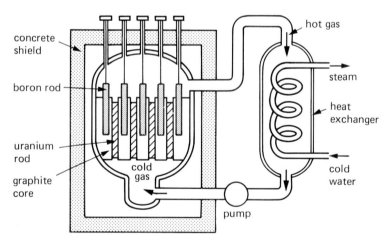

Fig. 75.8

down the fission neutrons; fission of uranium 235 occurs more readily with slow than with fast neutrons. Carbon dioxide gas is pumped through the core and carries off heat to the heat exchanger where steam is produced. The concrete shield gives protection from γ rays and neutrons. The radioactive fission fragments must be removed periodically if the nuclear fuel is to be used efficiently.

4. What changes, if any, occur in the atomic number of a radioactive atom if the nucleus emits **(a)** an α particle, **(b)** a β particle, **(c)** a γ ray? (*E.A.*)

5. Atomic bombs and atomic reactors both provide large quantities of energy through chain reactions.
   (a) Explain what is meant by a chain reaction.
   (b) Explain how the reaction, which is violent in the case of a bomb, is slowed down in the reactor. (*E.A.*)

check list
p.291

76 Electronics

Electronics is being used more and more in our homes, factories, offices, banks, shops and hospitals. The development of semiconductor devices such as transistors and integrated circuits ('chips') has given us, among other things, pocket calculators that are also clocks and musical instruments, Fig. 76.1a, heart pacemakers, Fig. 76.1b, microcomputers, robots, TV games, machines for teaching spelling and arithmetic and even for recognizing signatures.

Semiconductors

Semiconductors, of which silicon and germanium are the two best known, are insulators if they are very pure, especially at low temperatures. However, their conductivity can be greatly increased by adding tiny, but controlled amounts of certain other substances (called 'impurities') by a process known as 'doping'. They can then be used to make diodes, transistors and integrated circuits. Two types of semiconductor material are obtained in this way.

In *n-type*, silicon is doped with phosphorus atoms which increase the number of free *negative* electrons that are able to move through the material.

In *p-type* silicon, boron atoms are used for doping. They create gaps, called *positive* 'holes', in the material and conduction occurs by electrons jumping from one hole to another. It is just as if positive holes were moving in the opposite direction and for that reason we usually consider that conduction in a *p*-type semiconductor is due to positive holes.

Junction diode

A diode is a two-terminal device which lets current pass through it in one direction only. One can be made by doping a crystal of pure silicon (or germanium) so that there is a region of *p*-type material in contact with a region of *n*-type material. The boundary between them is called the *junction*. The connection to the *p*-side is the *anode* (A) and that to the *n*-side is the *cathode* (C), Fig. 76.2a. A diode and its symbol are shown in Fig. 76.2b.

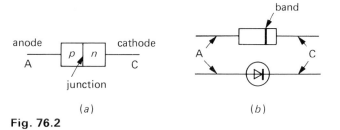

Fig. 76.2

(a) **Forward bias.** If a supply is connected as in Fig. 76.3a so that the *p*-type is positive and the *n*-type negative, the positive holes drift from *p*- to *n*-material across the junction while free negative electrons go from *n*- to *p*-material. This happens because the positive terminal of the supply repels positive holes in the *p*-type towards the junction, while the negative terminal of the supply does the same to the free negative electrons in the *n*-type. The diode conducts, it has a very low resistance and is forward biased.

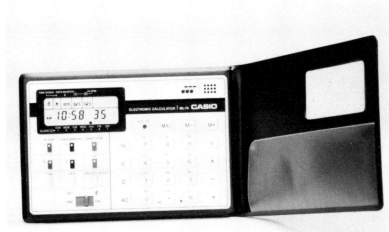

Fig. 76.1   (a)   (b)

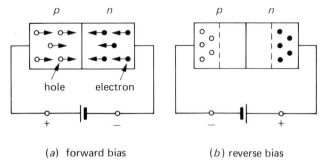

(a) forward bias   (b) reverse bias

Fig. 76.3

(b) Reverse bias. If the supply voltage is applied the other way round, Fig. 76.3b, the holes and electrons are attracted by the negative and positive terminals respectively of the supply and move to opposite ends of the diode, away from the junction so that there is no flow of charge across it. The region around the junction loses its charges (holes and electrons) and becomes an insulator. The diode hardly conducts, it has a very high resistance and is said to be reverse biased.

To sum up, a diode has an 'easy' and a 'difficult' direction for current to pass. Its one-way action can be shown with the circuit of Fig. 76.4 in which the lamp only lights when the diode is forward biased.

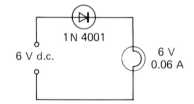

Fig. 76.4

(c) Characteristic curve. This was investigated in Topic 60 and shows how the current I through a diode varies with the p.d. V across it. The circuit is given again in Fig. 76.5a. Altering the rheostat changes V and if the corresponding values of I are noted, a graph like that in Fig. 76.5b for a typical silicon diode can be plotted. Reverse bias is obtained by reversing the battery connections.

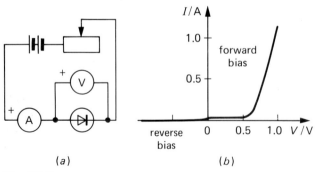

(a)   (b)

Fig. 76.5

222

The curve shows that conduction starts when the forward bias value of V is about 0.6 V (0.1 V for a germanium diode). Thereafter, a very small change in V causes a sudden, large increase in I, which must not exceed the rated forward current (1A for 1N4001). In reverse bias, which also has a maximum value (50 V for 1N4001), I is almost zero.

(d) Uses. Junction diodes are used

(i) as rectifiers to change a.c. to d.c., and

(ii) to prevent damage to a circuit by a reversed power supply. In Fig. 76.6a the battery is correctly connected to the circuit; it forward biases the diode

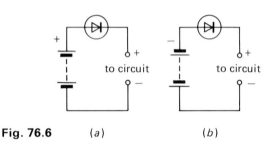

Fig. 76.6   (a)   (b)

and current passes. In Fig. 76.6b the battery is incorrectly connected but it reverse biases the diode so no current flows. The circuit is thus protected against wrong polarity connection of the supply.

Rectification and smoothing

The conversion of a.c. to steady d.c. is often necessary for electronic equipment such as a radio, which is supplied by the a.c. mains but requires d.c.

(a) Half-wave rectification. In the circuit of Fig. 76.7 the diode removes the negative half-cycles of the a.c. input to give a varying but one-way (direct) p.d. across the 'load' R requiring a d.c. supply. In practice R would be a piece of electronic equipment.

Fig. 76.7

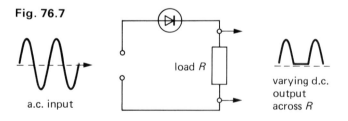

If the Y-input terminals of a C.R.O. are connected first across the a.c. input, then across R, the output waveform is seen to be the positive half-cycles of the input. Hence 'half-wave' rectification.

(b) Smoothing. The 'humps' in the varying d.c. from a rectifier can be smoothed to give a much steadier

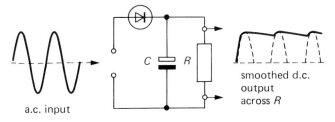

Fig. 76.8

d.c. by connecting a large capacitor C (electrolytic) across R as in Fig. 76.8.

On the positive half-cycles of the a.c. input the diode conducts and current passes through R *and* also into C to charge it up. On the negative half-cycles when the diode is reverse biased and non-conducting, C partly discharges through R. The charge stored in C acts as a reservoir and maintains the current in R. A steadier p.d. is thus produced across R as the output waveform shows. There is still a 'ripple' on the output which reveals itself as 'mains hum'.

(c) Full-wave rectification. In this case both half-cycles of the a.c. to be rectified are used. In the bridge rectifier circuit of Fig. 76.9, the current follows the

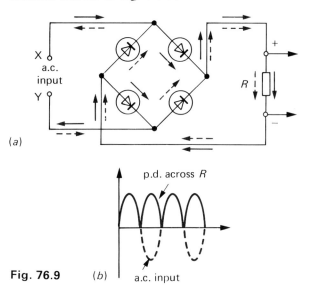

Fig. 76.9 (b)

solid arrows when X is positive and Y negative and the broken arrows on the negative half-cycles when the polarities of X and Y are reversed. During both half-cycles, current passes through R and in the *same* direction, giving a p.d. like that shown. It can be smoothed by connecting a large capacitor across R.

(d) Power supply unit. If a transformer with its primary connected to the a.c. mains supplies the unit, the live wire from the mains must have a switch and fuse. The iron core of the transformer must also be earthed to prevent shock in case the insulation of the windings breaks down.

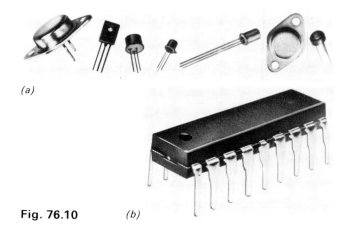

(a)

Fig. 76.10 *(b)*

Transistors

Transistors are the small, three terminal, semiconductor devices which have revolutionized electronics. They are made as separate devices, like those in Fig. 76.10*a* in their cases, and also as parts of integrated circuits (ICs) where many thousands may be packed on a 'chip' of silicon, Fig. 76.10*b*.

Transistors act as fast *switches* and as *amplifiers*.

(a) Action. The simplified structure of the *n-p-n* junction transistor, the commonest type, is shown in Fig. 76.11 with its symbol. It consists of two *p-n* junctions (in effect two diodes back-to-back) arranged as a sandwich with a thin *p*-type 'filling', the *base* between two thicker slices of *n*-type material, called the *collector* and the *emitter*. The arrow on the symbol gives the direction in which conventional (positive) current would flow; electron flow is in the opposite direction.

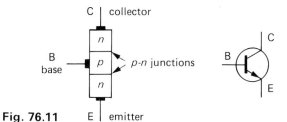

Fig. 76.11

There are two current paths through a transistor. One is the *base-emitter path* and the other is the *collector-emitter* (via base) *path*. The transistor's value arises from the fact that it can link circuits connected to each path so that the current in one controls that in the other.

If a p.d. (e.g. $+6$ V) is applied across an *n-p-n* transistor so that the collector becomes positive with respect to the emitter, the base being unconnected, the base-collector *p-n* junction is reverse biased

223

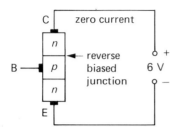

Fig. 76.12

(since the + of the supply goes to the *n*-type collector). Current cannot pass through the transistor, Fig. 76.12.

If the base-emitter junction is now forward biased by applying a p.d. V_{BE} of about +0.6 V (for a silicon transistor, +0.1 V for germanium), Fig. 76.13,

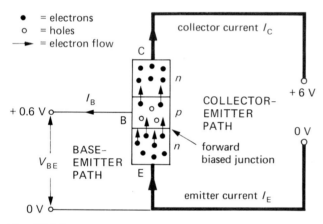

Fig. 76.13

electrons flow from the *n*-type emitter across the junction (as they would in a diode) into the *p*-type base. Their loss is made good by electrons entering the emitter from the external circuit to form the emitter current I_E.

In the base only a small proportion of the electrons from the emitter combine with holes because the base is very thin and is lightly doped. The loss of holes which does occur is made good by some flowing to the base from the base power supply. This creates a small base current I_B. Most electrons pass through the base under the attraction of the positive collector. They cross the base-collector junction and become the collector current I_C in the external circuit.

If we regard I_B as the *input* current and I_C as the *output* current, a transistor acts

(i) as a *switch* in which I_B turns on and controls I_C, i.e. if $I_B = 0$, $I_C = 0$ and I_B starts to flow when $V_{BE} \approx$ +0.6 V (the 'turn-on' p.d.); thereafter I_C increases when I_B does,

(ii) as a *current amplifier* since I_C is greater than I_B and a small change in I_B causes a larger change in I_C.

224

(b) Current gain. Typically I_C is 10 to 1000 times greater than I_B depending on the transistor. The *current gain* is an important property of a transistor and is defined by

$$current\ gain = \frac{I_C}{I_B}$$

For example, if $I_C = 5$ mA and $I_B = 50\ \mu A = 0.05$ mA the current gain $= 5/0.05 = 100$.

Since the current leaving equals that entering,

$$I_E = I_B + I_C$$

(c) Further points. For an *n-p-n* transistor the collector and base must be positive with respect to the emitter.

The input (base-emitter) and output (collector-emitter) circuits in Fig. 76.13 have a common connection at the emitter. The transistor is said to be in *common-emitter connection*.

In general, silicon is preferred to germanium for transistors, one reason being that silicon can work at higher temperatures (175°C compared with 75°C).

Transistors are damaged if they get too hot due, for example, to carrying too large a collector current. Having a *heat sink*, Fig. 76.14, on the transistor helps to prevent overheating, the excess heat being lost from the black plastic cooling fins.

Fig. 76.14

Other semiconductor devices

(a) Light dependent resistor (L.D.R.). Its action depends on the fact that the resistance of certain semiconductors such as cadmium sulphide, decreases as the intensity of the light falling on them increases. For example, it might be several megohms in the dark but only a few kilohms in daylight.

An L.D.R., its symbol and a circuit which shows its action is given in Fig. 76.15a,b. When light from a lamp falls on the 'window' of the L.D.R., its resistance decreases and the increased current is large enough to bring on the bulb.

L.D.R.s are used in photographic exposure meters.

(b) Thermistor. It contains semiconducting metallic oxides whose resistance decreases markedly when

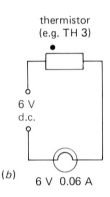

(a)

Fig. 76.15

(b)

L.D.R.
(e.g. ORP 12)

6 V
d.c.

6 V 0.06 A

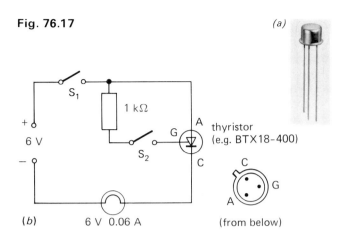

Fig. 76.17

(a)

S_1

1 kΩ

+
6 V
−

S_2

G
A
C

thyristor
(e.g. BTX18–400)

C

G

A

(b)
6 V 0.06 A

(from below)

the temperature rises either due to heating the thermistor directly or to passing a current through it.

One is shown in Fig. 76.16a,b with its symbol and a circuit to demonstrate its action. Heating the thermistor with a match lights the bulb. A thermistor in series with a meter marked in °C can measure temperatures in, for example, the range −5 to 70°C.

(a)

Fig. 76.16

(b)

thermistor
(e.g. TH 3)

6 V
d.c.

6 V 0.06 A

(c) Thyristor or silicon controlled rectifier (S.C.R.). This is a silicon diode with a third connection called the *gate*. When forward biased it does not conduct until a small gate current enters it. Conduction continues even if the gate current becomes zero; it stops only if the battery to the diode itself is disconnected or if the diode current falls below a certain value.

A thyristor, its symbol and a circuit which shows its action when S_1 and S_2 are closed, is given in Fig. 76.17a,b.

(d) Light emitting diode (L.E.D.). An L.E.D., shown in Fig. 76.18a,b, with its symbol, is a junction diode made from the semiconductor gallium arsenide phosphide. When *forward biased* it conducts and emits red, yellow or green light. No light is emitted on reverse bias which, if it exceeds 5 V, may damage the L.E.D.

In use, an L.E.D. must have a suitable resistor R in series with it to limit the current, which typically

may be 10 mA (0.01 A). The value of R depends on the supply p.d. For example, in Fig. 76.18b it is 5 V and if we take the p.d. across a forward biased L.E.D. as 2 V then R is given by

$$R = \frac{(\text{supply p.d.} - 2)\ V}{0.01\ A} = \frac{(5 - 2)\ V}{0.01\ A} = 300\ \Omega$$

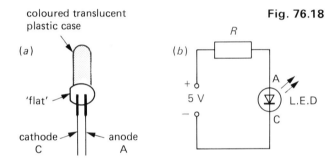

coloured translucent
plastic case

Fig. 76.18

(a)

'flat'

cathode
C

anode
A

(b)

R

+
5 V
−

A

L.E.D

C

L.E.D.s are used as indicator lamps in radio receivers and other electronic equipment. Many calculators, clocks, cash registers and measuring instruments have seven-segment red or green numerical displays, Fig. 76.19a. Each segment is an L.E.D. and depending on which have a p.d. across them, the display lights up the numbers 0 to 9, as in Fig. 76.19b.

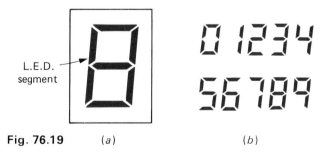

L.E.D.
segment

01234
56789

Fig. 76.19   *(a)*   *(b)*

L.E.D.s are small, reliable, have a long life, their operating speed is high and their current requirement is very modest.

(e) Photodiode. This is a normal junction diode having a 'window' through which light can enter. It works in reverse bias and the tiny reverse current increases with the light intensity. Photodiodes are used as fast counters which produce a current pulse every time a beam of light is interrupted. Its symbol is the same as for an L.E.D. but with the arrows reversed.

Transistor as a switch

(a) Advantages. Transistors have many advantages over other electrically operated switches such as relays and reed switches. They are small, cheap, reliable, have no moving parts, their life is almost indefinite (in well-designed circuits) and they can switch on and off millions of times a second.

Digital electronics is concerned with the use of transistors as switches in devices such as computers, pocket calculators, digital watches, weighing machines and increasingly in telecommunications.

(b) 'On' and 'off' states. A transistor is considered to be 'off' when the collector current is zero or very small. It is 'on' when the collector current is much larger. The resistance of the collector-emitter path is large when the transistor is 'off' (as it is for an ordinary mechanical switch) and small (ideally it should be zero) when it is 'on'.

To switch a transistor 'on' requires the base voltage (and therefore the base current) to exceed a certain minimum value ($+0.6$ V for silicon).

(c) Basic switching circuits. Two are shown in Fig. 76.20a,b. In both, only one battery is used to obtain the positive voltages for the collector and base of the transistor. The 'on' state is shown by the lamp in the collector circuit becoming fully lit.

Fig. 76.20

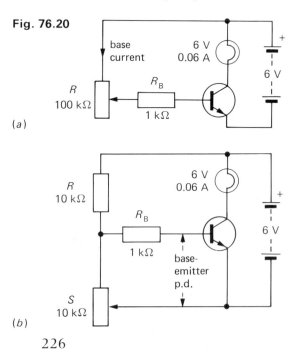

(a)

(b)

226

Rheostat control is used in (a) where the base-emitter junction is forward biased to 'switch-on' by reducing R until the *base current* is large enough to make the collector current light the lamp. (The base resistor R_B is *essential* in case R is made zero and results in $+6$ V from the battery being applied directly to the base. This would produce very large base and collector currents and destroy the transistor by overheating.)

Potential divider control is used in (b). Here 'switch-on' is obtained by adjusting the variable resistor S until the p.d. across S (which is the base-emitter p.d. and depends on the value of S compared with that of R) exceeds $+0.6$ V or so.

Note. In a potential divider the p.d.s across the resistors are in the ratio of their resistances. For example, in Fig. 76.20b if $R = 10$ kΩ and $S = 5$ kΩ then p.d. across $R = V_R = 4$ V and p.d. across $S = V_S = 2$ V, i.e. $V_R/V_S = 4V/2V = 2/1$ and $V_R + V_S = 6$ V = p.d. across R and S in series where $R/S = 10$ k$\Omega/5$ $\Omega = 2/1$.

In general

$$\frac{V_R}{V_S} = \frac{R}{S} \quad \text{and} \quad V_S = (V_R + V_S)S/(R+S)$$

Also see question 3 on page 233.

Experiment: transistor switching circuits

The components can be mounted on a circuit board, e.g. an S-DeC, Fig. 76.21a. Fig. 76.21b shows how to lengthen transistor leads and also how to make connections (without soldering) to parts that have 'tags', e.g. variable resistors.

(a)

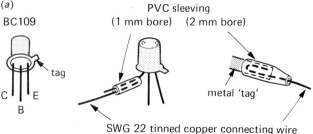

BC109   PVC sleeving
(1 mm bore)   (2 mm bore)

metal 'tag'

SWG 22 tinned copper connecting wire held in contact by sleeving

Fig. 76.21 (b)

In many alarm circuits, devices such as L.D.R.s, thermistors and microphones are used in potential divider arrangements to detect small changes of light intensity, temperature and sound level respectively. These changes then cause a transistor to switch on and activate an alarm, e.g. a lamp or bell.

(a) Light-operated. In the circuit of Fig. 76.22 the L.D.R. is part of a potential divider. The lamp comes on when the L.D.R. is shielded due to more of the battery p.d. being dropped across the increased resistance of the L.D.R. (i.e. more than 0.6 V) and less across R. In the dark, the base-emitter p.d. increases as does the base current and so also the collector current.

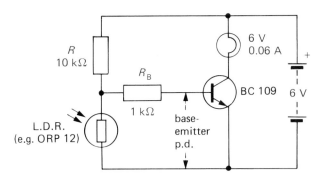

Fig. 76.22

If the L.D.R. and R are interchanged the lamp goes off in the dark and the circuit could act as a light-operated intruder alarm.

(b) Temperature-operated. In the high-temperature alarm circuit of Fig. 76.23 a thermistor and a resistor R form a potential divider across the 6 V supply.

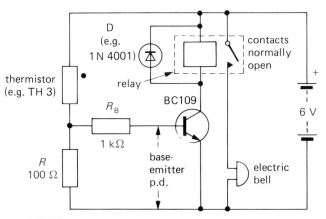

Fig. 76.23

When the temperature of the thermistor rises, its resistance decreases and a larger fraction of the 6 V supply is dropped across R, i.e. the base-emitter p.d. increases. When it exceeds 0.6 V or so the transistor switches on and collector current (too small to ring

the bell directly) goes through the relay coil. The relay contacts close, enabling the bell to obtain directly from the 6 V supply, the larger current it needs.

The diode D protects the transistor from damage by the large e.m.f. induced in the relay coil (due to its inductance) when the collector current falls to zero at switch off. The diode is forward biased by the induced e.m.f. (which tries to maintain the current through the relay coil) and, because of its low forward resistance (e.g. 1 Ω), offers an easy path for the current produced. To the 6 V supply the diode is reverse biased and its high resistance does not short-circuit the relay coil when the transistor is on.

If the thermistor and R are interchanged, the circuit could act as a frost-warning device.

(c) Time-operated. In the circuit of Fig. 76.24 when S_1 and S_2 are closed, the lamp is on and the transistor is off because the base-emitter p.d. is zero (due to S_2 short-circuiting C and stopping it charging up). If S_2 is opened, C starts to charge through R and, *after a certain time*, the base-emitter p.d. exceeds 0.6 V causing the transistor to switch on. This operates the relay whose contacts open and switch off the lamp. The *time delay* between opening S_2 and the lamp going off increases if either C or R are increased.

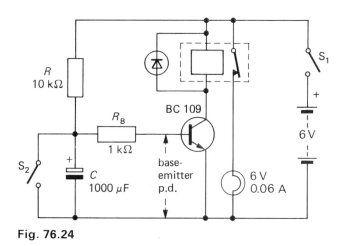

Fig. 76.24

The circuit is reset by opening S_1 and closing S_2 to let C discharge. It could be used as a timer to control a lamp in a photographic dark room.

(d) Sound-operated with latching. The variable resistor R in Fig. 76.25 is adjusted so that the transistor switches 'on' only when someone speaks into the microphone. The emitter current then provides the gate current which triggers the thyristor and allows current to flow through the lamp. The lamp stays on till the 6 V supply is disconnected, i.e. the thyristor acts as a latching switch.

227

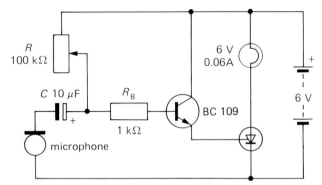

Fig. 76.25

The capacitor C stops d.c. from the battery passing via R through the microphone and upsetting the operation of the transistor. But it allows the a.c. produced in the microphone by the sound to pass to the base.

Using a relay and bell instead of a lamp (as in Fig. 76.23) the circuit could form the basis of a sound-operated intruder alarm.

Multivibrator circuits

Multivibrators are switching circuits containing two transistors and when one is 'on' the other is 'off'. There are three types.

(a) Bistable. The circuit is given in Fig. 76.26. If L_1 comes on when it is first connected then Tr 1 is 'on' (i.e. large collector current) and Tr 2 is 'off' (near zero collector current). The circuit stays in this state, i.e. it *latches* on to it and *remembers* it and is a *memory* circuit.

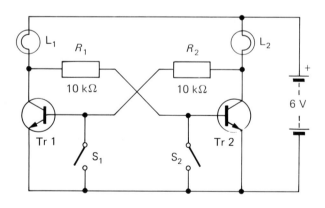

$L_1 = L_2 = 6 \text{ V } 0.06 \text{ A} : \text{Tr 1} = \text{Tr 2} = \text{BC 109}$

Fig. 76.26

However, if S_1 is momentarily closed, this connects the base of Tr 1 to 0 V, i.e. it is no longer forward biased and so switches 'off'. L_1 therefore goes off and L_2 lights up because Tr 2 comes 'on' due to its base

228

now being connected via R_1 and L_1 to $+6$ V, i.e. it is forward biased. The circuit will stay in this second stable state (hence bistable) until S_2 is closed momentarily, when L_1 lights again.

By adding a few more components to a bistable, a circuit is obtained which changes from one stable state to the other every time a switching pulse is applied to an input called the 'trigger'. The pulse can be obtained either by connecting and disconnecting 'trigger' to $+6$ V or from an astable multivibrator (see below). Each lamp then only comes on every second trigger pulse, i.e. at *half* the frequency of the trigger pulses. The bistable is a *divide-by-two* circuit and many joined together make a binary counter. Bistables are used as dividers in digital watches.

(b) Astable. A circuit is shown in Fig. 76.27. In this case Tr 1 and Tr 2 switch 'on' and 'off' automatically causing L_1 and L_2 to flash in turn. The action is due to C_1 charging and discharging through R_1 and C_2 doing the same through R_2. The circuit is used for the flashing lights warning of road works.

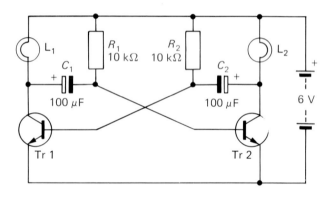

$L_1 = L_2 = 6 \text{ V } 0.06 \text{ A} : \text{Tr 1} = \text{Tr 2} = \text{BC 109}$

Fig. 76.27

The flashing rate depends on the values of $C_1 \times R_1$ and $C_2 \times R_2$. For example if C_1 and C_2 are replaced by 0.1 μF capacitors and L_2 by a headphone (or small loudspeaker), the transistors switch 'on' and 'off' so quickly that L_1 seems to be on all the time and an audible note (i.e. a string of very rapid 'clicks') is heard in the headphone.

An astable has no stable states but switches from one state to the other of its own accord at a rate determined by the circuit components. One of its uses is as an *oscillator* to produce the notes in an electronic organ. Another use is as a timer or clock to control and operate an electronic system.

(c) Monostable, Fig. 76.28. It has one stable state in which it normally rests with Tr 1 'on' and Tr 2 'off'.

It can be switched to its unstable state for a certain time, depending on the values of C_1 and R_1, if the base of Tr 1 is connected momentarily to 0 V by closing S_1. During this time L_2 is alight.

A monostable can be used as a *timer* to bring on a light for a known time.

All three types of multivibrator are made as ICs.

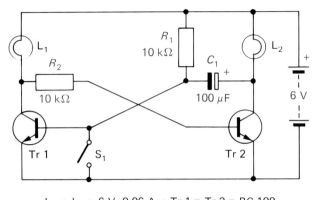

$L_1 = L_2 = 6$ V 0.06 A : Tr 1 = Tr 2 = BC 109

Fig. 76.28

Logic gates

Logic gates are switching circuits used in computers and other digital electronic systems. They 'open' and give a 'high' output voltage, i.e. a signal (e.g. 5 V), depending on the combination of voltages at their inputs, of which there is usually more than one.

There are six basic types, all made from transistors in integrated circuit form. The behaviour of each is described by a *truth table* showing what the output is for all possible inputs. 'High' (e.g. 5 V) and 'low' (e.g. near 0 V) outputs and inputs are represented by 1 and 0 respectively and are referred to as *logic levels* 1 and 0.

(a) NOT gate or inverter. This is the simplest gate, with one input and one output. It produces a 'high' output if the input is 'low', i.e. NOT high and vice-versa. Whatever the input, the gate inverts it. The symbol and truth table are given in Fig. 76.29.

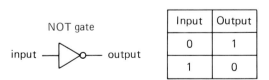

Input	Output
0	1
1	0

Fig. 76.29

(b) OR, NOR, AND, NAND gates. All these have two or more inputs and one output. The truth tables and symbols for 2-input gates are shown in Fig. 76.30. Try to remember the following.

OR : output is 1 if input A *OR* input B *OR* both are 1

NOR : output is 1 if neither input A *NOR* input B is 1

AND : output is 1 if input A *AND* input B are 1

NAND : output is 1 if input A *AND* input B are *NOT* both 1

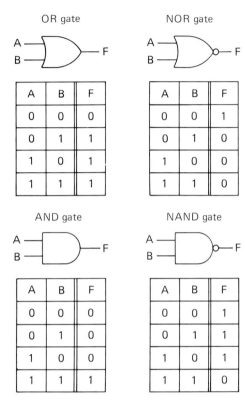

A	B	F
0	0	0
0	1	1
1	0	1
1	1	1

A	B	F
0	0	1
0	1	0
1	0	0
1	1	0

A	B	F
0	0	0
0	1	0
1	0	0
1	1	1

A	B	F
0	0	1
0	1	1
1	0	1
1	1	0

Fig. 76.30

Note that the outputs of the NOR and NAND gates are those of the OR and AND gates respectively but inverted. They have a small circle at the output end of their symbols to show this inversion.

(c) Exclusive OR gate. It gives a 'high' output when either input is 'high' but not when both are 'high'. Unlike the ordinary OR gate (sometimes called the *inclusive* OR gate), it excludes the case of both inputs being 'high' for a 'high' output. It is also called the *difference* gate because the output is 'high' when the inputs are different.

The symbol and truth table are shown in Fig. 76.31.

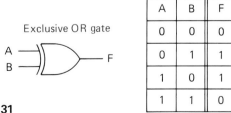

A	B	F
0	0	0
0	1	1
1	0	1
1	1	0

Fig. 76.31

229

(d) Testing logic gates. The truth tables for the various gates can be conveniently checked by having the logic gate IC mounted on a small board with sockets for the power supply, inputs A and B and output F, Fig. 76.32. A 'high' input (i.e. logic level 1) is obtained by connecting the input socket to

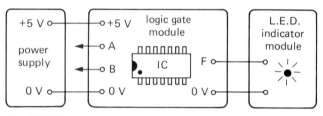

Fig. 76.32

the positive of the power supply, e.g. +5 V and a 'low' one (i.e. logic level 0) to 0 V.

The output can be detected using an indicator module containing an L.E.D. which lights up for a '1' and stays off for a '0'.

(e) Using gates to build logic circuits. Fig. 76.33 shows a circuit in which an L.D.R. and a thermistor are used with an AND gate to produce a switch which is both light- and temperature-dependent. For example, it could be used to ring a bell (via a relay) if the temperature of a *hot* radiator fell during the *day*.

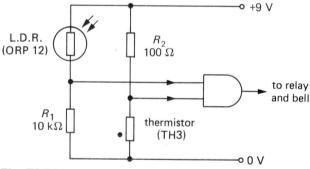

Fig. 76.33

The input from the potential divider formed by the L.D.R. and R_1 is a '1' in *daylight* (since the L.D.R. resistance is then small compared with 10 kΩ). The input from the R_2-thermistor potential divider is a '1' when it is *cold* (since the resistance of the thermistor is then several times greater than 100 Ω). The AND gate 'opens' (because both its inputs are 1's) and gives a '1' output which rings the bell. When the radiator is hot, the input from the thermistor is a '0' (because its resistance is now much less than 100 Ω) and the output from the AND gate is 0. The bell remains silent.

230

Transistor as an amplifier

This is the other main use for transistors and is what *analogue electronics* (e.g. radio and television) is chiefly concerned with.

In a transistor amplifier a small a.c. signal, due, for example, to sound falling on a microphone, is applied as the *input* to its base-emitter circuit. This causes small changes of the steady base current which result in much larger changes of collector current. These produce as the *output*, an amplified version of the original sound in an earphone in the collector circuit. For the amplification to be distortionless, the output waveform must be an exact copy of the input waveform and is achieved only if the base current changes about a suitable steady value.

In the simple circuit of Fig. 76.34a the steady base current is decided by the resistance of R_B which should be about 100 times greater than the resistance of the earphone. Capacitor C stops d.c. from the battery passing through the microphone (and upsetting the biasing) but allows the a.c. signal from the microphone to pass to the base.

Fig. 76.34

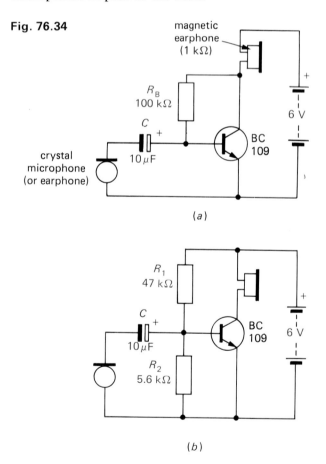

(a)

(b)

An alternative arrangement for forward biasing the base-emitter junction to the required value using a potential divider R_1–R_2 is shown in Fig. 76.34b.

Practical amplifiers have several transistors with the output of one fed to the input of the next. This increases the amplification.

Operational amplifier

Operational amplifiers (op amps) were originally designed to solve mathematical equations electronically by performing operations such as addition. Today, in IC form, they have many uses, two of the most important being **(a)** as high-gain d.c. and a.c. voltage amplifiers and **(b)** as switches. A typical one contains about twenty transistors.

An op amp has one output and two inputs, called the *inverting* input (marked $-$) and the *non-inverting* input (marked $+$), as shown on its symbol in Fig. 76.35. It operates from a dual power supply giving equal positive and negative d.c. voltages in the range ± 5 V to ± 15 V. The centre point on the power supply is common to the input and output circuits and is taken as 0 V.

Fig. 76.35

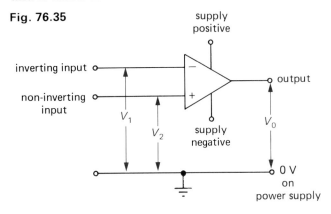

If a small *positive* voltage V_1 is applied to the inverting input ($-$), with the non-inverting input ($+$) connected to 0 V so that $V_2 = 0$, the output is an amplified *negative* voltage V_0. On the other hand if V_1 is negative, V_0 is positive. In both cases V_0 is of opposite sign to V_1, i.e. it is inverted and in antiphase with V_1.

If a small *positive* voltage V_2 is applied to the non-inverting input, with the inverting input connected to 0 V so that $V_1 = 0$, the output in this case is an amplified *positive* voltage V_0. Similarly if V_2 is negative, V_0 is negative. In both cases V_0 has the same sign as V_2, i.e. it is not inverted and is in phase with V_2.

Op amp voltage amplifier

In this role op amps almost always use *negative feedback*, i.e. part of the output is fed back to the external input so that it is in antiphase with it. The

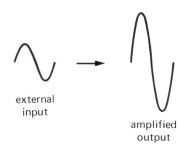

(a) no negative feedback

external input → amplified output

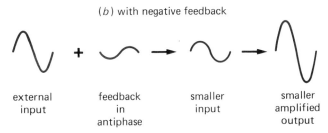

(b) with negative feedback

external input + feedback in antiphase → smaller input → smaller amplified output

Fig. 76.36

feedback subtracts from the input, thereby reducing it and the output, as shown in Fig. 76.36. The *voltage gain A* of the amplifier, defined by

$$A = \frac{output\ voltage}{input\ voltage}$$

is thus reduced but this is more than compensated for by the gain becoming *accurately predictable* and *more constant over a wider range of input frequencies*.

In the basic circuit of Fig. 76.37 for an *inverting* op amp voltage amplifier, the non-inverting terminal is held at 0 V and the input voltage V_i (a.c. or d.c.) to be amplified, is applied via resistor R_i to the inverting terminal. The output voltage V_0 is therefore in antiphase with the input. The *feedback resistor* R_f, by feeding back a certain fraction (depending on the value of R_f) of the output to the *inverting* terminal, ensures that the feedback is negative.

Fig. 76.37

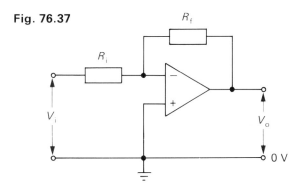

N.B. Power supply connections not shown

231

It can be proved that the gain is given by

$$A = \frac{V_0}{V_i} = -\frac{R_f}{R_i} \qquad (1)$$

For example, if $R_f = 100$ kΩ and $R_i = 10$ kΩ, $A = -10$. The negative sign shows that V_0 is negative if V_i is positive and vice-versa.

From equation (1) we see that A depends only on the values of the two resistors R_f and R_i (which can be known accurately) and not on the particular op amp used. Also note that the maximum value of the output voltage cannot exceed the range between the positive and negative of the power supply, i.e. 18 V peak-to-peak on a ± 9 V supply.

To check the gain equation experimentally, V_i (at about 1 kHz) can be obtained from a signal generator and measured, along with V_0, using a CRO as an a.c. voltmeter (Topic 73), with the time base on so that the waveforms of V_i and V_0 are seen to be undistorted as in Fig. 76.38a.

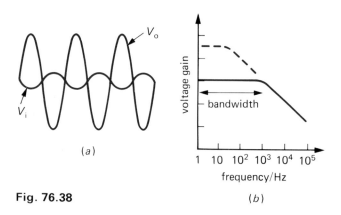

Fig. 76.38

To show that the gain decreases as the frequency of V_i increases, the signal generator should be set at different frequencies up to 100 kHz and a graph plotted of *gain* against *frequency* using a frequency scale that increases by powers of ten, as in Fig. 76.38b (to accommodate the wide range of values). The frequency range over which the gain is nearly constant is called the *bandwidth* of the amplifier; it decreases if the gain is increased (by making R_f/R_i greater), as shown by the dotted graph.

Op amp voltage comparator

When both inputs of an op amp are used at the same time, the amplified output voltage V_0 is the *difference* between the two input voltages V_1 and V_2. But because of its very high gain (e.g. up to 10^5 for d.c. and low frequency a.c. without negative feedback), if the difference between V_1 and V_2 is greater than

232

about 0.1 mV, then in theory V_0 exceeds the power supply voltage but in practice it cannot. Hence for values of V_1 and V_2 which are more than this difference, V_0 remains constant at the supply voltage (e.g. $+9$ V or -9 V on a ± 9 V supply) and the op amp is said to be *saturated*. In this condition it can be used as a *switch*.

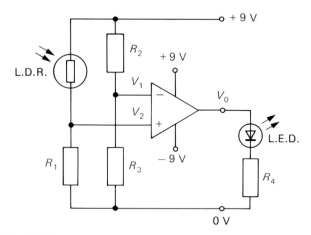

Fig. 76.39

For example, in the light-operated alarm circuit of Fig. 76.39, the inputs V_1 and V_2 are supplied by potential dividers and the op amp compares them. In the dark, the resistance of the L.D.R. is much greater than R_1, making V_2 less than V_1, the 'reference' voltage (set by the values of R_2 and R_3), the difference being large enough to saturate the op amp. Since V_1 (at the inverting input) is positive, V_0 will be negative and about -9 V. If light falls on the L.D.R., its resistance decreases and increases V_2. When V_2 (the non-inverting input) exceeds V_1, the op amp switches to its other saturated state with V_0 about $+9$ V. This positive voltage lights the L.E.D. (i.e. the alarm).

Counting in binary

Normally we count on the decimal or scale of ten system. Electronic counting is done on the *binary* or scale of two system using the numbers 0 and 1, called 'bits' (from *binary digits*). The table opposite shows 0 to 15 in binary, the least significant bit being at the extreme right.

Bits 0 and 1 are represented electrically by 'high' and 'low' voltages respectively, e.g. by 5 V and 0 V. L.E.D.s can be used to detect the voltage carried by a wire; if it is 'high' the L.E.D. lights, if it is 'low' it doesn't. The simple circuit in Fig. 76.40 shows the principle for a 4-bit number (0101 = 5).

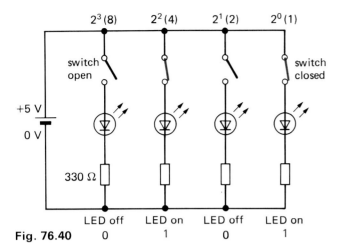

2^3 (8)
2^2 (4)
2^1 (2)
2^0 (1)
switch open
switch closed
+5 V
0 V
330 Ω
LED off
0
LED on
1
LED off
0
LED on
1

Fig. 76.40

Binary				Decimal
2^3 (8)	2^2 (4)	2^1 (2)	2^0 (1)	
0	0	0	0	0
0	0	0	1	1
0	0	1	0	2
0	0	1	1	3
0	1	0	0	4
0	1	0	1	5
0	1	1	0	6
0	1	1	1	7
1	0	0	0	8
1	0	0	1	9
1	0	1	0	10
1	0	1	1	11
1	1	0	0	12
1	1	0	1	13
1	1	1	0	14
1	1	1	1	15

Q Questions

1. Fig. 76.41*a* shows a lamp, a semiconductor diode and a cell connected in series. The lamp lights when the diode is connected in this direction. Say what happens to each of the lamps in *b*, *c* and *d*. Give reasons for your answer. (*S.E.*)

D
L
(a)
D_1
L_2
L_1
D_2
(b)
E_1
D
L_1
E_2
L_2
(c)
E_1
D_1
L_1
E_2
D_2
L_2
(d)

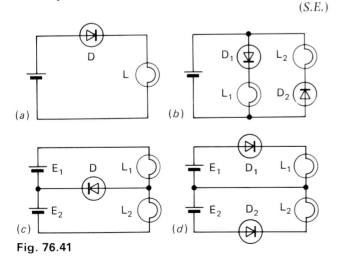

Fig. 76.41

2. Fig. 76.42 shows a circuit demonstrating an important use of a transistor. State the action in the circuit when the resistance of the variable resistor is decreased. State the importance of this action. (*E.A.*)

M
M
+
–
–
+

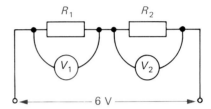

Fig. 76.42

3. What are the readings on the high resistance voltmeters V_1 and V_2 in the potential divider circuit of Fig. 76.43 if **(a)** $R_1 = R_2 = 10$ kΩ, **(b)** $R_1 = 10$ kΩ, $R_2 = 50$ kΩ and **(c)** $R_1 = 20$ kΩ and $R_2 = 10$ kΩ?

R_1
R_2
V_1
V_2
6 V

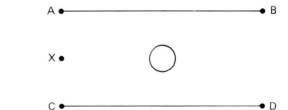

Fig. 76.43

4. Fig. 76.44 shows an incomplete electronic circuit.

(a) Copy the diagram and make the following additions.

(i) In the circle draw the symbol for an *n-p-n* transistor.

(ii) Add the symbol for a battery, correctly connected between B and D.

A
B
X
C
D

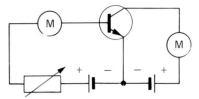

Fig. 76.44

(iii) Draw the symbol for a thermistor between points A and X.

(iv) Draw the symbol for a variable resistor between points X and C.

(v) Draw the symbol for a lamp in the collector circuit.

(vi) Connect the base and emitter of the transistor to the correct places in the circuit.

(b) As the circuit is arranged the lamp will not light. If there is a rise in temperature the lamp will light. Explain this.

(c) A simple rearrangement of this circuit would make the lamp light when the temperature falls. Explain what changes should be made, and how they will affect the working of the circuit.

(d) In the original circuit, instead of a variable resistor,

a fixed resistor could have been used between points X and C. Explain why it might have been considered desirable to use a variable resistor for these two circuits. (J.M.B./A.L./N.W.16+)

5. A simple moisture warning circuit is shown in Fig. 76.45 in which the moisture detector is two closely-spaced copper rods.

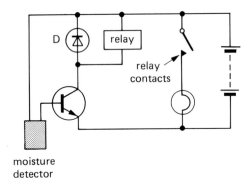

moisture
detector

Fig. 76.45

(a) Describe how the circuit works when the detector gets wet.
(b) Warning lamps are often placed in the collector circuit of a transistor. Why is a relay used here?
(c) If the relay operates with a current of 30 mA and the current gain of the transistor is 60, what base current is needed for the relay to work?
(d) What is the function of D?

6. (a) Name the three types of multivibrator circuit.
(b) State what each type does.
(c) Give one use for each type.

7. The combined truth tables for five logic gates A, B, C, D, E are given below. State what kind of gate each one is.

Inputs		Outputs				
		A	B	C	D	E
0	0	0	0	0	1	1
0	1	0	1	1	1	0
1	0	0	1	1	1	0
1	1	1	1	0	0	0

8. What do the symbols represent in Fig. 76.46a,b,c,d,e,f?

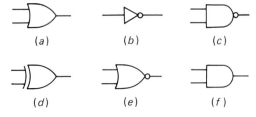

(a)　　　(b)　　　(c)

(d)　　　(e)　　　(f)

Fig. 76.46

check list p.292

9. Write down the truth table for the logic circuit in Fig. 76.47.

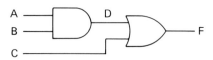

Fig. 76.47

10. (a) Draw the circuit for a single stage common emitter amplifier which includes a microphone, an earphone, a resistor and a capacitor.
(b) What is the function of each of the components mentioned in (a)?

11. (a) State two important uses of an op amp.
(b) Explain the terms 'inverting' and 'non-inverting' inputs with reference to an op amp.
(c) Draw the symbol for an op amp.

12. (a) Explain the term negative feedback.
(b) Give two reasons why op amp voltage amplifiers use negative feedback.

13. The circuit of Fig. 76.48 is for an inverting op amp voltage amplifier. What is
(a) its voltage gain,
(b) the output voltage if the input is an alternating voltage of peak value (i) ± 1 V, (ii) ± 2 V, and the supply voltage is ± 15 V?

Fig. 76.48

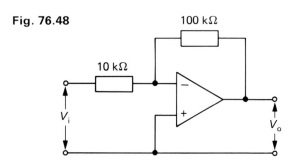

14. How does the voltage gain of an op amp vary with the frequency of the input?

15. (a) How is an op amp used to compare voltages?
(b) Voltages V_1 and V_2 with waveforms like those in Fig. 76.49 are applied simultaneously to the inverting and non-inverting inputs respectively of an op amp on a ± 6 V supply. Copy Fig. 76.48 and draw below it the waveform of the output voltage V_0.

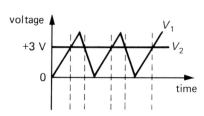

Fig. 76.49

77 Radio and television

Radio waves

Radio waves are electromagnetic radiation (pp. 43 and 44) which vary in wavelength from a few mm to several km. Typical values are given below.

Waves →	Long	Medium	Short	v.h.f.	u.h.f.	Micro
Wave-length	1500 m	300 m	30 m	3 m	30 cm	3 cm
Fre-quency	200 kHz	1 MHz	10 MHz	10^2 MHz	10^3 MHz	10^4 MHz
Use		Radio		T.V.		Radar

They can travel in the three ways shown in Fig. 77.1. *Ground waves* follow the earth's surface and have a limited range, being greatest (1500 km) for long waves and only a few km for v.h.f. (very high frequency) waves. *Sky waves* are bounced back from the ionosphere (p. 44) and are used in long-distance communication by long, medium and short waves.

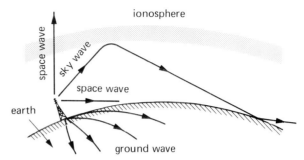

Fig. 77.1

Space waves travel in straight lines and can penetrate the ionosphere. v.h.f., u.h.f. (ultra high frequency) and microwaves travel in this way.

Fig. 77.2 shows the Goonhilly Downs, Cornwall, aerial for receiving microwave signals relayed across the Atlantic by a synchronous satellite (i.e. one which travels round the earth in 24 hours and appears to be at rest).

Fig. 77.2

Outline of radio

Radio waves are emitted by aerials when a.c. flows in them but the length of the aerial must be comparable with the wavelength of the wave produced for the radiation to be appreciable. A 50 Hz a.c. corresponds to a wavelength of 6×10^6 m (since $v = f\lambda$ where $v = 3 \times 10^8$ m and $f = 50$ Hz).

Alternating currents with frequencies below about 20 kHz are called *audio frequency* (a.f.) currents; those with frequencies greater than this are *radio frequency* (r.f.) currents. Therefore, so that aerials are not too large, they are supplied with r.f. currents. However, speech and music generate a.f. currents and so some way of combining a.f. with r.f. is required if they are to be sent over a distance.

(a) Transmitter. In this an *oscillator* produces an r.f. current which would cause an aerial connected to it to send out an electromagnetic wave, called a *carrier wave*, of constant amplitude and having the same frequency as the r.f. current. If a normal receiver picked up such a signal nothing would be heard. The r.f. signal has to be modified or *modulated* so that it 'carries' the a.f. This is done in various ways.

In *amplitude modulation* (a.m.) the amplitude of the r.f. is varied so that it depends on the a.f. current from the microphone, the process occurring in a *modulator*. This type of transmitter is used for medium and long-wave broadcasting in Britain: Fig. 77.3a is a block diagram for one.

In *frequency modulation* (f.m.), which is used in v.h.f. broadcasts, the frequency of the carrier is altered at a

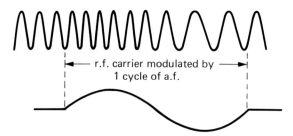

Fig. 77.4

rate equal to the frequency of the a.f. but the amplitude remains constant, Fig. 77.4. This kind of signal is fairly free from electrical interference.

(b) Receiver. A block diagram of a simple a.m. receiver is shown in Fig. 77.3b. The *tuning circuit* selects the wanted signal from the *aerial*. The *detector* (or demodulator) separates the a.f. (speech or music) from the r.f. carrier. The *amplifier* then boosts the a.f. which produces sound in the loudspeaker.

Electrical oscillations

The production of r.f. currents is impossible with mechanical generators but is readily achieved electrically.

(a) Oscillatory circuit. It contains a capacitor C and a coil L, Fig. 77.5a. If C is charged and then discharges through L, current flows backwards and forwards round the circuit, alternately discharging and charging C, first one way then the other. The frequency of the a.c. produced depends on C and L: the smaller they are the higher is the frequency and each circuit has a *natural frequency* of oscillation. Due

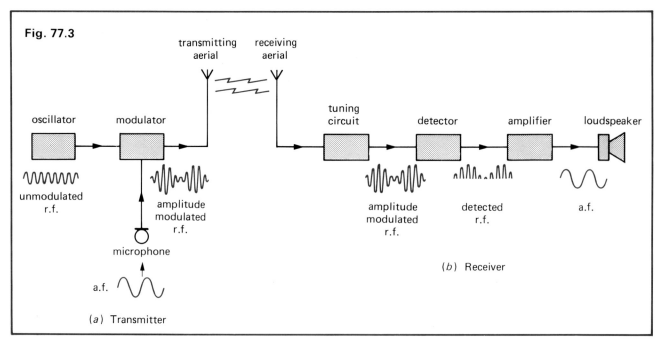

Fig. 77.3

transmitting aerial   receiving aerial

oscillator   modulator   tuning circuit   detector   amplifier   loudspeaker

unmodulated r.f.   amplitude modulated r.f.   amplitude modulated r.f.   detected r.f.   a.f.

microphone

a.f.

(a) Transmitter

(b) Receiver

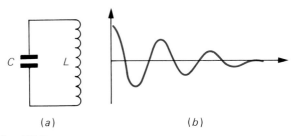

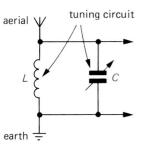

Fig. 77.5

Fig. 77.8   earth

to the resistance of the coil, the current eventually stops and a decaying or damped oscillation is obtained, Fig. 77.5*b*.

Slow damped oscillations (about 2 Hz) can be shown using the circuit of Fig. 77.6 with the C.R.O. on its slowest time base speed.

Fig. 77.6

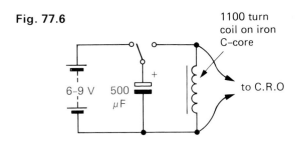

(b) Transistor oscillator. To obtain oscillations that do not die away, energy has to be fed into the *L-C* circuit at the right time. This can be done using a transistor as in Fig. 77.7. The coil L_1, by being near L, feeds the energy needed into the input (base-emitter) circuit of the transistor.

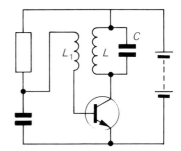

Fig. 77.7

Tuning circuit

Different transmitting stations send out radio waves of different frequencies. Each induces a signal in the aerial of a radio receiver but if the aerial is connected to an *L-C* circuit, Fig. 77.8, only the signal whose frequency equals the natural frequency of the *L-C* circuit is selected. If the value of C is varied different stations can be 'tuned in'.

This is an example of *electrical resonance* (p. 50).

Detection

In *detection* or *demodulation* the a.f. which was 'added' to the r.f. in the transmitter is recovered in the receiver.

Suppose the amplitude modulated r.f. signal V of Fig. 77.9*a* is applied to the detector circuit of Fig. 77.9*d*. The diode produces rectified pulses of r.f. current I, Fig. 77.9*b*. These charge up C during the positive half-cycles as well as flowing through the earphone. During the negative half-cycles when the diode is non-conducting, C *partly* discharges through the earphone.

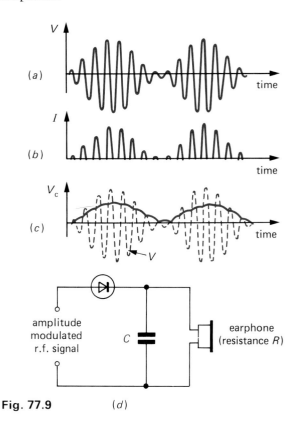

Fig. 77.9   (*d*)

The p.d. V_c across C (and the earphone) varies as in Fig. 77.9*c* if C and R have the correct values. Apart from the slight r.f. ripple V_c has the same frequency and shape as the modulating a.f. and produces the original sound in the earphone.

237

Fig. 77.10

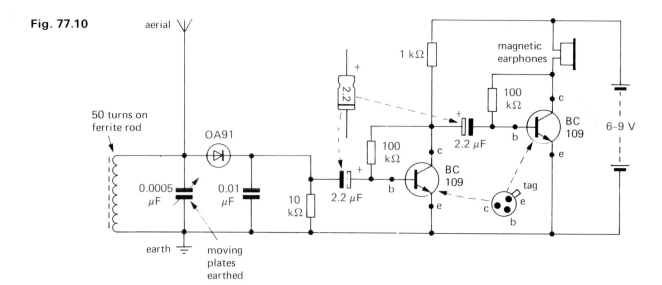

Germanium diodes are used as detectors. If a loud-speaker is to be operated, the detector is followed by several stages of a.f. amplification. A suitable resistor then replaces the earphone.

⋀⋀⋀ Experiment: simple radio receiver

Connect the circuit of Fig. 77.10 on an S-DeC, Fig. 77.11a. The transistor leads can be lengthened and connections made to the 'tags' on the variable capacitor as in Fig. 76.21b.

The resistor colour code is given on p. 241.

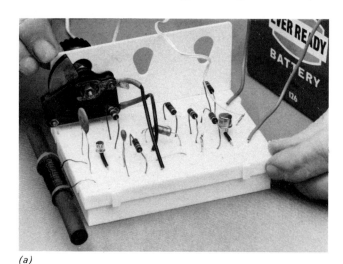

(a)

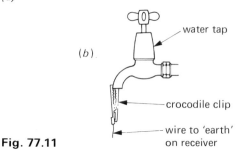

Fig. 77.11

For an aerial, support a length of wire (e.g. 10 m) as high as you can; making an earth connection to a water tap will improve reception, Fig. 77.11b.

You should be able to tune in one or two stations (depending on your location) by altering the variable capacitor.

Television

(a) Black and white. A TV receiver is basically a C.R.O. with two time bases. The horizontal or *line time base* acts as in the C.R.O. The vertical or *frame time base* operates at the same time and draws the spot at a much slower rate down to the bottom of the screen and then returns it almost at once to the top. The spot thus 'draws' a series of parallel lines of light (625) which cover the screen, Fig. 77.12.

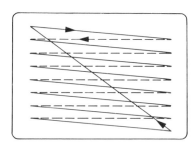

Fig. 77.12

A picture is produced by the incoming signal altering the number of electrons which travel from the electron gun to the screen. The greater the number the brighter the spot. The brightness of the spot varies from white through grey to black as it sweeps across the screen. A complete picture appears every 1/25 s, but because of the persistence of vision we see the picture as continuous. If each picture is just

Fig. 77.13

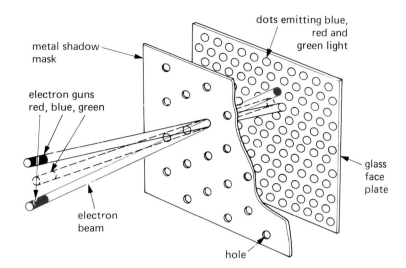

metal shadow mask

electron guns red, blue, green

electron beam

hole

dots emitting blue, red and green light

glass face plate

slightly different from its predecessor, the resultant effect is that of a 'movie' and not a sequence of 'stills'.

(b) Colour. One type of colour TV has three electron guns and the screen is coated with about a million tiny light-emitting 'dots' arranged in triangles. One 'dot' in each triangle emits red light when hit by electrons, another green light and the third blue light.

As the three electron beams scan the screen, an accurately placed 'shadow mask' consisting of a perforated metal plate with about one-third of a million holes ensures that each beam strikes only dots of one 'colour', e.g. electrons from the 'red' gun strike only 'red' dots, Fig. 77.13.

When a triangle of dots is struck it may be that the red and green electron beams are intense but not the blue. The triangle will emit red and green light strongly and appear yellowish (see p. 27). The triangles of dots are struck in turn, and since the dots are so small and the scanning so fast, we see a continuous colour picture.

Q Questions

1. **(a) (i)** Draw a labelled diagram of the electromagnetic spectrum showing how radio waves fit in with other types of electromagnetic radiation. What do all these radiations have in common?
   (ii) A radio station broadcasts on a frequency of 100 MHz. Assuming the speed of light to be 3×10^8 m/s what will be the wavelength of the radio frequency carrier wave from this station?
   (b) What is meant by the terms **(i)** audio frequencies, **(ii)** radio frequencies? Illustrate your answer by reference to sound waves and carrier waves.(*N.W.*)

2. Fig. 77.14 shows the circuit of a simple radio receiver.
   (a) State the names of the parts labelled A, B, C,
   (b) What is the purpose of the part labelled D?

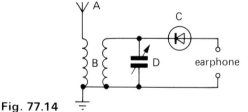

earphone

Fig. 77.14

78 Additional questions

Core level (all students)

Electrons

1. (a) How are cathode rays (i) produced in a cathode ray tube, (ii) made to travel along the tube to the screen?
(b) State three ways in which cathode rays differ from light rays.
(c) Give three uses of cathode ray tubes.

Radioactivity

2. Explain what is meant by *half-life, background radiation*.

At a certain instant the corrected count-rate registered on a detector placed close to an α-particle emitter is 200 per second and this falls to 50 per second in 12 minutes. Determine the half-life of the source. (*L.*)

Atomic structure

3. Which of the following statements most correctly represents the important deductions from experiments on the scattering of alpha particles by matter?

A Alpha particles are nuclei of helium atoms
B Atoms are composed of protons, neutrons and electrons
C Atoms consist of positively and negatively charged matter
D Electrons move in orbits round the nucleus of atoms
E Atoms have a small nucleus carrying a positive charge (*J.M.B.*)

4. The following represents part of a radioactive series in which the chemical symbols have been replaced by letters.

$$\overset{(i)}{} \qquad \overset{(ii)}{} \qquad \overset{(iii)}{}$$
$${}^{234}_{90}A \longrightarrow {}^{234}_{91}B \longrightarrow {}^{234}_{92}C \longrightarrow {}^{230}_{90}D$$

Name the particles emitted in the three changes.

Write down the two letters which represent isotopes. (*W.*)

Electronics

5. (a) Draw a diagram to show the construction of a junction diode.
(b) Explain the differences between a conductor and a semiconductor.
(c) Why are some semiconductors called 'p' type and others 'n' type?
(d) A puzzle box, Fig. 78.1, contains two lamps and other simple components connected so that, when terminal T_1 is connected to the positive pole of a cell, lamp L_1 lights, but when terminal T_2 is connected to the positive, lamp L_2 lights. Suggest what is in the puzzle box and how the connections are made.

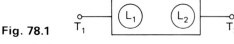

Fig. 78.1

6. (a) Sketch a graph to show what is meant by half-wave rectification.
(b) Sketch a graph to show what is meant by full-wave rectification.
(c) Draw a diagram of a *bridge circuit* using four semiconductor diodes which would produce full-wave rectification. Mark on your diagram where the input voltage is applied and where the output voltage is produced.
(d) What is the purpose of a smoothing capacitor?
(*E.M.*)

7. An *n-p-n* transistor circuit is shown in Fig. 78.2 with suitable meters M_1 and M_2.
(a) Why is it called a 'common emitter circuit'?
(b) Which meter measures the base current?
(c) Which junction is forward biased?
(d) How would the readings on M_1 and M_2 compare?
(e) What happens to the readings on (i) M_1, (ii) M_2 when R_1 is reduced?
(f) Why is R_2 included in the circuit?
(g) Redraw the circuit so that only one battery is required to supply the transistor.

Fig. 78.2

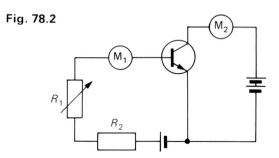

8. Fig. 78.3 is a diagram of a simple safety circuit.
(a) (i) Name component A. (ii) Name component B.
(b) (i) What is the name of the terminal of component B to which component A is connected? (ii) Why is component A usually needed in this part of this type of circuit?
(c) (i) How many contacts must be closed to operate this circuit? (ii) Explain exactly what happens in the circuit when the required number of contacts is closed.

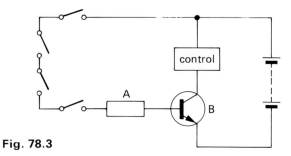

Fig. 78.3

Fig. 78.4 is a diagram of a simple security circuit.

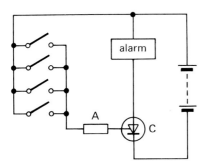

Fig. 78.4

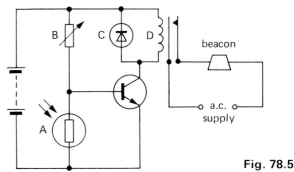

Fig. 78.5

(d) (i) Component C is known by two names, in addition to its abbreviation. Give BOTH full names. **(ii)** What is the name of the terminal of component C to which component A is connected?

(e) (i) How many contacts have to be closed to operate this circuit? **(ii)** Explain exactly what happens when the required number of contacts is closed.

(f) (i) Explain why component C (Fig. 78.4) is more useful than component B (Fig. 78.3) in a warning circuit. **(ii)** Explain how the circuit in Fig. 78.4 may be returned to its normal state after operation.

(J.M.B./A.L./N.W.16+)

9. Fig. 78.5 is designed to switch on the beacon automatically when daylight fades.

(a) (i) What is the full name of component A? **(ii)** What is the abbreviation of its name? **(iii)** What changes the value of this component? **(iv)** Explain fully how the properties of the component change.

(b) Name component B.

(c) Component D is the coil of a reed switch: **(i)** Why has this coil no metal core? **(ii)** What is the function of the coil?

(d) (i) Explain how the transistor circuit works, mentioning the function of each component. **(ii)** Explain how the transistor circuit controls the operation of the beacon.

(e) The beacon operates on a.c. Explain why this supply cannot be used for the transistor circuit.

(J.M.B./A.L./N.W.16+)

Radio and television

10. The circuit of a very simple radio receiver is shown in Fig. 78.6. It can be divided into three parts along the dotted lines.

(a) Name part X and state what it does.

(b) Name part Y and state what each of the *three* components does.

(c) Name part Z and state its function.

(d) Draw three diagrams to show the colour coding on the 1 kΩ, 10 kΩ and 100 kΩ resistors. (See the Resistor Colour Code below.)

Fig. 78.6

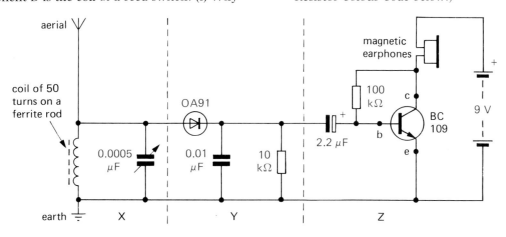

Resistor Colour Code

Black	0	Green	5
Brown	1	Blue	6
Red	2	Violet	7
Orange	3	Grey	8
Yellow	4	White	9

— 1st figure (e.g. yellow = 4)
— 2nd figure (e.g. violet = 7)
— no. of zeros (e.g. red = 00)
— tolerance

value = 4700 Ω = 4.7 kΩ

4th band = tolerance: gold = ±5%, silver = ±10%, no colour = ±20%

11. (a) Explain the meaning of the terms *audio frequency* and *radio frequency.*
(b) What is a synchronous communications satellite? Explain why such satellites are used to transmit television signals across the Atlantic ocean.
(c) Radio signals travel at a speed of 300 000 km/s. How long would it take a signal from the USA to reach this country if the total path was 6000 km?

(E.A.)

Further level (higher grade students)

Electrons

12. (a) Outline very briefly
    (i) how Thomson measured the specific charge e/m for the electron, and
    (ii) how Millikan measured the electronic charge e.
(b) Calculate e/m for an electron if $e = 1.6 \times 10^{-19}$ C and $m = 9.1 \times 10^{-31}$ kg.

13. (a) When the Y-amp gain control on a C.R.O. is set on 2 V/div, an a.c. input produces a vertical line 10 divisions long. What is (i) the peak voltage, (ii) the r.m.s. voltage, of the input?
(b) What is the frequency of an alternating p.d. which is applied to the Y-plates of a C.R.O. and produces five complete waves covering 10 horizontal divisions of the screen when the time base setting is 10 ms/div?

Radioactivity

14. Describe simple experiments (details of instruments are not expected) to show the difference between α and β radiation when (a) passed through a magnetic field, and (b) appropriate sheets of material are placed in their paths.

What are the essential differences in nature and properties between the above types of radiation and γ radiation?

State, with a reason, an essential precaution which is necessary when using equipment known to emit γ radiation or X-rays.

A radioactive substance has a half-life of 1 minute. By what factor would you expect the level of activity to drop after 4 minutes?

(S.)

Other topics

79 Materials and structures

About materials

(a) Classification. There are only about one hundred basic substances or elements from which the vast range of materials is made. It is useful to classify common materials into *metals* (e.g. iron, gold), *natural non-metals* (e.g. wood, stone) and *synthetic* (man-made) *materials* (e.g. nylon, concrete). The members of each group have certain distinguishing features. What are they?

(b) Uses. Different materials are used for different jobs, the choice depending, among other things, on their properties. There are good reasons why concrete is used for constructing large buildings, wood for furniture, glass for windows, aluminium for saucepans, polythene for washing-up bowls, copper for electrical cables, cotton and nylon for clothes, steel for car bodies and rubber for elastic bands. Your knowledge of physics should enable you to justify all these choices (see question 2 at the end of the topic).

While the twentieth century has seen the arrival of plastics and composite materials like fibreglass, metals still predominate because of their strength and ductility.

(c) Reserves. Although there is an abundance of many materials, the reserves (like those of oil and coal, p. 133) are limited. It is estimated that depending on the rate of use, new finds and economic factors, present known reserves of many metals, e.g. copper, platinum, gold, will not last for more than 50 years or so. Also, some countries are better off than others, e.g. 30% of all mercury is in Spain and 75% of all chromium in South Africa.

Experiments: stretching a wire, a rubber band and a polythene strip

In each case you can obtain a graph showing how the extension varies with the stretching (tensile) force.

(a) Copper wire. One arrangement was shown on p. 65, Fig. 23.6. *Safety glasses should be worn to protect the eyes* in case the wire snaps.

(b) Rubber band. A possible method is given in Fig. 79.1. In case the band breaks, *have some padding to protect the floor* and also *ensure your feet are not under the weights.*

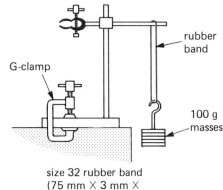

Fig. 79.1 size 32 rubber band (75 mm × 3 mm × 1 mm unstretched)

(c) Polythene strip, Fig. 79.2. Take the same safety precautions as in (b). Use a cleanly cut strip with no ragged edges.

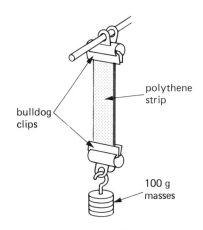

Fig. 79.2 polythene strip 500 gauge (20 cm × 1 cm × 0.1 mm)

Investigations with beams, pillars and bridges

These are meant to encourage you to try your own methods of investigation.

(a) Bending beams. Apply increasing loads to 'beams' (strips or sheets) of various materials, e.g. metals (mild steel, aluminium, lead), cardboard, wood, expanded polystyrene, polythene etc., as in Fig. 79.3.

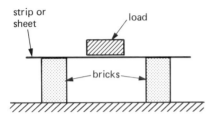

Fig. 79.3

Write a brief report on each material stating whether it behaved elastically (i.e. recovered its original size and shape) and, if possible, under what conditions it was permamently deformed (i.e. broken).

(b) Compressing a pillar. Apply increasing loads to pillars made from Plasticine (of different lengths and diameters), a short paper cylinder, a long paper cylinder and a corrugated paper cylinder, Fig. 79.4.

Write a short report on how each pillar behaves.

Using sheets of paper of the same size (and not too much Sellotape) find what shape and size of pillar can support the greatest load.

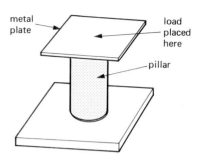

Fig. 79.4

(c) Building a bridge. Design and construct a model road bridge using for example, drinking straws or Meccano strips. Its minimum length should be at least twice the length of the longest piece of strip or straw used. It must be able to bear a load 50 times its own weight placed anywhere on it.

Q Questions

1. Which of the following materials are **(a)** metals, **(b)** natural non-metals and **(c)** synthetics?

    Oil, copper, wood, Perspex, wool, titanium, granite, glass, carbon, coal, brass, PVC.

2. Select *ten* common materials (other than those mentioned in paragraph **(b)** of 'About materials'), give one use for each and why it is used for that purpose.

check list p.292

80 Insulation of buildings: *U*-values

Energy losses

The inside of a building can only be kept at a steady temperature above that outside by heating it at a rate which equals the rate it is losing energy. The loss occurs mainly by conduction through the walls, roof, floors and windows. For a typical house in the U.K. where no special precautions have been taken, the contribution each of these makes to the total loss is shown in the first table. The substantial reduction of this loss which can be achieved, especially by wall and roof insulation, is shown in the second table.

	Percentage of total energy loss due to			
walls	roof	floors	windows	draughts
35	25	15	10	15

Percentage of each loss saved by				
insulating		carpets on floors	double glazing	draught excluders
walls	roof			
65	80	≈ 30	50	≈ 60
Percentage of total loss saved = 60				

U-value

This is a term used by heating engineers and is defined as follows.

The U-value for a specified heat conductor (e.g. a single-glazed window) *is the heat energy lost per second through it per square metre when there is a temperature difference of 1°C between its surfaces.*

The rate of heat energy loss (i.e. the heat energy lost per second) through a conductor can therefore be calculated knowing its *U*-value, from

rate of energy loss = U-value × surface area × temp. difference.

Some *U*-values in joules per second per square metre per °C (i.e. $W/(m^2 °C)$) are given in the table. Note that good insulators have low *U*-values.

	U-value
Double brick wall with air cavity	1.7
Double brick wall with 75 mm thick cavity insulation	0.6
Tiled roof without insulation	2.2
Tiled roof with 75 mm thick roof insulation	0.45
Single-glazed 6 mm thick glass window	5.6
Double-glazed window with 20 mm air gap	2.9

✎ Worked example

Calculate the rate of heat energy loss through the four double brick air cavity walls of a building measuring 10 m by 10 m along the ground. The walls are 5 m high and the temperatures outside and inside the building are 3°C and 18°C respectively. U-value of walls = $2.0 \text{ W}/(m^2 °C)$.

Surface area of one wall = 10 m × 5 m = 50 m²
All walls have the same surface area
∴ Total surface area of four walls = 4 × 50 = 200 m²
Temperature difference = 18 − 3 = 15°C

Rate of heat energy loss from walls of building
= *U*-value × total surface area × temperature difference

$$= \left(2.0 \frac{W}{m^2 °C}\right) \times (200 \text{ m}^2) \times (15 °C) =$$

$$(2.0 \times 200 \times 15) \left(\frac{W}{m^2 °C} \times m^2 \times °C\right)$$

$$= 6000 \text{ W} = 6000 \text{ J/s} = 6 \text{ kW}$$

Note: a 6 kW heater would be needed to maintain a temperature of 18°C inside the building.

▣ Question

1. (a) How much heat energy is lost in an hour through a window measuring 2.0 m by 2.5 m when the inside and outside temperatures are 18°C and −2°C respectively if the window is (i) single-glazed, *U*-value $6.0 \text{ W}/(m^2 °C)$, (ii) double-glazed, *U*-value $3.0 \text{ W}/(m^2 °C)$?

(b) What power of heater is required in case (ii) to maintain this temperature if there are no ventilation losses?

check list
p.292

81 Resistivity; cells and recharging

Resistivity

Experiments show that the resistance R of a wire of a given material is

(i) directly proportional to its length l, i.e. $R \propto l$

(ii) inversely proportional to its cross-section area A, i.e. $R \propto 1/A$ (doubling A, halves R).

Combining these two statements, we get

$$R \propto l \times \frac{1}{A}$$

This can be written as an equation if we insert a constant

$$\therefore \quad R = \rho \frac{l}{A}$$

where ρ is a constant, called the *resistivity* of the material. If we put $l = 1$ m and $A = 1$ m^2 in the equation, then $\rho = R$.

The resistivity of a material is numerically equal to the resistance of a 1 m length of it of cross-section area 1 m^2.

The unit of resistivity is obtained by rearranging the equation to give $\rho = RA/l$. The units of the right-hand side (i.e. those of ρ) are ohm $\times$ metre^2/metre, i.e. ohm metre (Ω m).

Knowing ρ for a material, the resistance of any sample of it may be calculated. The wide range of resistivities for different materials at room temperature is shown below. The resistivities of metals increase at higher temperatures, for most other materials they decrease as the temperature rises.

	Metals		Alloy	Semiconductors		Insulator
Copper	Aluminium	Constantan	Germanium	Silicon	Polythene
1.7×10^{-8}	2.7×10^{-8}	4.9×10^{-7}	6.0×10^{-1}	2.3×10^{3}	10^{16}

✎ Worked example

Calculate the resistance of a copper wire 1 km long and 0.5 mm diameter if the resistivity of copper is $1.7 \times 10^{-8} \ \Omega$ m.

Converting all units to metres, we get

length $l = 1$ km $= 1000$ m $= 10^3$ m
diameter $d = 0.5$ mm $= 0.5 \times 10^{-3}$ m

If r is the radius of the wire, the cross-section area $A = \pi r^2 = \pi (d/2)^2 = (\pi/4)d^2$

$$\therefore \quad A = \frac{\pi}{4}(0.5 \times 10^{-3})^2 \ \text{m}^2 \approx 0.20 \times 10^{-6} \ \text{m}^2$$

$$R = \rho \frac{l}{A} = \frac{(1.7 \times 10^{-8} \ \Omega \ \text{m}) \times (10^3 \ \text{m})}{0.20 \times 10^{-6} \ \text{m}^2} = 85 \ \Omega$$

New types of electric cell

(a) Fuel cells. In one type, hydrogen (the fuel) and oxygen combine to form water and generate an e.m.f. between two porous metal electrodes in an electrolyte, Fig. 81.1. The electrodes also act as catalysts to speed up the reaction.

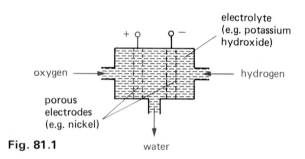

Fig. 81.1

Fuel cells are used in spacecraft where small, light power supplies are essential: the water produced provides a drinking supply.

(b) Photovoltaic cells. These change light into electrical energy directly. Types designed as solar cells, Fig. 81.2, are made of silicon. They produce about 0.5 V per cell in full sunlight with a maximum current of 35 mA or so per cm^2 of cell and an efficiency of 10%. Panels of solar cells are used in artificial satellites to power electronic equipment.

Fig. 81.2

More about secondary cells

(a) Recharging an accumulator. It is advisable to recharge an accumulator once a month using a circuit like that in Fig. 81.3. The supply must be d.c.

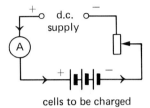

Fig. 81.3   cells to be charged

of greater e.m.f. than that of the cells to be recharged. Note that the + of the supply goes to the + of the cells and that the current is adjusted by the rheostat to the recommended recharging value usually shown on the cells.

(b) Capacity of a cell. This states how long a cell can supply current and is measured in *ampere-hours* (A h) for a 10-hour discharge time. A 30 A h cell would sustain 3 A for 10 hours, but while 1 A would be supplied for more than 30 hours, 6 A would not flow for 5 hours.

Q Questions

1. What is the resistance of a wire of length 300 m and cross-section area 1.0 mm^2 made of material of resistivity 1.0×10^{-7} Ω m?

2. Calculate the resistance of an aluminium cable of length 10 km and diameter 2.0 mm if the resistivity of aluminium is 2.7×10^{-8} Ω m.

3. What length of an alloy wire of resistivity 5.0×10^{-7} Ω m and diameter 0.50 mm is required to make a 5.0 Ω resistor?

check list
p.292

Core level revision questions
(for all students)

Light and sight

1. The image of a window on the screen of a pinhole camera is

    A virtual, inverted and larger
    B virtual, upright and smaller
    C real, upright and larger
    D real, inverted and larger
    E real, inverted and smaller

2. In Fig. C1 the completely dark region is

    A PQ   **B** PR   **C** QR   **D** QS   **E** RS

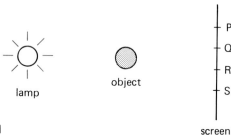

Fig. C1   screen

3. In Fig. C2 a ray of light is shown reflected at a plane mirror. What is
    (a) the angle of incidence,
    (b) the angle the reflected ray makes *with the mirror*?

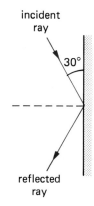

Fig. C2

4. In Fig. C3 at which of the points A to E will the observer at X see the image of the object in the plane mirror?

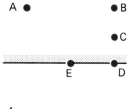

Fig. C3

248

5. In Fig. C4 a ray of light IO changes direction as it enters glass from air.
    (a) What name is given to this effect?
    (b) Which line is the normal?
    (c) Is the ray bent towards or away from the normal in the glass?
    (d) What is the value of the angle of incidence in air?
    (e) What is the value of the angle of refraction in glass?

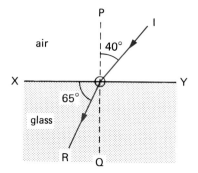

Fig. C4

6. In Fig. C5 which of the rays A to E is most likely to represent the ray emerging from the parallel-sided sheet of glass?

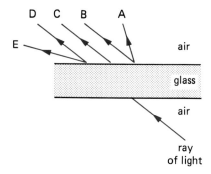

Fig. C5

7. In Fig. C6 which diagram shows the correct path of the ray through the prism?

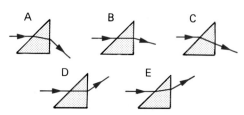

Fig. C6

8. If the critical angle for glass is 42°, what diagram A to E in Fig. C7 does *not* represent the behaviour of a ray of light falling on a glass-air boundary?

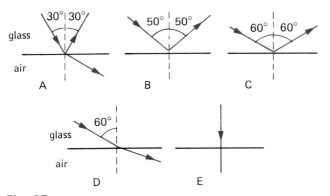

Fig. C7

9. A ray of light is shown passing through the upper prism of a prismatic periscope in Fig. C8. In which position A to E must the lower prism be placed for someone at X to use it?

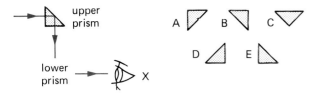

Fig. C8

10. A convex lens produces a magnified, inverted image of an object if the distance of the object from the lens is

    A greater than two focal lengths
    B equal to two focal lengths
    C between one and two focal lengths
    D equal to one focal length
    E less than one focal length

11. The eye forms an image on the retina which

    1 is inverted
    2 is focused by the lens changing shape
    3 has its brightness controlled by the size of the pupil

Which statement(s) is (are) correct?

    A 1, 2, 3   **B** 1, 2   **C** 2, 3   **D** 1   **E** 3

12. In Fig. C9 a narrow beam of white light is shown passing through a glass prism and forming a spectrum on a screen.
    (a) What is the effect called?
    (b) Which colour of light appears at **(i)** A, **(ii)** B?

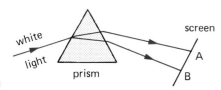

Fig. C9

13. Which pair of colours A to E when added together give white light?

    A yellow and blue   **B** red and blue
    C green and yellow   **D** red and green
    E green and blue

14. To focus a lens camera on a nearer object is

    A the distance between the lens and film increased
    B the distance between the lens and film decreased
    C the aperture made larger
    D the aperture made smaller
    E the shutter speed changed?

15. When using a magnifying glass to see a small object

    1 an upright image is seen
    2 the object should be less than one focal length away
    3 a real image is seen

Which statement(s) is (are) correct?

    A 1, 2, 3   **B** 1, 2   **C** 2, 3   **D** 1   **E** 3

Waves and sound

16. In the transverse wave shown in Fig. C10 distances are in centimetres. Which pair of entries A to E is correct?

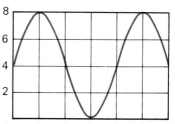

Fig. C10

	A	B	C	D	E
Amplitude	2	4	4	8	8
Wavelength	4	4	8	8	12

17. The lines in Fig. C11 are the crests of straight ripples produced in a ripple tank by a wave generator.
    (a) What is the wavelength of the ripples if there are 6 complete waves in a distance of 30 cm?
    (b) If the wave generator makes 4 vibrations per second what is the frequency of the ripples?
    (c) What is the equation connecting the wavelength (λ), the frequency (f) and the speed (v) of the ripples?
    (d) Calculate the speed of the ripples.

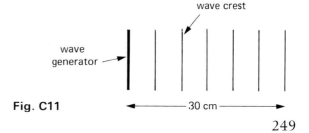

Fig. C11

249

18. When water waves go from deep to shallow water, the changes (if any) in its speed, wavelength and frequency are

	Speed	Wavelength	Frequency
A	greater	greater	the same
B	greater	less	less
C	the same	less	greater
D	less	the same	less
E	less	less	the same

19. When the straight water waves in Fig. C12 pass through the narrow gap in the barrier they are diffracted. What changes (if any) occur in

    (a) the shape of the waves,
    (b) the speed of the waves,
    (c) the wavelength?

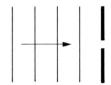

Fig. C12

20. (a) When two sets of waves arrive at the same point at the same time, under what conditions do they **(i)** completely cancel out, **(ii)** produce a larger wave?
(b) If A and B in Fig. C13 are two dippers vibrating together (in phase) in a ripple tank, what happens **(i)** at P if PA = PB, **(ii)** at Q if QB − QA = half a wavelength, **(iii)** at R if RB − RA = one wavelength?

Fig. C13

21. Sometimes 'light + light = darkness'. This behaviour of light is called

    A reflection   B refraction   C dispersion
    D interference   E deviation

22. Which one of the following is not electromagnetic waves?

    A infrared   B gamma rays   C ultraviolet
    D X-rays   E sound

23. Compared to radio waves, the wavelength, frequency and speed of light are

	Wavelength	Frequency	Speed
A	greater	smaller	the same
B	greater	smaller	greater
C	greater	greater	the same
D	smaller	greater	smaller
E	smaller	greater	the same

250

24. The wave travelling along the spring in Fig. C14 is produced by someone moving end X of the spring to and fro in the directions shown by the arrows.
    (a) Is the wave longitudinal or transverse?
    (b) What is the region called where the coils of the spring are **(i)** closer together, **(ii)** farther apart, than normal?

Fig. C14

25. Sound and light waves

    A travel through a vacuum
    B travel as longitudinal waves
    C travel with the same speed in air
    D can be diffracted
    E have similar wavelengths

26. The signal sent out by a sonar echo sounder in a ship is received back from the sea bed directly below the ship, Fig. C15, 2 seconds later. The speed of sound in sea water, in m/s, is

    A 750   B 1500   C 2250   D 3000
    E 3750

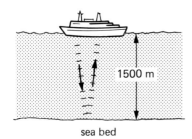

sea bed

Fig. C15

27. If a note played on a piano has the same pitch as one played on a guitar, they have the same

    A frequency   B amplitude   C quality
    D loudness   E harmonics

28. The waveforms of two notes P and Q are shown in Fig. C16. Which one of the statements A to E is true?

    A P has a higher pitch than Q and is not so loud
    B P has a higher pitch than Q and is louder
    C P and Q have the same pitch and loudness
    D P has a lower pitch than Q and is not so loud
    E P has a lower pitch than Q and is louder

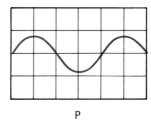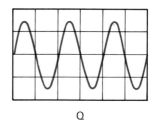

             P   Q

Fig. C16

29. The loudness of the note produced by a vibrating wire can be increased by

    A increasing the length of the wire
    B decreasing the length of the wire
    C increasing the amplitude of vibration
    D increasing the tension of the wire
    E decreasing the tension of the wire

30. Examples of transverse waves are

    1 water waves in a ripple tank
    2 all electromagnetic waves
    3 sound waves

Which statement(s) is (are) correct?

    A 1, 2, 3   B 1, 2   C 2, 3   D 1   E 3

31. Notes of the same pitch played on the guitar and the clarinet sound different because

    1 different overtones are present in each case
    2 the fundamental frequencies are different
    3 a guitar string vibrates transversely and the air column in a clarinet vibrates longitudinally.

Which reason(s) is (are) correct?

    A 1, 2, 3   B 1, 2   C 2, 3   D 1   E 3

Matter and molecules

32. The basic SI units of mass, length and time are

	Mass	Length	Time
A	kilogram	kilometre	second
B	gram	centimetre	minute
C	kilogram	centimetre	second
D	gram	centimetre	second
E	kilogram	metre	second

33. Density can be calculated from the expression

    A mass/volume   B mass × volume
    C volume/mass   D weight/area
    E area × weight

34. A cube of density 1000 kg/m^3 and side of length 2 m has a mass in kg of

    A 2000   B 4000   C 8000   D 16 000
    E 32 000

35. Which of the following properties are the same for an object on earth and on the moon?

    1 weight   2 mass   3 density

Use the answer code

    A 1, 2, 3   B 1, 2   C 2, 3   D 1   E 3

36. The spring in Fig. C17 stretches from 10 cm to 22 cm when a force of 4 N is applied. If it obeys Hooke's law, its total length in cm when a force of 6 N is applied is

    A 28   B 42   C 50   D 56   E 100

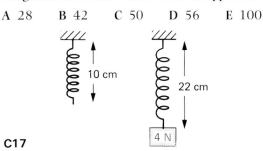

Fig. C17

37. Which one of the following statements is *not* true?

    A The molecules in a solid vibrate about a fixed position
    B The molecules in a liquid are arranged in a regular pattern
    C The molecules in a gas exert negligibly small forces on each other, except during collisions
    D The densities of most liquids are about 1000 times greater than those of gases because liquid molecules are much closer together than gas molecules
    E The molecules of a gas occupy all the space available.

38. In the oil-film experiment to estimate the size of an oil molecule

    1 the oil film is assumed to be one molecule thick
    2 the size of a molecule is found by dividing the volume of the oil drop by the area of the film
    3 the oil must be coloured

Which statement(s) is (are) correct?

    A 1, 2, 3   B 1, 2   C 2, 3   D 1   E 3

39. Diffusion occurs more quickly in a gas than in a liquid because

    A molecules in a gas have more frequent collisions than molecules in a liquid
    B gas molecules are larger
    C gas molecules move randomly
    D on average, molecules in a gas are further apart than molecules in a liquid
    E heat is needed to cause diffusion in a liquid.

Forces and pressure

40. The metre rule in Fig. C18 is pivoted at its centre. If it balances, the mass of M is given by the equation

    A $M + 50 = 40 + 100$ B $M \times 40 = 100 \times 50$
    C $M/50 = 100/40$ D $M/50 = 40/100$
    E $M \times 50 = 100 \times 40$

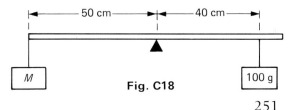

Fig. C18

251

41. The uniform beam of weight 30 N and length 4 m is hinged at end A, as in Fig. C19. The force F, in N, which must be applied vertically upwards at a distance of 1 m from end B so that the beam is horizontal is

    A 60   B 50   C 40   D 30   E 20

Fig. C19

42. The resultant of a force of 5 N acting at right angles to a force of 12 N at a point 0, Fig. C20, is

    A 5 N   B 7 N   C 12 N   D 13 N   E 17 N

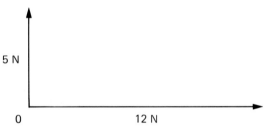

Fig. C20

43. The work done by a force is

    1 calculated by multiplying the force by the distance moved in the direction of the force
    2 measured in joules
    3 the amount of the energy changed.

Which statement(s) is (are) correct?

    A 1, 2, 3   B 1, 2   C 2, 3   D 1   E 3

44. The main energy change occurring in the device named is

Device	Main energy change
1 electric lamp	electrical to heat and light
2 battery	chemical to electrical
3 pile driver	k.e. to p.e.

Which statement(s) is (are) correct?

    A 1, 2, 3   B 1, 2   C 2, 3   D 1   E 3

45. If a lift of mass 200 kg is raised by an electric motor through a height of 15 m in 20 s,

    1 the weight of the lift is 2000 N
    2 the useful work done is 30 000 J
    3 the power output of the motor is 1.5 kW

Which statement(s) is (are) correct?

    A 1, 2, 3   B 1, 2   C 2, 3   D 1   E 3

46. The efficiency of a machine which raises a load of 200 N through 2 m when an effort of 100 N moves 8 m is

    A 0.5%   B 5%   C 50%   D 60%   E 80%

252

47. How much work is done when
    (a) a weight of 10 N is raised 2 m vertically and 3 m horizontally at the same time,
    (b) the trolley of mass 10 kg is pushed up to the platform at the top of the slope in Fig. C21?

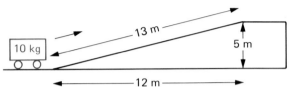

Fig. C21

48. Which one of the following statements is *not* true?

    A Pressure is the force acting on unit area
    B Pressure is calculated from force/area
    C The SI unit of pressure is the pascal (Pa) which equals 1 newton per square metre (1 N/m^2)
    D The greater the area over which a force acts the greater is the pressure
    E Force = pressure × area

49. Which of the following will damage a wood-block floor that can withstand a pressure of 2000 kPa (2000 kN/m^2)?

    1 A block weighing 2000 kN standing on an area of 2 m^2
    2 An elephant weighing 200 kN standing on an area of 0.2 m^2
    3 A girl of weight 0.5 kN wearing stiletto-heeled shoes standing on an area of 0.0002 m^2

Use the answer code

    A 1, 2, 3   B 1, 2   C 2, 3   D 1   E 3

50. The pressure at a point in a liquid

    1 increases as the depth increases
    2 increases if the density of the liquid increases
    3 is greater vertically than horizontally

Which statement(s) is (are) correct?

    A 1, 2, 3   B 1, 2   C 2, 3   D 1   E 3

51. A mercury manometer is connected to a gas supply as shown in Fig. C22. The pressure of the gas supply exceeds atmospheric pressure by the pressure exerted by a column of mercury of length

    A 15 mm   B 20 mm   C 25 mm
    D 30 mm   E 45 mm

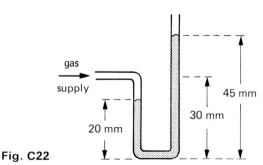

Fig. C22

52. Some air is trapped above the mercury in the tube in Fig. C23. If atmospheric pressure is 750 mmHg, the pressure exerted by the trapped air in mmHg is

A 30   B 80   C 720   D 770   E 800

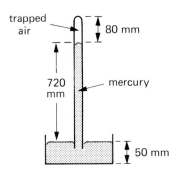

Fig. C23

53. If the piston in Fig. C24 is pulled out of the cylinder from position X to position Y, without changing the temperature of the air enclosed, the air pressure in the cylinder is

A reduced to a quarter   B reduced to a third
C the same   D trebled   E quadrupled

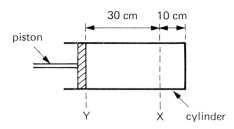

Fig. C24

54. Archimedes' principle states, 'When a body is wholly or partly submerged in a fluid, the upthrust on the body equals

A the mass of the body
B the mass of fluid displaced
C the weight of fluid displaced
D the weight of the body
E the volume of the body.'

55. An object weighs 200 N in air and 120 N when totally submerged in a liquid of density 800 kg/m³. What is

(a) the upthrust on the object
(b) the weight of liquid displaced
(c) the mass of liquid displaced
(d) the volume of liquid displaced
(e) the volume of the object

Motion and energy

56. The speeds of a car travelling on a straight road are given below at successive intervals of 1 second.

Time (s)	0	1	2	3	4
Speed (m/s)	0	2	4	6	8

The car travels

1 with an average velocity of 4 m/s
2 16 m in 4 s
3 with a uniform acceleration of 2 m/s²

Which statement(s) is (are) correct?

A 1, 2, 3   B 1, 2   C 2, 3   D 1   E 3

57. If a train travelling at 10 m/s starts to accelerate at 1 m/s² for 15 s on a straight track, its final velocity in m/s is

A 5   B 10   C 15   D 20   E 25

58. The equally-spaced dots on the ticker tape in Fig. C25 are made by a vibrator of frequency 50 Hz. The speed of the tape in m/s is

A 0.04   B 0.1   C 0.2   D 0.4   E 1

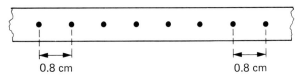

Fig. C25

59. The velocity-time graph for a short trip by a girl cyclist is shown in Fig. C26.

(a) How far does she travel during the trip?
(b) What is her average velocity for the trip?
(c) What is her initial acceleration?

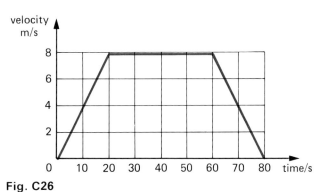

Fig. C26

60. If the acceleration due to gravity is 10 m/s², an object falling from rest will

1 fall with a constant speed of 10 m/s
2 fall 10 m every second
3 have a speed of 20 m/s after 2 seconds

Which statement(s) is (are) correct?

A 1, 2, 3   B 1, 2   C 2, 3   D 1   E 3

61. Which one of the velocity-time graphs in Fig. C27 represents an object falling freely in a vacuum?

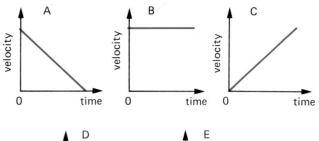

Fig. C27

62. An unbalanced force of 50 N acts on a mass of 5 kg. The acceleration of the mass in m/s² will be

A 0.1   B 10   C 45   D 50   E 250

63. A force of 20 N pulls a block of mass 2 kg along a horizontal bench and is opposed by a constant frictional force of 12 N, Fig. C28. The acceleration of the block in m/s² is

A 40   B 18   C 10   D 4   E 0.1

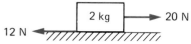

Fig. C28

64. A 3 kg mass falls with its terminal velocity. Which combination A to E gives its weight, the air resistance and the resultant force acting on it?

	Weight	Air resistance	Resultant force
A	0.3 N down	zero	zero
B	3 N down	3 N up	3 N up
C	10 N down	10 N up	10 N down
D	30 N down	30 N up	zero
E	300 N down	zero	300 N down

65. A trolley of mass 3.0 kg moving at 4.0 m/s collides with, and remains attached to a stationary trolley of mass 1.0 kg, Fig. C29. Their combined momentum in kg m/s, after the collision is

A 3.0   B 4.0   C 7.0   D 8.0   E 12

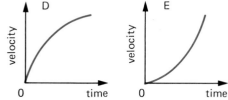

Fig. C29

254

66. Which one of the following statements is *not* correct?

A Kinetic energy (k.e.) is the energy a body has because of its motion
B Potential energy (p.e.) is the energy a body has because of its position or condition
C k.e. and p.e. are vector quantities
D k.e. is calculated in J from $\frac{1}{2}mv^2$ where m is the mass in kg and v is the speed in m/s
E p.e. is calculated in J from mgh where m is the mass in kg, h is the height in m and g equals 10 N/kg

67. A stone of mass 2 kg is dropped from a height of 4 m. Neglecting air resistance, the k.e. of the stone in joules just before it hits the ground is

A 6   B 8   C 16   D 80   E 160

68. An object of mass 2 kg is fired vertically upwards with a k.e. of 100 J. Neglecting air resistance, which of the numbers in A to E below is

(a) the velocity in m/s with which it is fired,
(b) the height in m to which it will rise?

A 5   B 10   C 20   D 100   E 200

69. An object has k.e. of 10 J at a certain instant. If it is acted on by an opposing force of 5 N, which of the numbers A to E below is the further distance it travels in metres before coming to rest?

A 2   B 5   C 10   D 20   E 50

70. A boy whirls a ball at the end of a string round his head in a horizontal circle, centre 0. If he lets go of the string when the ball is at X, Fig. C30, the ball flies off in the direction

A 1   B 2   C 3   D 4   E 5

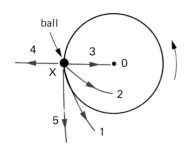

Fig. C30

Heat and energy

71. Which one of the following statements is *not* true?

A Temperature tells us how hot an object is
B Temperature is measured by a thermometer which uses some property of matter (e.g. the expansion of mercury) that changes continuously with temperature
C Heat flows naturally from an object at a lower temperature to one at a higher temperature
D The molecules of an object move faster when its temperature rises
E Temperature is measured in °C, heat is measured in joules.

72. When a metal bar is cooled the

    A length and density decrease but mass is the same
    B length and mass decrease but density is the same
    C length is the same but mass and density decrease
    D length decreases, density increases but mass is the same
    E length and density increase but mass is the same

Which one of these statements is correct?

73. When a pond freezes over in winter, ice only forms at its surface because

    1 water has its maximum density at 4 °C
    2 ice floats on water
    3 ice is a good conductor of heat

Which statement(s) is (are) correct?

    A 1, 2, 3   B 1, 2   C 2, 3   D 1   E 3

74. (a) Convert the following temperatures to kelvin (i) 7 °C, (ii) 25 °C, (iii) −23 °C
    (b) Convert the following temperatures to °C (i) 300 K, (ii) 573 K, (iii) 200 K.

75. The pressure exerted by a gas in a container

    1 is due to the molecules of the gas bombarding the walls of the container
    2 decreases if the gas is cooled
    3 increases if the volume of the container increases

Which statement(s) is (are) correct?

    A 1, 2, 3   B 1, 2   C 2, 3   D 1   E 3

76. From the five graphs A to E in Fig. C31 choose the one which, in each case, gives the most correct relationship between the quantities y and x given below.

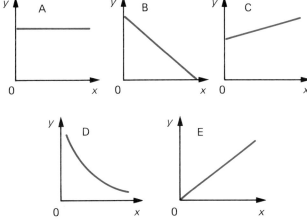

Fig. C31

	y	x
(a)	Volume of a fixed mass of gas at a constant pressure	Temperature in K
(b)	Pressure of a fixed mass of gas at a constant volume	Temperature in °C
(c)	Pressure of a fixed mass of gas at constant temperature	Volume of gas
(d)	Pressure of a fixed mass of gas at constant temperature	1/(Volume of gas)

77. When equal masses of water and paraffin are supplied with heat at the same rate, the temperature of the paraffin rises faster because paraffin has a

    A smaller density
    B lower boiling point
    C smaller specific heat capacity
    D greater specific heat capacity
    E lower melting point

78. The specific heat capacity (c) of a substance is

    1 the quantity of heat needed to raise the temperature of 1 kg by 1 °C
    2 calculated using the equation $Q = m \times \Delta\theta \times c$ where $\Delta\theta$ (in °C or K) is the temperature rise when mass m (in kg) is supplied with quantity of heat Q (in J)
    3 1000 J/(kg °C) if 4000 J of heat raise the temperature of 0.5 kg from 20 °C to 28 °C.

Which statement(s) is (are) correct?

    A 1, 2, 3   B 1, 2   C 2, 3   D 1   E 3

79. A drink is cooled more by ice at 0 °C than by the same mass of water at 0 °C because ice

    A floats on the drink
    B has a smaller specific heat capacity
    C gives out latent heat to the drink as it melts
    D absorbs latent heat from the drink to melt
    E is a solid

80. The specific latent heat of vaporization (l_v) of a liquid is

    1 the quantity of heat needed to change 1 kg from liquid to vapour without change of temperature
    2 calculated from the equation $Q = m \times l_v$ where the units of Q and l_v are J and J/kg respectively
    3 2×10^6 J/kg if a 2 kW heating element after bringing it to its boiling point, boils off 0.1 kg in 100 s

Which statement(s) is (are) correct?

    A 1, 2, 3   B 1, 2   C 2, 3   D 1   E 3

81. Which one of the following statements is *not* true?

    A Salt melts the ice on an icy road because salt solution has a lower freezing point than pure water
    B The temperature of pure melting ice is 0 K
    C Vegetables cook quicker in a steam pressure cooker because increased pressure raises the boiling point of water
    D Ether spilt on the hand feels colder than water at the same temperature because ether evaporates more rapidly
    E When a liquid evaporates its temperature falls because the fastest molecules leave the surface and the average k.e. of those left is less.

82. Which of the following statements is/are true?

    1 In cold weather the wooden handle of a saucepan feels warmer than the metal pan *because* wood is a better conductor of heat

    2 Convection occurs when there is a change of density in the parts of a fluid

    3 Conduction and convection cannot occur in a vacuum

    A 1, 2, 3   B 1, 2   C 2, 3   D 1   E 3

83. Which one of the following statements is *not* true?

    A Energy from the sun reaches the earth by radiation only

    B A dull black surface is a good absorber of radiation

    C A shiny white surface is a good emitter of radiation

    D The best heat insulation is provided by a vacuum

    E A vacuum flask is designed to reduce heat loss or gain by conduction, convection and radiation.

Electricity and magnetism

84. Which one of the following is *not* true?

    A The needle of a magnetic compass can be made from steel

    B The attraction of an iron nail by a magnet is an example of induced magnetism

    C Heating a magnet weakens it

    D The magnetic field pattern between the N poles of two bar magnets is shown in Fig. C32

    E The law of magnetic poles states that: like poles attract, unlike poles repel.

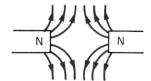

Fig. C32

85. If a positively charged acetate strip is brought near the cap of a positively charged electroscope and then removed without touching it, the leaf is deflected

    A less and returns to its original deflection

    B less and stays in that position

    C more and stays in that position

    D more and returns to its original position

    E more and then falls to zero

Which one of these statements is correct?

256

86. If the two uncharged metal spheres R and S on insulating stands, Fig. C33, are separated while the negatively charged polythene strip is held near R, the charges on R and S are

Fig. C33

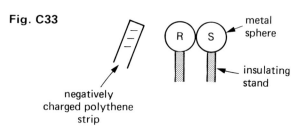

	A	B	C	D	E
R	+	−	−	+	zero
S	+	−	+	−	zero

87. In the circuit of Fig. C34 calculate
    (a) the total resistance,
    (b) the current in each resistor,
    (c) the p.d. across each resistor.

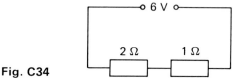

Fig. C34

88. Repeat Question 87 for the circuit of Fig. C35.

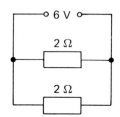

Fig. C35

89. Which circuit in Fig. C36 could be used to find the resistance of the lamp?

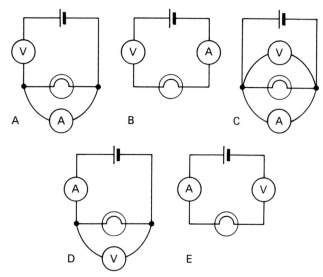

Fig. C36

90. The graphs in Fig. C37 show how the current *I* varies with the p.d. *V* applied to four devices. In which case does
  (a) the device obey Ohm's law,
  (b) the resistance decrease as the p.d. increases,
  (c) the resistance increase as the p.d. increases,
  (d) the device conduct in only one direction?

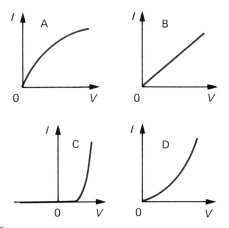

Fig. C37

91. An electric kettle for use on a 240 V supply is rated at 3000 W. The minimum current the cable supplying it should be able to carry for safe working is

  A 2 A   B 5 A   C 10 A   D 15 A   E 30 A

92. Which one of the following statements is *not* true?

  A In a house circuit lamps are wired in parallel
  B Switches, fuses and circuit breakers are placed in the neutral wire
  C An electric fire has its earth wire connected to the metal case to prevent the user receiving a shock
  D When connecting a three-core cable to a 13 A three-pin plug the *brown* wire goes to the *live* pin
  E The cost of operating three 100 W lamps for 10 hours at 5p per unit is 15p.

93. Which one of the following statements about a lead-acid cell is *not* true?

  A It can produce a large direct current
  B It has an e.m.f. of 2.0 V
  C It can be recharged by alternating current
  D It has a low internal resistance
  E A car battery consists of six in series.

94. The diagrams in Fig. C38 represent magnetic fields caused by current-carrying conductors. Which one is due to
  (a) a long straight wire,
  (b) a circular coil,
  (c) a solenoid?

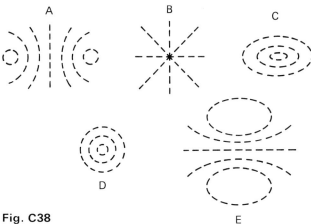

Fig. C38

95. Which of the following statements is/are true?

  1 An electromagnet consists of a coil of wire wound on a soft iron core
  2 The strength of the magnetic field produced by an electromagnet increases if the strength of the current and/or the number of turns of wire is increased
  3 In Fig. C39 when the switch is closed the gap G increases.

  A 1, 2, 3   B 1, 2   C 2, 3   D 1   E 3

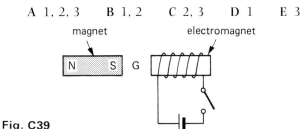

Fig. C39

96. Which one of the following statements is *not* true?

  A If a current is passed through the wire XY in Fig. C40*a*, a vertically upwards force acts on it
  B If a current is passed through the wire PQ in Fig. C40*b*, it does not experience a force
  C If a current is passed through the coil in Fig. C41 it rotates clockwise
  D If the coil in Fig. C41 had more turns and carried a larger current, the turning effect would be greater
  E In a moving coil loudspeaker a coil moves between the poles of a strong magnet.

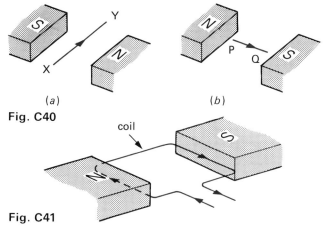

Fig. C40

(a)   *(b)*

Fig. C41

97. Which statement(s) is (are) correct?

   1 An ammeter is connected in series in a circuit and a voltmeter in parallel
   2 An ammeter has a high resistance
   3 A voltmeter has a low resistance

   A 1, 2, 3   B 1, 2   C 2, 3   D 1   E 3

98. Which one of the following statements is *not* true when a magnet is pushed N pole first into a coil, Fig. C42?

   A An e.m.f. is induced in the coil and causes a current through the galvanometer
   B The induced e.m.f. increases if the magnet is pushed in faster and/or the coil has more turns
   C Mechanical energy is changed to electrical energy
   D The coil tends to move to the right because the induced current makes face X a N pole which is repelled by the N pole of the magnet
   E The effect produced is called electrostatic induction.

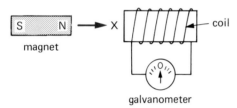

Fig. C42

99. Which of the following statements is/are true?

   1 A transformer changes an alternating p.d. from one value to another according to the equation $V_s/V_p = N_s/N_p$
   2 Transformers are used to send electrical energy by cable over long distances at high voltage and low current to cut heat loss
   3 The number of turns on the primary of a transformer which has 200 turns on the secondary and is designed to deliver 12 V from the 240 V mains supply, Fig. C43, is 5000.

   A 1, 2, 3   B 1, 2   C 2, 3   D 1   E 3

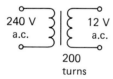

Fig. C43

100. Which of the units A to E could be used to measure (a) electric charge, (b) electric current, (c) p.d., (d) energy, (e) power?

   A ampere   B joule   C volt   D watt
   E coulomb

258

Electrons and atoms

101. Which of the following statements about the deflection of the beam of electrons by the p.d. between the plates P and Q in Fig. C44 is/are true?

   1 Plate P is negative
   2 The deflection would be greater if the p.d. was greater
   3 The deflection would be greater if the electrons were moving slower.

   A 1, 2, 3   B 1, 2   C 2, 3   D 1   E 3

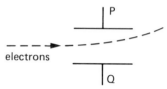

Fig. C44

102. Which of the patterns in Fig. C45 could appear on the screen of a C.R.O. if an alternating p.d. is applied to the Y-plates with the time base (a) off, (b) on?

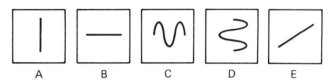

Fig. C45

103. Identify the radiations X, Y and Z from Fig. C46.

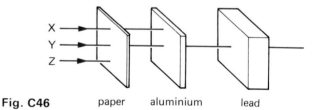

Fig. C46   paper   aluminium   lead

	A	B	C	D	E
X	alpha	beta	gamma	gamma	beta
Y	beta	alpha	alpha	beta	gamma
Z	gamma	gamma	beta	alpha	alpha

104. The radioactive substance whose decay curve is given in Fig. C47 has a half-life in minutes of

   A 1   B 2   C 3   D 4   E 5

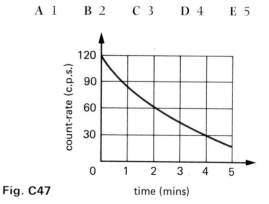

Fig. C47   time (mins)

105. A radioactive source which has a half-life of 1 hour gives a count-rate of 100 c.p.s. at the start of an experiment and 25 c.p.s. at the end. The time taken by the experiment was, in hours,

    A 1   B 2   C 3   D 4   E 5

106. Which one of the following statements is *not* true?

    A An atom consists of a tiny nucleus surrounded by orbiting electrons

    B The nucleus contains protons and neutrons, called nucleons, in equal numbers

    C A proton has a positive charge, a neutron is uncharged and their mass is about the same

    D An electron has a negative charge of the same size as the charge on a proton but it has a much smaller mass

    E The number of electrons equals the number of protons in a normal atom.

107. A lithium atom has a nucleon (mass) number of 7 and a proton (atomic) number of 3.

    1 Its symbol is $^{7}_{4}\text{Li}$

    2 It contains 3 protons, 4 neutrons and 3 electrons

    3 One of its isotopes has 3 protons, 3 neutrons and 3 electrons

Which statement(s) is (are) correct?

    A 1, 2, 3   B 1, 2   C 2, 3   D 1   E 3

108. Which symbol A to E is used in equations for nuclear reactions to represent (**a**) an alpha particle, (**b**) a beta particle, (**c**) a neutron, (**d**) an electron?

    A $^{0}_{-1}\text{e}$ B $^{1}_{0}\text{n}$ C $^{4}_{2}\text{He}$ D $^{1}_{-1}\text{e}$ E $^{1}_{1}\text{n}$

109. (**a**) Radon $^{220}_{86}\text{Rn}$ decays by emitting an alpha particle to form an element whose symbol is

    A $^{216}_{85}\text{At}$ B $^{216}_{86}\text{Rn}$ C $^{218}_{84}\text{Po}$ D $^{216}_{84}\text{Po}$
    E $^{217}_{85}\text{At}$

    (**b**) Thorium $^{234}_{90}\text{Th}$ decays by emitting a beta particle to form an element whose symbol is

    A $^{235}_{90}\text{Th}$ B $^{230}_{89}\text{Ac}$ C $^{234}_{89}\text{Ac}$ D $^{232}_{88}\text{Ra}$
    E $^{234}_{91}\text{Pa}$

110. Which statement(s) is (are) true?

    1 A rectifier changes d.c. to a.c.

    2 A diode acts as a rectifier

    3 The circuit in Fig. C48 can be used to obtain the rectified output shown.

    A 1, 2, 3   B 1, 2   C 2, 3   D 1   E 3

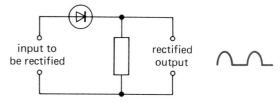

Fig. C48

111. Which one of the following statements about the *n-p-n* transistor circuit in Fig. C49 is *not* true?

    A The collector current I_C is zero until base current I_B flows

    B I_B is zero until the base-emitter p.d. V_{BE} is $+0.6$ V

    C A small I_B can switch-on and control a large I_C

    D When used as an amplifier the input is connected across B and E

    E X must be connected to supply $-$ and Y to the $+$

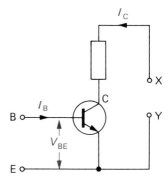

Fig. C49

Further level revision questions
(for higher grade students)

(A) SHORT TYPES (1 to 10 minutes each)

Light and sight

1. An optician's test card is fixed 80 cm behind the eyes of a patient, who looks into a plane mirror 300 cm in front of him, as shown in Fig. F1. The distance from his eyes to the image of the card is

   A 300 cm   **B** 380 cm   **C** 600 cm
   D 680 cm   **E** 760 cm   (L.)

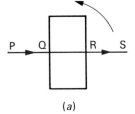

Fig. F1

2. (a) Why is a concave mirror used as a make-up mirror?
(b) Why is a convex mirror used as a driving mirror?
(c) Why is a concave parabolic mirror used as **(i)** a car headlamp reflector, **(ii)** a dish aerial?

3. Fig. F2a shows a ray of light PQ incident normally on a rectangular glass block, so that the emergent ray RS is neither displaced nor deviated from the original direction. The block is then turned in an anticlockwise direction to a new position. Which one of the diagrams in Fig. F2b best illustrates what happens to the emergent ray RS? (N.I.)

Fig. F2

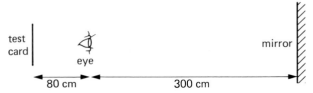

(a)

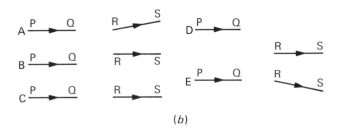

(b)

4. A ray of light is incident on the plane surface of a transparent material at such an angle that the reflected and refracted rays are at right angles to each other. Draw a diagram to illustrate this. (S. part qn.)

5. A 45° right-angled prism ABC, made of glass of critical angle 42° is used to turn light through 90°.
(a) Copy Fig. F3, continue the paths of the two rays until they emerge into the air again.
(b) Why does light not emerge from face AC?
(c) Using the two rays, explain what is meant by lateral inversion.

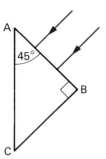

Fig. F3

6. A converging lens is used to form an image of a bright object on a white screen. If the lower half of the lens is covered with a sheet of metal, as shown in Fig. F4, then

   A the image disappears entirely
   B the lower half of the image disappears
   C the upper half of the image disappears
   D the brightness of the image is reduced
   E the image is unchanged.   (N.I.)

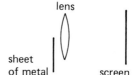

Fig. F4

7. A translucent white plastic bottle has green printing on it. An electric lamp with red glass is suspended inside the bottle and switched on in a darkened room. The green printing on the bottle will appear to be

   A black   **B** blue   **C** magenta   **D** red
   E yellow   (L.)

Waves and sound

8. State whether the following waves are transverse or longitudinal *and* progressive or stationary:
(a) the vibrations of a stretched wire (which is fixed at both ends) when plucked;
(b) the sound from the above wire to an observer;
(c) a ripple on a large pond;
(d) the vibrations of an air column in a long test tube. (S.)

260

9. Fig. F5 represents a plan view of a horizontal ripple tank. Two dippers, A and B, vibrate in-phase with the same frequency. At a point P constructive interference is observed, and at a nearby point Q destructive interference is observed.

Fig. F5

(a) On the same axes sketch two graphs of displacement against time for the vibration of the water at P, one for the waves from A, the other for the waves from B. On the same axes sketch a third graph showing the displacement of the water at P due to both sets of waves arriving at P together.

(b) In exactly the same way, sketch three graphs of displacement against time for the water at Q, one for waves from A, another for waves from B and the third for both sets of waves arriving at Q together.

(c) (i) State a relationship between the distances AP and BP.

(ii) State a relationship between the distances AQ and BQ. *(J.M.B.)*

10. Fig. F6 illustrates crests of circular wavefronts radiating from a point source O. If the time taken for a wavefront to travel from P to Q is 10 s, and the wavelength of the waves is 2 m, the speed of the waves, in m/s, is

    A 0.20 B 0.80 C 1.00 D 1.25 E 5.00

                                           (N.I.)

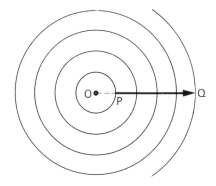

Fig. F6

11. Which of the following is (are) true both for light waves and for sound waves?

  1. They can travel through a vacuum.
  2. Speed of wave = frequency of wave × wavelength.
  3. They transfer energy from one place to another.

    A 1, 2, 3 B 1, 2 C 2, 3 D 1 E 3   *(L.)*

12. Fig. F7 shows a ray of sunlight incident upon a triangular glass prism, such that a spectrum is produced on the screen **W**.

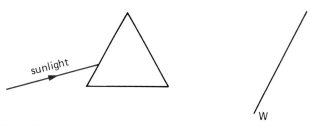

Fig. F7

(a) Copy the diagram, continue the ray to illustrate how the spectrum is formed on a white screen at **W**. Mark the violet end of the spectrum **V**, and the red end **R**.

(b) If the white screen is replaced with a pure red screen, describe carefully what you would observe on the screen, giving a reason for your answer.

(c) The refractive index of the glass is 1.50 for red light. Using your answer to (a), state whether you would expect the refractive index of glass for violet light to be greater than, equal to or less than 1.50. Explain how you deduced your answer.

(d) The glass prism transmits both some ultraviolet and some infrared radiation.

  (i) On the diagram, mark clearly with a **U** the region on the screen at which you would detect ultraviolet radiation, and with an **I** the region at which you would detect infrared radiation.

  (ii) State one method by which you could detect the presence of ultraviolet radiation.

  (iii) State whether frequency of ultraviolet radiation is greater than, equal to or less than that of infrared radiation. *(N.I. part qn.)*

13. Stationary waves are set up in a long thin cord. The points where the cord appears at rest are 0.15 m apart. What are these points called? If the frequency of the vibrations is 200 Hz, calculate the speed of waves along the cord. *(S.)*

14. In the double slits experiment using monochromatic light, Fig. F8, how would the fringe pattern be affected if

  (a) light of shorter wavelength was used,
  (b) the slit separation was decreased,
  (c) the slits were made narrower, and
  (d) the screen was moved closer?

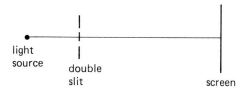

Fig. F8

15. A man standing between two cliffs X and Y at the point P fires a pistol, Fig. F9. If the speed of sound in air is 330 m/s, the first two echoes which he hears will be separated by a time interval, in seconds, of

    A 0.5 B 0.75 C 1.25 D 1.50 E 2.50 (N.I.)

Fig. F9

16. When prospecting for oil geologists may cause a small explosion at the surface of the earth. The distance to possible oil-bearing rocks can then be found by recording how long the sound wave reflected from the rocks, takes to return to the surface, Fig. F10.

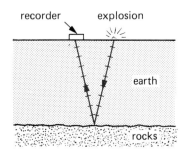

Fig. F10

In an exploration 3.6 s is recorded between the explosion and the return of the reflected wave. Taking the average speed of sound through the earth as 1800 m/s, the rock causing the reflection is at a depth in metres of

    A 1800×3.6 B $1800/3.6$

    C $\dfrac{1800 \times 3.6}{2}$ D $\dfrac{1800 \times 2}{3.6}$

    E $1800 \times 3.6 \times 2$

Matter and molecules

17. (a) The smallest division marked on a metre rule is 1 mm. A student measures a length with the ruler and records it as 0.835 m. Is he justified in giving three significant figures?
(b) The SI unit of density is
  A kg m   B kg/m^2 C kg m^3 D kg/m
  E kg/m^3

18. A solid block has dimensions 0.1 m × 0.5 m × 0.2 m and is made of material of density 9000 kg/m^3. It rests on a horizontal surface. Calculate:
    (i) the mass of the solid,
    (ii) the maximum pressure it can exert on the surface. (W.)

262

19. A force of 40 N is necessary to extend a spring by 20 mm. If the acceleration of free fall is 10 m/s^2, the extension, in mm, when a mass of 0.5 kg is attached to the lower end of the spring is

    A 0.20   B 1.00   C 2.50   D 5.00
    E 10.00   (N.I.)

20. The graph in Fig. F11 shows the displacement of a pendulum bob from its rest position as it varies with time.

From the graph, determine
    (i) the amplitude of the oscillation,
    (ii) the time for one complete oscillation,
    (iii) the distance of the bob from its rest position after 0.8 seconds.

On a copy of the diagram, draw the graph which represents a pendulum swinging with half the amplitude and twice the frequency. (W.)

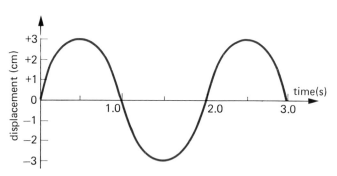

Fig. F11

Forces and pressure

21. A uniform rod has equally spaced graduation marks on each side of its mid point at which it is pivoted, Fig. F12. Four identical metal discs are placed on the left-hand side of the pivot at the three mark. Which one of the following arrangements of discs, placed on the right-hand side of the pivot, could be used to restore equilibrium?

    A 3 placed at the 2 mark and 1 at the 4 mark
    B 2 placed at the 2 mark and 2 at the 4 mark
    C 1 placed at the 4 mark and 3 at the 3 mark
    D 1 placed at the 2 mark and 3 at the 4 mark
    E 1 placed at the 1 mark and 3 at the 2 mark

                                        (N.I.)

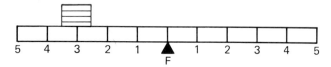

Fig. F12

22. Two parallel forces, of 2 N and 4 N respectively, act on a body through points P and Q as shown in Fig. F13.

Their resultant

   1 has a magnitude of 3 N
   2 acts through the mid-point of PQ
   3 is parallel to the two forces shown

Which statement(s) is (are) correct?

   A 1, 2, 3   B 1, 2   C 2, 3   D 1   E 3 (L)

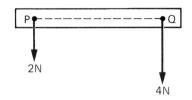

Fig. F13

23. Fig. F14 shows the horizontal forces exerted on a tree by two tractors in an attempt to pull it out of the ground.

Draw a diagram to a stated scale and use it to determine the magnitude and direction of the resultant force exerted on the tree by the two tractors.   (C.)

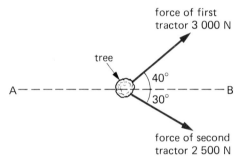

Fig. F14

24. In Fig. F15 two forces acting at a point O are represented in magnitude and direction by OX and OY. The third force required to maintain equilibrium is represented in magnitude and direction by

   A OZ B ZO C XY D YZ E XZ   (O.L.E.)

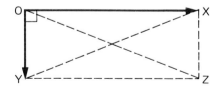

Fig. F15

25. An electric motor drives a machine which lifts a mass of 2 kg through the height of 6 m in 4 s at constant speed.
   (a) State how the work done by a force is calculated.
   (b) How much work is done in lifting the 2 kg mass?
   (c) How much power is used to lift the 2 kg mass?
   (d) If the electrical power input is 40 W, what is the efficiency?
   (e) State two causes of loss of efficiency.   (O.L.E.)

26. A load of 10 N is being raised slowly with the aid of the simple pulley system shown in Fig. F16. Each of the pulleys weighs 2 N.

If the system has no friction losses, the effort required is

   A 5 N   B 6 N   C 7 N   D 10 N   E 12 N
                                   (O.L.E.)

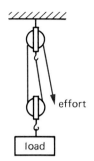

Fig. F16

27. Fig. F17 illustrates a mercury manometer (using a tube of uniform bore), connected to a gas container through a tap T, which is initially closed. Atmospheric pressure is 750 mm of mercury. On opening tap T, the mercury level in the left-hand arm

   A falls through 125 mm
   B rises through 125 mm
   C falls through 250 mm
   D rises through 250 mm
   E falls through 500 mm.   (N.I.)

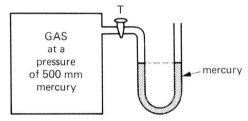

Fig. F17

28. Fig. F18 represents a simple hydraulic lift. Calculate the maximum load that can be lifted using a downward effort, as shown, of 5 N.   (W.)

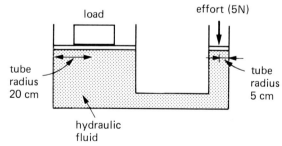

Fig. F18

29. (a) Copy Fig. F19.

   (i) Fig. F19*a* illustrates a simple mercury barometer. State what is contained in the region above the mercury column in the tube, and mark *clearly* the height which you would measure to determine atmospheric pressure.

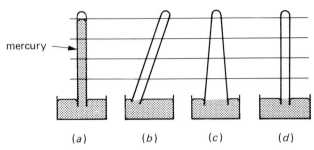

mercury →

(a)   (b)   (c)   (d)

Fig. F19

(ii) In Fig. F19*b* the tube has been tilted. Mark carefully on it the level of the mercury in the tube.
(iii) In Fig. F19*c*, a barometer has been made using a tube of non uniform bore. Mark carefully on it the level of the mercury in the tube.
(iv) In Fig. F19*d*, air has been introduced into the region above the mercury column so that it is at a pressure of $\frac{1}{4}$ of atmospheric pressure. Mark carefully on it the level of the mercury in the tube.

(b) On a particular day, atmospheric pressure at sea level was found to be 102 000 N/m². At the same time, the reading on a mercury barometer placed at the top of a mountain was 0.600 m. The acceleration of free fall is 10 m/s² at both sea level and the top of the mountain, and the density of mercury is 13 600 kg/m³.
(i) Determine the pressure of the atmosphere at the top of the mountain in N/m².
(ii) Calculate the difference in atmospheric pressure, in N/m², between the top of the mountain and sea level.
(iii) If the density of air is 1.2 kg/m³ between the top of the mountain and sea level, determine the height of the mountain top above sea level.   (*N.I.*)

30. An experiment was performed to find the connection between the pressure and volume of a fixed mass of gas when the temperature was kept constant. The graph in Fig. F20 shows the pressure (*p*) plotted against $\dfrac{1}{\text{volume}} \left(\dfrac{1}{V}\right)$

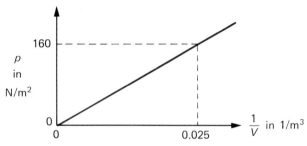

Fig. F20

(a) (i) What relationship between pressure and volume can you deduce from the above graph?
(ii) Give a reason for your answer.

264

(b) Using the kinetic theory of gases, explain
(i) how a gas exerts a pressure,
(ii) the variation in pressure when the volume is reduced.

(c) If the above experimental results show that when the pressure of the gas is 160 N/m² the corresponding value of $\dfrac{1}{V}$ is 0.025/m³, calculate the value of the volume of the gas when the pressure of the gas changes to 4800 N/m².   (*N.I.*)

31. (a) A cubic block of side 3.00 m and density 800 kg/m³, is attached to the base of a tank containing water of density 1000 kg/m³ by means of an inextensible cable of negligible mass, Fig. F21. The acceleration of free fall is 10 m/s².

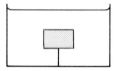

Fig. F21

(i) Determine the mass of the block.
(ii) Determine the weight of the block.
(iii) Determine the upthrust acting on the block.
(iv) Determine the tension in the cable.

(b) The cable is now released, and the block rises to the surface of the water, where it subsequently floats.
(i) Determine the upthrust on the block when it is floating.
(ii) Determine the volume of the block which is beneath the surface of the water while the block is floating.   (*N.I.*)

32. A hydrometer sinks to a greater depth in water than in a certain liquid L. From this it can be concluded that

1 the density of L is greater than that of water
2 the expansivity of L is smaller than that of water
3 the hydrometer displaces a greater mass of water than of L.

Which statement(s) is (are) correct?

A 1, 2, 3   B 1, 2   C 2, 3   D 1   E 3
          (*L.*)

Motion and energy

33. A radio transmitter directs pulses of waves towards a satellite from which reflections are received 10 milliseconds after transmission. If the speed of radio waves is 3×10^8 m/s, how far away is the satellite?   (*S.*)

34. The velocity-time graph of Fig. F22 illustrates the motion of a ball which has been projected vertically upwards from the surface of the moon. The weight of the ball on earth, where the acceleration of free fall is 10 m/s², is 2 N.

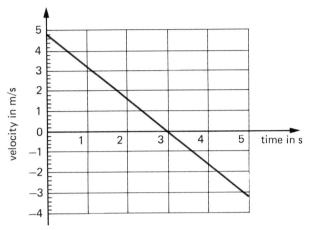

Fig. F22

(a) State why the velocity becomes negative (after 3 seconds).

(b) Determine a value for the acceleration of free fall on the moon, showing clearly how you obtain your answer.

(c) Determine the total distance travelled by the ball in the first 5 seconds, showing clearly how you obtain your answer.

(d) Determine the weight of the ball on the moon, showing clearly how you obtain your answer.

(e) If the ball were projected vertically upwards with the same velocity on earth, and friction is neglected, what difference, if any, would you expect to observe in a velocity-time graph (using the same scales) for the ball on earth, compared with that given above? Give a reason for your answer.   (N.I.)

35. Two metal spheres, of masses 4 kg and 1 kg, held at different heights above the ground, are released simultaneously, Fig. F23. If the acceleration of free fall is 10 m/s², then the difference in time, in seconds, at which they hit the ground is

   A 0   B 1   C 3   D 4   E 5   (N.I.)

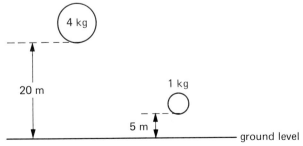

Fig. F23

36. A helicopter of mass 3000 kg rises vertically at a constant speed of 25 m/s. The acceleration of free fall is 10 m/s². The resultant force, in N, acting on the helicopter is

   A zero   B 30 000 downwards
   C 45 000 upwards   D 75 000 upwards
   E 105 000 upwards   (N.I.)

37. A hard ball, mass 0.50 kg, travels along a straight line with a speed 1.5 m/s. It hits a fixed sheet of metal and rebounds back along the same path with the same speed. Calculate the average force exerted on the metal during the collision if the ball remains in contact with the metal for 1.0 millisecond.   (O. and C.)

38. Water, which flows over a weir at a rate of 900 kg/s, takes 1.5 s to fall vertically into the stream below.

(a) (i) What is the speed with which the falling water hits the stream?
(ii) What is the height through which the water falls?

(b) Calculate (i) the weight of water falling over the weir in 5.0 s, (ii) the work which has been done on this weight of water when it hits the stream below the weir.

(c) Calculate the power of the falling water at the instant it hits the stream.

(d) State the energy transformations which occur as the water falls from the weir into the stream below.   (C.)

39. Copy and complete Fig. F24 by showing the two forces acting on a motor-cycle travelling at the appropriate speed round a banked curve. Show the direction in which the resultant of these forces must be acting. (W.)

Fig. F24

40. Fig. F25 shows a drawing pin which rests on the deck of a record player. The deck is horizontal and is rotating at a constant rate.

Which two of the following quantities associated with the pin are changed when the drawing pin has been carried from A to B:

speed, velocity, kinetic energy, potential energy, momentum?
                             (O. and C.)

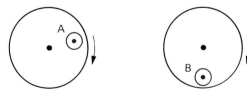

Fig. F25

Heat and energy

41. An uncalibrated thermometer is immersed in melting ice. The length of the thread of mercury is 35 mm. When the thermometer is immersed in steam from water boiling under a pressure of one standard atmosphere the length of the thread becomes 215 mm. What is the temperature when the length of the thread of mercury is 71 mm?

(O. and C.)

42. Two marks on a metal bar are 50.00 cm apart at a temperature of 293 K. This distance increases by 0.096 cm when the temperature rises to 373 K. Calculate the linear expansivity of the metal.　　　　　*(S.)*

43. Fig. F26 shows the variation of the density of water with temperature. It may be deduced from the graph that

　1　a given mass of water has a minimum volume at 4 °C
　2　between 0 °C and 4 °C convection will not take place in a vessel of water heated at the bottom
　3　expansion occurs when water freezes

Which statement(s) is (are) correct?

　A　1, 2, 3　　B　1, 2　　C　2, 3　　D　1　　E　3

(L.)

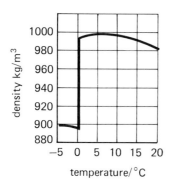

Fig. F26

44. An ideal gas, initially at a temperature of 27 °C, is heated until both pressure and volume are doubled. The final temperature, in °C, is

　A　6.75　　B　27　　C　108　　D　927　　E　1200

(N.I.)

45. A quantity of helium occupies a volume 3.0×10^{-3} m³ inside a high pressure gas cylinder. The pressure of the helium is 1.5×10^{7} Pa (N/m²) when the cylinder is stored at 7 °C. What volume will the helium occupy if the contents of the cylinder are used to inflate a balloon to a pressure 1.0×10^{5} Pa (N/m²) when the temperature is 20 °C?　　　　　*(O. and C.)*

46. When a 2.0 kg block of metal at 600 °C is immersed in water at its boiling point (100 °C), 0.4 kg of steam is produced. Assuming that there are no heat losses to the surroundings and that the specific latent heat of vaporization of water is 2.3×10^{6} J/kg, the specific heat capacity of the metal, in J/(kg °C) is

　A　7.67×10^{2}　　B　9.20×10^{2}　　C　7.67×10^{4}
　D　9.20×10^{4}　　E　2.30×10^{8}　　　　*(N.I.)*

47. A well-lagged calorimeter contains 0.10 kg of water and has an electric immersion heater in the liquid. A second identical lagged calorimeter contains 0.10 kg of paraffin and an identical heater. Both fluids are initially at 20 °C. The heaters are switched on at the same instant. If the paraffin reaches a temperature of 30 °C in 10 minutes, how long will it take for the water to attain this temperature? Assume that the thermal capacity of the calorimeter and any heat losses are negligible. (Take the specific heat capacity of paraffin as 2100 J/(kg °C) and that of water as 4200 J/(kg °C).)　　*(O. and C.)*

48. A small quantity, 0.010 kg, of water at 17 °C is added to a larger mass of ice at 0 °C contained in a vacuum flask. Calculate the greatest mass of ice which can be melted. (Take the specific heat capacity of water as 4200 J/(kg °C) and the specific latent heat of fusion of ice as 340×10^{3} J/kg.)　　　　　*(O. and C.)*

49. A heater of power 200 W was used to keep liquid boiling in a vessel. During a period of 50 seconds, the mass of liquid decreased by 0.005 kg. Assuming that all the heat energy was given to the liquid, the specific latent heat of vaporization of the liquid, in J/kg, is

　A　50　　B　800　　C　1.0×10^{4}　　D　4.0×10^{4}
　E　2.0×10^{6}　　　　　　*(N.I.)*

50. Answer each part of this question in terms of the *kinetic theory of matter.*
　(a) A sample of liquid at its boiling point is to be converted to vapour at the same temperature. Why must heat energy be supplied to bring about this change?
　(b) When all the liquid has been converted into vapour at the boiling point, the vapour fills a container at atmospheric pressure. Explain why the volume of the vapour is much greater than the original volume of the liquid.
　(c) Why does the vapour exert a pressure on its container?
　(d) If the temperature of the container is raised, why does the pressure of the vapour inside the container rise? (Assume that the volume of the container does not change.)　　　　　*(C.)*

Electricity and magnetism

51. (a) When a positively charged rod is brought near to a negatively charged electroscope and is then removed without touching, the leaf

A rises more and stays at that deflection
B rises more and returns to its previous deflection
C falls and stays at that deflection
D falls and returns to its previous deflection

(b) If the electric field between two charged conducting spheres X and Y is as shown in Fig. F27, then

A X is positive and Y is negative
B X is negative and Y is positive
C X is positive and Y is positive
D X is negative and Y is negative

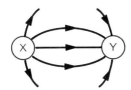

Fig. F27

52. What is the electric charge on a cloud from which an average current of 1500 A flows during a flash of lightning to earth lasting 2 s?

53. When a 5 ohm resistance is connected across the terminals of a 20 V battery of negligible internal resistance, the number of coulombs passing through the resistance in 10 seconds is

A 0.4　B 2.5　C 10.0　D 40.0　E 1000.0
(N.I.)

54. (a) A wire was connected to a battery and it was found that the energy converted into heat was 30 joules when 20 coulombs of charge flowed through the wire in 5 seconds. Calculate
(i) the potential difference between the ends of the wire;
(ii) the current flowing through the wire;
(iii) the resistance of the wire;
(iv) the average power developed in the wire.

(b) If the current in the wire were doubled and all the energy were released as heat in the wire, how much heat would be produced in the wire in 5 seconds?
(J.M.B.)

55. A 9 V battery is composed of six 1.5 V cells, each of internal resistance 0.2 Ω, connected in series. The greatest current that can be obtained from such a battery is

A 0.3 A　B 1.3 A　C 1.5 A
D 7.5 A　E 9.2 A　(O.L.E.)

56. The graph in Fig. F28 shows how the current I through a tungsten filament lamp varies with the voltage V across it.
(a) What is the voltage when the current is 0.2 amperes?

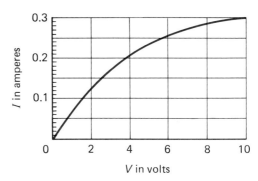

Fig. F28

(b) What is the resistance of the lamp when I is 0.2 amperes?
(c) What is the increase in current when V is increased from 2 V to 6 V?
(d) Does the resistance of the filament increase or decrease as V increases from 2 V to 6 V? What causes the change?
(e) Draw a diagram of the circuit you would use to take readings to plot this graph.　(O.L.E.)

57. An electric motor takes a current of 5.0 A from a 4.0 V supply. Calculate the power input to the motor.

The motor lifts a weight of 50 N through a vertical height of 3.0 m in 10 s. Calculate the average useful work done per second by the motor in lifting this weight.

Suggest a reason for the difference between this quantity and the power input to the motor.　(C.)

58. Two coils X and Y, having the same dimensions and number of turns, are wound around a piece of soft iron as in Fig. F29. A steady direct current passes through X and Y. Which one of the following statements is correct?

A End P becomes a North pole, while end Q becomes a South pole.
B End P becomes a South pole, while end Q becomes a North pole.
C Both end P and end Q become North poles.
D Both end P and end Q become South poles.
E Neither end P nor end Q become poles, since the effects of the two coils cancel out.　(N.I.)

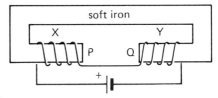

Fig. F29

59. A galvanometer can be converted into a voltmeter by connecting

A a large resistance in series with it
B a small resistance in parallel with it
C a large resistance in parallel with it
D a small resistance in series with it

267

60. Fig. F30 shows four resistors connected to an accumulator of e.m.f. 2.0 V and negligible internal resistance. The resistances of the individual resistors are shown on the diagram. Calculate the total current flowing from the accumulator. The points B and D are now joined by a wire. Indicate, with a reason, the direction in which you would expect current to flow in the wire BD.   (C.)

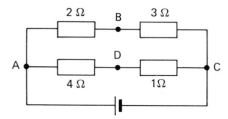

Fig. F30

61. When S is closed in Fig. F31, the readings on *V* and *A* change from

	V	*A*		*V*	*A*
A	2 V to 4 V	0.5 A to 1 A	C	4 V to 2 V	1 A to 2 A
B	2 V to 4 V	1 A to 2 A	D	4 V to 2 V	0.5 A to 1 A

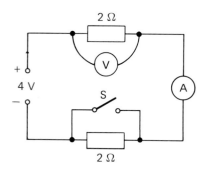

Fig. F31

62. Two coils P and Q are wound on a soft iron former as in Fig. F32. When a steady current passes through coil P, end N of the coil behaves as a magnetic north pole, while end S behaves as a magnetic south pole. Key K is initially open, and is then closed. Which one of the following statements concerning the circuit with coil Q is correct after the key is closed?

    A A current will flow momentarily from X to Y through G.
    B A current will flow momentarily from Y to X through G.
    C A steady current will flow from X to Y through G.
    D A steady current will flow from Y to X through G.
    E No current will flow through G.   (N.I.)

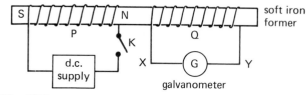

Fig. F32

63. Fig. F33 shows an incomplete diagram of a simple a.c. generator,

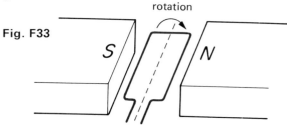

Fig. F33

    (a) Copy and complete the diagram so that the following are shown and labelled
        (i) the slip rings
        (ii) the brushes
        (iii) the direction of the induced current when the coil is at the position shown and the generator is connected to an external circuit.
    (b) Sketch a graph showing the variation of output current with time. Label the axes.
    (c) State for what positions of the coil the output voltage is **(i)** maximum, **(ii)** zero, giving a reason for your answer in each case.
    (d) State two methods by which the output voltage of an a.c. generator can be increased.
    (e) (i) A 3 V (effective value) output from an a.c. generator is applied to the primary coil of a transformer containing 30 turns. If the secondary coil contains 10 turns, determine the output voltage (effective value).
        (ii) If the generator is capable of supplying 6 W, calculate the greatest output current (effective value) which can be taken from the secondary windings.   (N.I.)

64. Fig. F34 shows how the output voltage from an a.c. generator varies with time. The frequency of the output, in Hz, is

    A 0.01   **B** 0.10   **C** 10.00   **D** 100.00
    E 200.00   (N.I.)

Fig. F34

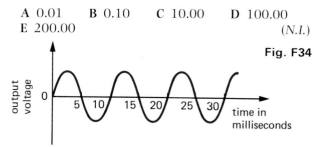

65. The circuit in Fig. F35 is for a transformer supplying a lamp. If the transformer is 100% efficient the resistance of the lamp is

    A 1.5 Ω   **B** 3 Ω   **C** 6 Ω   **D** 12 Ω
    E 24 Ω

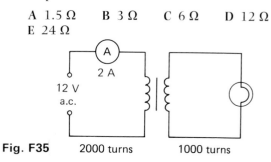

Fig. F35   2000 turns   1000 turns

Electrons and atoms

Questions 66 to 68

The diagrams in Fig. F36 represent five possible paths taken by atomic particles (initially travelling to the right) after they enter a magnetic field. Crosses (+) indicate a magnetic field directed into the page, while arrows pointing up indicate a magnetic field in that direction.

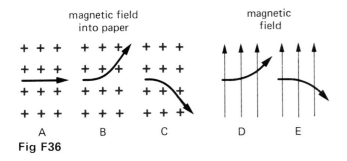

Fig F36

For each of the following particles, select the diagram which corresponds most closely to the expected path taken. Each diagram may be used once, more than once, or not at all.

66. electron.   **67.** α particle.   **68.** neutron. (*N.I.*)

69. (a) What is meant by the *specific charge* of a particle?
   (b) The specific charge of an electron is 1.76×10^{11} C/kg and the specific charge of a proton is 9.6×10^7 C/kg. Assuming that both particles carry the same quantity of charge, calculate the mass of a proton compared with the mass of an electron.

(*J.M.B. part qn.*)

70. The waveform in Fig. F37a is displayed on a C.R.O. A student then alters *two* controls and obtains the waveform in Fig. F37b. The two controls adjusted were

   A time base and Y-shift   B X-shift and Y gain
   C Y-gain and time base   D focus and X-gain
   E X-gain and Y-gain

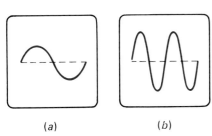

(a)   (b)

Fig. F37

71. A radioactive source emits α, β and γ radiations. A suitable detector is first placed very close to the source and then moved about 10 cm away. Explain any differences in count rate that you would expect. What further effects would be obtained by your placing **(a)** a very thin sheet of aluminium, **(b)** a thick sheet of lead, between source and detector? (*S.*)

72. The radiation from a radioactive source is arranged so that it passes through a postcard and then at right angles to a strong magnetic field. A detector shows the presence of radiation straight through *and* to one side only of the original direction. Which types of radiation are being detected?

   A α particles and β particles
   B α particles and γ rays
   C β particles and γ rays
   D α particles and neutrons
   E γ rays and neutrons (*O.L.E.*)

73. (a) Explain what is meant by **(i)** atomic mass number and **(ii)** atomic number.
   (b) (i) The isotope $^{238}_{92}U$ decays by alpha-emission to an isotope of thorium (Th). Compare the $^{238}_{92}U$ and thorium nuclei, explaining the changes which have occurred in the uranium nucleus.
   (ii) The thorium nucleus decays by beta-emission to an isotope of protactinium (Pa). Compare the thorium and Pa nuclei, accounting for the changes you describe. (*J.M.B.*)

74. A radioactive sample has a half-life of 20 minutes, and at a certain time a detector records 120 counts per second. Calculate the count-rate recorded by the detector one hour later.

The sample emits α particles. State briefly the nature of these particles. (*C.*)

75. (a) A nitrogen nuclide is written as $^{14}_{7}N$. What information about the structure of the nitrogen atom can be deduced from this symbol of the nuclide?
   (b) A nuclide whose symbol is $^{16}_{7}N$ is an isotope of nitrogen. In what way is an atom of this type of nitrogen different from the atom in **(a)**?
   (c) The nuclide $^{16}_{7}N$ decays to become an oxygen nuclide by emitting an electron. Write down an equation to show this process.
   (d) The half-life of the nuclide $^{16}_{7}N$ is 7.3 s. What does this mean? A sample of this type of nitrogen is observed for 29.2 s. Calculate the fraction of the original radioactive isotope remaining after this time.

(*C.*)

76. A neutral atom has a mass number of 7 and an atomic number of 3. State **(a)** the number of protons in the nucleus; **(b)** the number of electrons surrounding the nucleus. How many neutrons are present and where are they situated? (*S.*)

77. Fig. F38 represents a certain atom which is made up of protons (P), neutrons (N), and electrons (e).

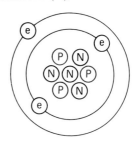

Fig. F38

269

(a) What is meant by the *mass number* of an atom and what is its value in this case?

(b) What is meant by the *atomic number* of an atom and what is its value in this case?

(c) What is the charge: (i) on the nucleus of this atom; (ii) on the atom as whole?

(d) How do protons, neutrons and electrons compare in mass?

(e) What happens to the atomic number of an atom when it emits: (i) an α particle; (ii) a β particle?

(O.L.E.)

Questions 78 to 81

    A joule
    B metre per second squared
    C newton
    D kilogram metre per second
    E watt

Which of the units listed above could be used to measure

78. the tension in a stretched spring?

79. the rate at which a machine is doing work?

80. the change of momentum of a billiard ball when it bounces off the cushion of a billiard table?

81. the change in the potential energy of a falling parachutist? (L.)

(B) LONG QUESTIONS (20 to 25 minutes each)

Light and sight

82. (a) The paths of two rays of light from a coin on the bottom of a swimming pool full of water are shown in Fig. F39. Copy the diagram and draw the paths of the rays which reach the eye E of someone standing at the side of the pool. Add construction lines to show where the image of the coin appears to be.

Fig. F39

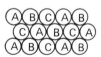

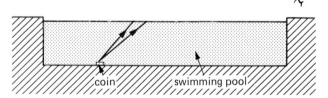

On your diagram draw the paths of two other rays from the coin, one falling on the water surface at an angle of incidence equal to the critical angle of water (49°) and the other at an angle of incidence greater than the critical angle.

(b) How would you find the refractive index by the real and apparent depth of a Perspex block?
In such an experiment the real depth was 14 cm and the apparent depth 10 cm. What value does this give for the refractive index of Perspex?

(c) Draw a diagram to show how a 45° right-angled prism can turn light through 180° by total internal reflection.

83. Explain, with the aid of ray diagrams, how a lens, of focal length 100 mm, can be used

(a) in a simple camera to photograph a distant object,

(b) in a slide projector to project the image of a transparency ('slide') on to a screen,

(c) as a magnifying glass to examine the transparency.

In each case show clearly where the photographic film or transparency should be placed in relation to the focal plane of the lens, and state the type of image being formed by the lens. The details of the other parts of the instruments are not required.

When the lens is used as a magnifying glass, the final image produced has a linear magnification of 2.5. Determine, by using a ray diagram drawn to scale, or otherwise, the distance from the lens to the object.

(O. and C.)

84. (a) (i) Draw a ray diagram to show how a pure spectrum of white light can be projected onto a screen using a prism and any other necessary apparatus.
(ii) Describe and explain the appearance of a red-coloured pen as it is moved through the white-light spectrum.

(b) If the screen of a working colour television set is closely observed it may be seen that it is made up of a lattice of regularly arranged minute spots of light. Each spot of light emits one of the three primary colours and the lattice of these three coloured spots A, B and C is regularly arranged as in Fig. F40. Any colour can therefore be formed on the screen by various combinations of one, two or three spots emitting their individual colours.

Fig. F40

(i) Name the three primary colours used in the screen lattice.
(ii) Explain how the television screen can appear magenta using this lattice.
(iii) Describe and account for the appearance of a yellow coloured screen when it is viewed through a red filter.

(c) A student interested in astronomy attempts to identify various constellations of stars in the night sky with the aid of a book which he illuminates using a powerful electric torch. Describe the two changes in the optical system of his eye as he looks from the book to the stars and explain why these changes occur.

(L.)

Waves and sound

85. Write down four properties common to all electro-magnetic waves.

Describe how you would make use of a ripple tank to show
(a) reflection of straight (plane) waves by a concave boundary, and
(b) diffraction of water waves.
Sketch the wave pattern you would expect to observe in each case.

A straight vibrator, vibrating at 7 Hz, causes water ripples to travel across the surface of a shallow tank. The waves travel a distance of 35 cm in 1.25 seconds. Calculate the wavelength of the waves. (*W.*)

86. Two identical dippers A and B vibrate in a ripple tank in phase with each other and with the same constant frequency and amplitude and produce waves in the tank.

Fig. F41

OA = OB

The disturbance of the water at points along the line MN, shown in Fig. F41, is observed. At Q and S the disturbance is greatest, whereas at P and R the water is at rest. R is the only position between Q and S where the water is not disturbed.
(i) Explain why the disturbance at Q is as described above.
(ii) Why is there no disturbance at R?

For double slit fringes (Young's slit fringes) observed with light,

$$\frac{\text{separation of fringes}}{\text{wavelength}} = \frac{\text{distance from slits to screen}}{\text{distance between the slits}}$$

Make use of this formula in answering (iii) and (iv) below.

The frequency at which the dippers are driven is slowly increased until the position of maximum disturbance, originally at S, has moved to R.
(iii) Calculate the new wavelength if the original wavelength was 10 mm.
(iv) Describe how the separation of the dippers A and B could now be changed to restore the position of maximum disturbance to S.
(v) The mechanism driving the dippers is altered so that they now vibrate half a cycle out of phase with each other. The frequency of vibration and the separation AB are the same as in parts (i) and (ii) of the question. At which of the points P, Q, R and S will the water be at rest? (*O. and C.*)

87. (a) Explain what is meant by interference of light. Draw a diagram to show why a diffraction grating illuminated with light of a single wavelength can produce constructive interference at more than one angle to the direct beam.
(b) Draw a diagram to show the appearance on a screen of the first-order and second-order spectra produced when white light falls normally on a diffraction grating.
(c) How would the diffraction spectrum of a sodium lamp differ from that of a normal electric filament lamp, and what would the differences indicate?
(*O.L.E. part qn.*)

Forces and pressure

88. (a) Describe an experiment to show that for a body in equilibrium the sum of the clockwise moments about a point is equal to the sum of the anticlockwise moments about the same point.

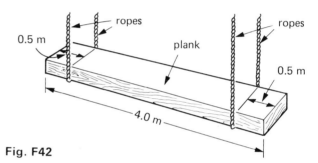

Fig. F42

(b) A painter stands on a uniform plank 4.0 m long and of mass 30 kg. The plank is suspended horizontally from vertical ropes attached 0.5 m from each end as shown in Fig. F42. The mass of the painter is 80 kg. Calculate the tensions in the ropes when the painter is 1.0 m from the centre of the platform. State briefly (no calculation required) how you would expect the tensions in the ropes to vary as the painter moves along the plank.

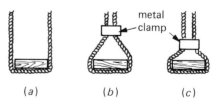

Fig. F43

Fig. F43*a* shows an end view of the plank. Fig. F43*b* and *c* show two other possible ways in which the plank may be supported by the ropes. State, giving your reasons, whether or not you would expect
(i) the tensions to be the same in the vertical parts of the rope in each diagram and
(ii) the tensions to be the same in the sloping parts of the rope in *b* and *c*. (*L.*)

271

89. State Archimedes' principle and describe an experiment you would perform to verify it.

Explain why a smaller volume of a block of wood is submerged when the block floats in water than when it floats in oil.

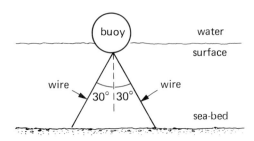

Fig. F44

A buoy of mass 100 kg is tethered to the sea-bed by two wires, each making an angle of 30° to the vertical, Fig. F44. 0.5 m³ of water is displaced by the buoy. Calculate
  (i) the weight of sea water, of density 1100 kg/m³, displaced by the buoy,
  (ii) the upward force exerted on the buoy by the water,
  (iii) the tensions in the wires. (*O. and C.*)

Motion and energy

90. (a) Explain what is meant by the acceleration of free fall *g*.
  (b) Describe carefully a laboratory experiment to obtain, as accurately as possible, a value for *g* using a free fall method.
  (c) A helicopter which is ascending at a steady speed of 20 m/s releases a parcel which takes 10 s to reach the earth. Initially upon release the parcel moves upwards. (In the following calculations take *g* to be 10 m/s². Show your calculations clearly.)
  (i) Find the time necessary for the parcel to reach its maximum height.
  (ii) What is the time taken for the parcel to reach the ground from the maximum height?
  (iii) What is the velocity of the parcel when it strikes the ground?
  (iv) What is the maximum height above the ground of the parcel?
  (v) What is the height of the helicopter at the instant the parcel is dropped?
  (vi) What is the height of the helicopter at the instant the parcel strikes the ground? (*N.I.*)

91. (a) Describe an experiment to show how the force applied to a body is related to the acceleration it produces. Sketch the apparatus you would use, state the observations you would make and show how you

would use these observations to obtain the relation between force and acceleration.
  (b) A stationary gun of mass 50 tonnes fires a shell of mass 100 kg with a velocity of 600 m/s. Calculate the initial velocity of recoil of the gun (1 tonne = 1000 kg).
  (c) Explain the difference between the mass and the weight of a body. State how these quantities change when the same object has its mass and weight measured, first on the earth and then on the moon.
  (*J.M.B.*)

Heat and energy

92. (a) Explain how the transfer of heat energy from a warm body to a cooler body by *radiation* differs from the transfer of heat energy by *convection*.
  (b) A solar heating system consists of a unit which may be mounted on the sloping roof of a house to produce warm water which is stored in a tank.

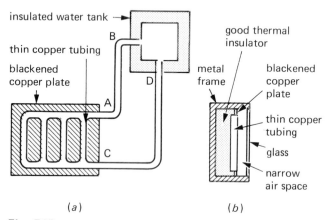

Fig. F45

Fig. F45*a* shows such a unit which consists of thin-walled copper tube which is embedded, partially, in a copper plate. The surface of the copper is blackened. This is mounted within a metal frame, as indicated in Fig. F45*b*. Sunlight irradiates the copper through a glass window and water is warmed and stored in the insulated water tank. The space behind the plate and tubes is filled with a good thermal insulator.
  (i) Why is it necessary to place the insulated water tank at a higher level than the unit? Which of the two pipes AB or CD carries the warmer water?
  (ii) Why are the tubes and plate made of copper? Why is the tube thin-walled?
  (iii) Explain why the unit is more effective if the surface of the tubes and plate is blackened rather than shiny.
  (iv) The purpose of the glass plate and narrow airspace is to reduce energy loss from the copper tubes and plate to the atmosphere. Explain *two* ways by which energy loss is minimized in this design.
  (*O. and C.*)

272

Electricity and magnetism

93. In modern house-wiring a 'square-pin' plug is used which has three pins and a fuse. List the colour codes for the wiring of the three pins. State which lead includes the fuse.

Describe the purpose and action of the fuse and suggest an appropriate fuse value for the following appliances when used on 250 V mains:

    (i) a 3 kW electric kettle,
    (ii) a 1 kW electric iron,
    (iii) a 400 W television set.

Draw a circuit diagram showing how two switches (e.g. hall and landing) can be used to control one light. (*W.*)

94. A mains supply is stated to be '240 V 50 Hz a.c.'. With the aid of a labelled sketch graph, explain the meaning of a.c. and 50 Hz.

Describe with the aid of a labelled diagram a transformer which could be used to operate a lamp marked 12 V 36 W from the 240 V mains supply. Assuming the transformer to be ideal (i.e. 100% efficient), calculate

    (i) the number of turns required in the secondary coil if there are 4000 turns in the primary,
    (ii) the current in the primary when the lamp is supplied with its normal operating current,
    (iii) the energy given out in 300 s by the lamp under normal operating conditions. (*C.*)

Electrons and atoms

95. An evacuated cathode ray tube has a circular flat screen of area 0.04 m². Calculate the force due to air pressure on the screen if atmospheric pressure is 10^5 pascal.

Give reasons why such a tube should be evacuated. If the potential difference between anode and cathode is 8 kV and the electron beam current is 1.5 mA, calculate the energy supplied per second to this beam. Describe the changes in energy when the electrons hit the screen.

    (*S.*)

96. (a) In Fig. F46 a beam of electrons is shown travelling in an evacuated glass tube between two parallel metal plates P and Q towards a fluorescent screen where it produces a spot of light at O.
    (i) Why must the tube be evacuated?
    (ii) What happens to the spot if an electric field is created between P and Q by applying a steady d.c. voltage across them with P positive?

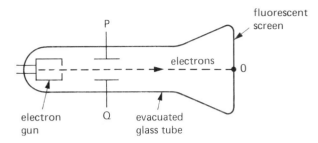

Fig. F46

    (iii) What happens to the spot if a strong magnetic field is applied directed vertically downwards from P to Q?
    (iv) What happens to the spot if the speed of the electrons is increased (by increasing the electron gun voltage) whilst the electric field is applied?

(b) When a correctly adjusted C.R.O. is connected across *R* in Fig. F47, which of the following waveforms is obtained?

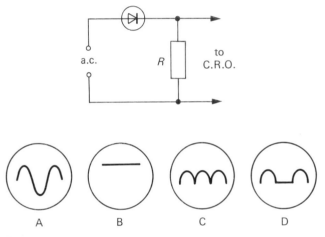

Fig. F47

97. (a) State *six* advantages which transistors have as switches.
    (b) When is a transistor considered to be **(i)** off, **(ii)** on, if it is used as as switch?
    (c) Draw the two basic transistor switching circuits and explain how each works.

98. (a) Draw a simple radio receiver circuit which has a tuning stage, a detector and an amplifier.
    (b) Explain the terms *modulation* and *detection*.
    (c) In a television receiver what is the function of **(i)** the line time base, **(ii)** the frame time base?

99. The circuit in Fig. F48 when completed is to be used as an early-morning alarm which rings a bell when it gets light.

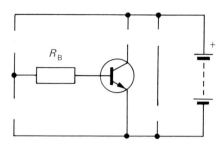

Fig. F48

(a) Copy and complete the circuit by adding in the correct gaps an L.D.R., a variable resistor, a relay, a diode and an electric bell.

(b) Explain how the circuit works.

(c) What is the purpose of (i) R_B, (ii) the diode?

(d) What is the advantage of using a variable resistor?

Mathematics for physics

USE THIS SECTION AS THE NEED ARISES

Solving physics problems

When tackling physics problems using mathematical equations it is suggested that you *do not substitute numerical values until you have obtained the expression in symbols which gives the answer*. That is, work in symbols until you have solved the problem and only then insert the numbers in the expression to get the final result.

This has two advantages. First, it reduces the chance of arithmetic (and copying down) errors. Second, you write less since a symbol is usually a single letter whereas a numerical value is often a string of figures.

Adopting this 'symbolic' procedure frequently requires you to change round an equation first. The next two sections and the questions that follow them are intended to give you practice in doing this and then substituting numerical values to get the answer.

Equations—type 1

In the equation $x = a/b$, the subject is x. To change it we *multiply or divide both sides* of the equation by the same quantity.

(i) To change the subject to a. We have

$$x = \frac{a}{b}$$

If we multiply both sides by b the equation will still be true.

$$\therefore \; x \times b = \frac{a}{b} \times b$$

The b's on the right-hand side cancel

$$\therefore \; b \times x = \frac{a}{\not{b}} \times \not{b} = a$$

$$\therefore \quad a = b \times x$$

(ii) To change the subject to b. We have

$$x = \frac{a}{b}$$

Multiplying both sides by b as before, we get

$$a = b \times x$$

Divide both sides by x

$$\therefore \; \frac{a}{x} = \frac{b \times x}{x} = \frac{b \times \not{x}}{\not{x}} = b \qquad \therefore \; b = \frac{a}{x}$$

Now try the following questions using these ideas.

Q Questions

1. What is the value of x if

(a) $2x = 6$ (b) $3x = 15$ (c) $3x = 8$

(d) $\dfrac{x}{2} = 10$ (e) $\dfrac{x}{3} = 4$ (f) $\dfrac{2x}{3} = 4$

(g) $\dfrac{4}{x} = 2$ (h) $\dfrac{9}{x} = 3$ (i) $\dfrac{x}{6} = \dfrac{4}{3}$

2. Change the subject to

(a) f in $v = f\lambda$ (b) λ in $v = f\lambda$
(c) I in $V = IR$ (d) R in $V = IR$
(e) m in $d = \dfrac{m}{V}$ (f) V in $d = \dfrac{m}{V}$
(g) s in $v = \dfrac{s}{t}$ (h) t in $v = \dfrac{s}{t}$
(i) a in $F = ma$ (j) I in $Q = It$
(k) Q in $C = \dfrac{Q}{V}$ (l) A in $P = \dfrac{F}{A}$
(m) p_1 in $p_1V_1 = p_2V_2$ (n) a in $x = abc$
(o) t in $W = ItV$ (p) $\Delta\theta$ in $Q = mc\Delta\theta$

3. Change the subject to

(a) I^2 in $P = I^2R$ (b) I in $P = I^2R$
(c) a in $s = \frac{1}{2}at^2$ (d) t^2 in $s = \frac{1}{2}at^2$
(e) t in $s = \frac{1}{2}at^2$ (f) v in $\frac{1}{2}mv^2 = mgh$
(g) y in $\lambda = \dfrac{ay}{D}$ (h) ρ in $R = \dfrac{\rho l}{A}$
(i) V_1 in $\dfrac{V_1}{T_1} = \dfrac{V_2}{T_2}$ (j) T_1 in $\dfrac{V_1}{T_1} = \dfrac{V_2}{T_2}$
(k) V_2 in $\dfrac{p_1V_1}{T_1} = \dfrac{p_2V_2}{T_2}$ (l) T_2 in $\dfrac{p_1V_1}{T_1} = \dfrac{p_2V_2}{T_2}$

4. By replacing (substituting) find the value of v in $v = f\lambda$ if

(a) $f = 5$ and $\lambda = 2$ (b) $f = 3.4$ and $\lambda = 10$
(c) $f = \frac{1}{4}$ and $\lambda = 8/3$ (d) $f = 3/5$ and $\lambda = 1/6$
(e) $f = 100$ and $\lambda = 0.1$ (f) $f = 3 \times 10^5$ and $\lambda = 10^3$

5. By changing the subject and replacing find

(a) f in $v = f\lambda$ if $v = 3.0 \times 10^8$ and $\lambda = 1.5 \times 10^3$
(b) h in $p = 10hd$ if $p = 10^5$ and $d = 10^3$
(c) a in $n = a/b$ if $n = 4/3$ and $b = 6$
(d) b in $n = a/b$ if $n = 1.5$ and $a = 3.0 \times 10^8$
(e) F in $p = F/A$ if $p = 100$ and $A = 0.2$

275

(f) s in $v = s/t$ if $v = 1500$ and $t = 0.2$

(g) t in $V = At$ if $V = 10^{-3}$ and $A = 10^4$

(h) V_2 in $p_1V_1 = p_2V_2$ if $p_1 = 10^5$, $V_1 = 30$ and $p_2 = 2 \times 10^5$

(i) t in $s = \frac{1}{2}gt^2$ if $s = 125$ and $g = 10$

(j) b in $x = abc$ if $x = 0.0016$, $a = 1$ and $c = 80$

(k) $\Delta\theta$ in $Q = mc\Delta\theta$ if $Q = 8000$, $m = 2$ and $c = 400$

(l) V_2 in $p_1V_1/T_1 = p_2V_2/T_2$ if $p_1 = 1$, $p_2 = 2$, $V_1 = 2$, $T_1 = 300$ and $T_2 = 500$

Equations—type 2

To change the subject in the equation $x = a + by$ we *add or subtract the same quantity from each side*. We may also have to divide or multiply as in type 1. Suppose we wish to change the subject to y in

$$x = a + by$$

Subtract a from both sides,

$$\therefore x - a = a + by - a = by$$

Divide both sides by b,

$$\therefore \frac{x-a}{b} = \frac{by}{b} = y \qquad \therefore y = \frac{x-a}{b}$$

Q Questions

6. What is the value of x if

(a) $x + 1 = 5$ **(b)** $2x + 3 = 7$ **(c)** $x - 2 = 3$

(d) $2(x - 3) = 10$ **(e)** $\frac{x}{2} - \frac{1}{3} = 0$ **(f)** $\frac{x}{3} + \frac{1}{4} = 0$

(g) $2x + \frac{5}{3} = 6$ **(h)** $7 - \frac{x}{4} = 11$ **(i)** $\frac{3}{x} + 2 = 5$

7. By changing the subject and replacing find the value of a in $v = u + at$ if

     (a) $v = 20$, $u = 10$ and $t = 2$

     (b) $v = 50$, $u = 20$ and $t = 0.5$

     (c) $v = 5/0.2$, $u = 2/0.2$ and $t = 0.2$

8. Change the subject in $v^2 = u^2 + 2as$ to a.

Proportion (or variation)

One of the most important mathematical operations in physics is finding the relation between two sets of measurements.

(a) Direct proportion. Suppose that in an experiment two sets of readings are obtained for the quantities x and y as in Table 1 (units omitted).

Table 1

x	1	2	3	4
y	2	4	6	8

276

We see that when x is doubled, y doubles; when x is trebled, y trebles; when x is halved, y halves and so on. There is a one-to-one correspondence between each value of x and the corresponding value of y.

We say that y is *directly proportional* to x, or y *varies directly* as x. In symbols

$$y \propto x$$

Also, the *ratio* of one to the other, e.g. y to x, is always the same, i.e. it has a constant value which in this case is 2. Hence

$$\frac{y}{x} = \text{a constant} = 2$$

The constant, called the *constant of proportionality* or *variation*, is given a symbol, e.g. k, and the relation (or law) between y and x is then summed up by the equation

$$\frac{y}{x} = k \qquad \text{or} \qquad y = kx$$

Notes. 1. In practice, because of inevitable experimental errors, the readings seldom show the relation so clearly as here.

2. If instead of using numerical values for x and y we use letters, e.g. x_1, x_2, x_3, etc., and y_1, y_2, y_3, etc., then we can also say

$$\frac{y_1}{x_1} = \frac{y_2}{x_2} = \frac{y_3}{x_3} = \dots\dots = k$$

or $y_1 = kx_1$, $y_2 = kx_2$, $y_3, = kx_3.\dots\dots$

(b) Inverse proportion. Two sets of readings for the quantities p and V are given in Table 2.

Table 2

p	3	4	6	12
V	4	3	2	1

There is again a one-to-one correspondence between each value of p and the corresponding value of V but when p is doubled, V is halved; when p is trebled, V has one-third its previous value and so on.

We say that V is *inversely proportional* to p, or V *varies inversely* as p, i.e.

$$V \propto \frac{1}{p}$$

Also, the *product* $p \times V$ is always the same ($= 12$) and we write

$$V = \frac{k}{p} \qquad \text{or} \qquad pV = k$$

where k is the constant of proportionality or variation and equals 12 in this case.

Using letters for values of p and V we can also say

$$p_1V_1 = p_2V_2 = p_3V_3 = \ldots \ldots \ldots = k$$

Graphs

Another useful way of finding the relation between two quantities is by a graph.

(a) Straight line graphs. When the readings in *Table 1* are used to plot a graph of y against x, a *continuous* line joining the points is *a straight line passing through the origin* O, Fig. 1. Such a graph shows there is direct

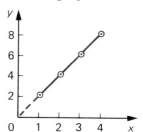

Fig. 1

proportionality between the quantities plotted, i.e. $y \propto x$. But note that the line must go through O.

A graph of p against V using the readings in *Table 2* is a curve, Fig. 2. However if we plot p against $1/V$ (or V

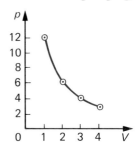

Fig. 2

against $1/p$) we get a straight line through the origin, showing that $p \propto 1/V$, Fig. 3 (or $V \propto 1/p$).

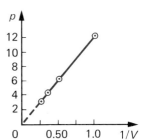

Fig. 3

p	V	$1/V$
3	4	0.25
4	3	0.33
6	2	0.50
12	1	1.0

(b) Slope or gradient. The slope or gradient of a straight-line graph equals the constant of proportionality. In Fig. 1, the slope is $y/x = 2$; in Fig. 3 it is $p/(1/V) = 12$.

In practice points plotted from actual measurements may not lie exactly on a straight line due to experimental errors. The 'best straight line' is then drawn 'through' them so that they are equally distributed about it. This automatically averages the results. Any points that are well off the line stand out and may be investigated further.

(c) Practical points.
(i) The axes should be labelled giving the quantities being plotted and their units, e.g. I/A meaning current in amperes.
(ii) If possible the origin of both scales should be on the paper and the scales chosen so that the points are spread out along the graph.
(iii) Mark the points ⊙ or ×.

Ｑ Questions

9. In an experiment different masses were hung from the end of a spring held in a stand and the extensions produced were as shown below.

Mass (g)	100	150	200	300	350	500	600
Extension (cm)	1.9	3.1	4.0	6.1	6.9	10.0	12.2

(a) Plot a graph of *extension* along the vertical (y) axis against *mass* along the horizontal (x) axis.
(b) What is the relation between *extension* and *mass*? Give a reason for your answer.

10. Pairs of readings of the quantities m and v are given below.

m	0.25	1.5	2.5	3.5
v	20	40	56	72

(a) Plot a graph of m along the vertical axis and v along the horizontal axis.
(b) Is m directly proportional to v? Explain your answer.
(c) Use the graph to find v when $m = 1$.

11. In an experiment to measure the density of copper the mass of a number of copper rivets was measured on a lever balance and the volume measured by displacement of water in a measuring cylinder. Several sets of readings were obtained by using different numbers of rivets. The readings obtained are shown below.

277

Reading no.	1	2	3	4	5	6	7	8	9	10
Mass (g)	18	30	36	60	42	54	66	72	78	90
Volume (cm^3)	2	3	4	4.5	5	6	7	8	9	10

(a) Draw a graph of these readings choosing a suitable scale (mass on vertical axis).

(b) Does your graph pass through the origin? Explain why you think it should or should not as the case may be.

(c) Did you take reading no. 4 into account when drawing your graph? Explain.

(d) The formula for the density of a substance is

$$\text{density} = \frac{\text{mass}}{\text{volume}}$$

By measuring the slope of the graph calculate the density of the copper used in this experiment. (Show clearly on your graph any measurements you have made in order to do this.)

(e) Why is it better to use your graph to calculate the density rather than just one pair of readings?

(W.M.)

12. The distance s (in metres) travelled by a car at various times t (in seconds) are shown below.

s	0	2	8	18	32	50
t	0	1	2	3	4	5

Draw graphs of (a) s against t, (b) s against t^2. What can you conclude?

Check lists of specific objectives

Light and sight

After studying *Topic 1: Light rays*, you should be able to

- give examples of effects which show that light travels in a straight line,
- explain the operation of a pinhole camera and draw ray diagrams to show the result of varying the object distance or the length of the camera,
- draw diagrams to show how shadows are formed using point and extended sources,
- use the terms 'umbra' and 'penumbra', and
- draw diagrams to show how solar and lunar eclipses occur.

After studying *Topic 2: Reflection of light*, you should be able to

- describe experiments to show that the angle of incidence equals the angle of reflection,
- state the laws of reflection and use them to solve problems,
- draw a ray diagram to show how a periscope works.

After studying *Topic 3: Plane mirrors*, you should be able to

- describe an experiment to show that the image in a plane mirror is as far behind the mirror as the object is in front and that the line joining the object and image is at right angles to the mirror,
- draw a diagram to explain the formation of a virtual image by a plane mirror,
- explain the term 'lateral inversion',
- explain how a plane mirror can improve the accuracy of measurements in physics, and
- state that when a plane mirror is rotated through a certain angle, the reflected ray turns through twice that angle.

After studying *Topic 4: Curved mirrors*, you should be able to

- draw diagrams to show the action of concave and convex mirrors on a parallel beam of light,
- explain why concave mirrors are used as reflectors in car headlamps,
- explain why concave mirrors are used as make-up and shaving mirrors, and
- explain why convex mirrors are used as driving mirrors.

After studying *Topic 5: Refraction of light*, you should be able to

- state what the term 'refraction' means,
- give examples of effects that show light can be refracted,
- describe experiments to study refraction,
- draw diagrams of the passage of light rays through rectangular blocks and recall that lateral displacement occurs for a parallel-sided block,
- recall that light is refracted because it changes speed when it enters another medium,
- recall the definition of refractive index as $n = c_{air}/c_{medium}$ and solve problems using it,
- describe experiments to find refractive index by the real and apparent depth method, and

- draw a diagram for the passage of a light ray through a prism.

After studying *Topic 6: Total internal reflection*, you should be able to

- explain with the aid of diagrams what is meant by 'critical angle' and 'total internal reflection',
- describe an experiment to find the critical angle of glass or Perspex,
- draw diagrams to show the action of totally reflecting prisms in periscopes and binoculars, and
- explain the action of light pipes.

After studying *Topic 7: Lenses*, you should be able to

- explain the action of a lens in terms of refraction by a number of small prisms,
- draw diagrams showing the effects of convex and concave lenses on a beam of parallel rays,
- recall the meaning of 'optical centre', 'principal axis', 'principal focus' and 'focal length',
- draw ray diagrams to show image formation by a convex lens,
- describe experiments to measure the focal length of a convex lens by the distant object and plane mirror methods,
- draw scale diagrams to solve problems on convex lenses, and
- recall the meaning of the term 'linear magnification'.

After studying *Topic 8: The eye*, you should be able to

- draw and label the main parts of the eye,
- explain how refraction at the cornea and lens produces an image on the retina,
- explain how the lens allows accommodation to occur,
- recall the meaning of 'short sight' and describe how it is corrected,
- recall the meaning of 'long sight' and describe how it is corrected, and
- explain the terms 'binocular vision' and 'persistence of vision'.

After studying *Topic 9: Colour*, you should be able to

- explain the terms 'spectrum' and 'dispersion',
- describe how a prism is used to produce a spectrum from white light,
- describe how the colours of the spectrum can be recombined to give white light,
- recall the factors affecting the colour of an object,
- recall the results of mixing coloured lights, and
- recall the results of mixing coloured pigments.

After studying *Topic 10: Simple optical instruments*, you should be able to

- describe with the aid of diagrams how a single lens is used (*a*) in a simple lens camera, (*b*) in a projector, (*c*) in a magnifying glass.

After studying *Topic 11: Microscopes and telescopes*, you should be able to

- describe with the aid of diagrams (*a*) a compound microscope, (*b*) a refracting astronomical telescope, (*c*) a reflecting astronomical telescope.

Waves and sound

After studying *Topic 13: Water waves*, you should be able to

- describe the production of pulses and progressive transverse waves on ropes, springs and ripple tanks,

- recall the meaning of 'wavelength', 'frequency', 'speed', 'amplitude' and 'phase',

- represent a transverse wave on a displacement–distance graph and extract information from it,

- recall the wave equation $v = f\lambda$ and use it to solve problems,

- describe experiments to show reflection of waves,

- recall that the angle of reflection equals the angle of incidence,

- draw diagrams for the reflection of straight wavefronts at plane and concave surfaces,

- describe experiments to show refraction of waves,

- recall that refraction at a straight boundary is due to change of wave speed but *not* of frequency,

- draw diagrams for the refraction of straight wavefronts at straight boundaries,

- explain the term 'diffraction',

- describe experiments to show diffraction of waves,

- draw diagrams for the diffraction of straight wavefronts at single slits of different widths,

- predict the effect of changing the wavelength or the size of the gap on diffraction of waves at a single slit,

- describe experiments to show interference of water waves using two point sources,

- draw interference patterns for waves from two point sources,

- explain interference as the superposition of crests or troughs at points where waves arrive 'in phase' or 'out of phase', and

- predict the effect on the interference pattern of changing the separation of the two sources or the wavelength of the waves.

After studying *Topic 14: Light waves*, you should be able to

- explain why diffraction of light is not normally observed,

- describe a simple Young's double-slit experiment to show light has a wave-like nature,

- recall that the colour of light depends on its frequency, that red light has a lower frequency (but longer wavelength) than blue light and that all colours travel at the same speed in air,

- describe a diffraction grating,

- explain with the aid of a diagram how a diffraction grating produces several orders of spectra, and

- explain briefly how a hologram is made.

After studying *Topic 15: Electromagnetic spectrum*, you should be able to

- recall the members of the electromagnetic spectrum,

- recall that all electromagnetic waves have the same speed in air and are progressive transverse waves, and

- distinguish between infrared radiation, ultraviolet radiation, radio waves and X-rays in terms of their wavelengths and other properties.

After studying *Topic 16: Sound waves*, you should be able to

- recall that sound is produced by vibrations,

- describe an experiment to show that sound is not transmitted through a vacuum,
- describe how sound travels in a medium as progressive longitudinal waves,
- recall the limits of audibility (i.e. the range of frequencies) for the normal human ear,
- explain echoes and reverberation,
- describe a simple method of estimating the speed of sound and recall its approximate value,
- solve problems using the speed of sound, e.g. thundercloud proximity,
- describe experiments to show diffraction and interference of sound waves, and
- recall some uses of ultrasonics.

After studying *Topic 17: Musical notes*, you should be able to
- use the terms 'pitch', 'loudness' and 'quality' (timbre) and connect them to wave properties,
- describe stationary (standing) waves and explain how they are produced in string instruments,
- recall the factors affecting the frequency of the note emitted by a vibrating string,
- describe a demonstration of resonance, and
- discuss noise pollution constructively.

Matter and molecules

After studying *Topic 19: Measurements*, you should be able to
- recall three basic quantities in physics,
- write a number in powers of ten (standard) form,
- recall the unit of length and the meaning of the prefixes 'kilo', 'centi', 'milli', 'micro', 'nano',
- use a ruler to measure length so as to eliminate errors due to parallax,
- give a result to an appropriate number of significant figures,
- measure regular and irregular areas,
- measure the volume of regular solids and liquids,
- recall the unit of mass and how mass is measured,
- recall the unit of time and how time is measured, and
- describe an experiment to find the period of a mass–spring system.

After studying *Topic 20: Density*, you should be able to
- define 'density' and perform calculations using $d = m/V$,
- describe experiments to measure the density of solids, liquids and air, and
- relate floating and sinking to density.

After studying *Topic 21: Weight and springs*, you should be able to
- recall that a force can cause a change in the motion, size or shape of a body,
- recall that the weight of a body is the force of gravity on it,
- recall the unit of force and how force is measured,
- describe experiments to study the relation between force and extension for springs,

- draw conclusions from force–extension graphs, and
- recall Hooke's law and solve problems using it.

After studying *Topic 22: Molecules*, you should be able to

- describe and explain an experiment to show Brownian motion,
- use the kinetic theory to explain the physical properties of solids, liquids and gases,
- describe an experiment to estimate the size of a molecule, and
- recall how crystals support the view that matter is made of small particles.

After studying *Topic 23: Properties of matter*, you should be able to

- describe and explain simple experiments on diffusion,
- state evidence for the existence of intermolecular forces, e.g. surface tension,
- recall the meaning of 'strength', 'stiffness', 'elasticity', 'ductility' and 'brittleness', and
- describe and explain what happens when a wire is stretched.

Forces and pressure

After studying *Topic 25: Moments and levers*, you should be able to

- define the moment of a force about a point,
- describe experiments to study turning effects on bodies in equilibrium,
- state the law of moments and use it to solve problems,
- explain the action of common tools and devices as levers,
- state the conditions for equilibrium when parallel forces act on a body, and
- describe the behaviour of beams and bridges when loaded.

After studying *Topic 26: Centres of gravity*, you should be able to

- recall that an object behaves as if its whole weight were concentrated at its centre of gravity,
- describe experiments to find the centre of gravity of an object, and
- connect the stability of an object to the position of its centre of gravity.

After studying *Topic 27: Adding forces*, you should be able to

- combine forces graphically to find their resultant using the parallelogram law,
- distinguish between vectors and scalars and give examples of each, and
- resolve a force into two components at right angles.

After studying *Topic 28: Work, energy, power*, you should be able to

- use the relation work $=$ force $\times$ distance moved, to calculate energy transfer,
- define the unit of work,
- identify various forms of energy,
- describe energy changes in given examples,

- state the principle of conservation of energy,
- recall that power is energy changed per unit time and solve problems,
- describe an experiment to measure your own power,
- recall that friction opposes motion between surfaces in contact, and
- recall that heat energy is produced when work is done against friction.

After studying *Topic 29: Machines*, you should be able to

- recognize a machine as a force-multiplier or a distance-multiplier,
- define the terms 'mechanical advantage' (M.A.), 'velocity ratio' (V.R.) and 'efficiency' and solve problems using them,
- describe an experiment to measure the efficiency of a pulley system, and
- recall the use and advantage of the wheel axle, inclined plane, screw and gears.

After studying *Topic 30: Pressure in liquids*, you should be able to

- define 'pressure' and recall the unit of pressure,
- connect the pressure in a fluid with its depth and density,
- recall that pressure is transmitted through a fluid and use it to explain the hydraulic jack and hydraulic car brakes, and
- use pressure $= 10$ *hd* to solve problems.

After studying *Topic 31: Atmospheric pressure*, you should be able to

- describe demonstrations of the effects of air pressure,
- describe uses of air pressure.
- describe how a Bourdon gauge or a U-tube manometer may be used to measure gas pressure,
- explain how a simple mercury barometer works,
- explain how an aneroid barometer works and is used as a weather glass and as an altimeter, and
- explain certain effects in aviation and diving in terms of pressure, e.g. ear 'popping', the 'bends',

After studying *Topic 32: Pumps and pressure*, you should be able to

- explain the action of a syringe,
- explain the action of a bicycle pump,
- describe an experiment to investigate the relation between the pressure and volume of a gas, and
- state Boyle's law and use it in calculations.

After studying *Topic 33: Floating, sinking and flying*, you should be able to

- state Archimedes' principle and use it to solve problems,
- state the principle of flotation and use it to solve problems,
- explain the behaviour of ships, submarines, balloons and airships,
- describe a hydrometer and what it does, and
- state Bernoulli's principle and use it to explain the action of a Bunsen burner, a spinning ball, an aerofoil and a yacht.

284

Motion and energy

After studying *Topic 35: Velocity and acceleration*, you should be able to

- explain the meaning of the terms 'speed', 'velocity', 'acceleration', 'displacement', and

- describe how velocity and acceleration may be measured using timers and tape charts.

After studying *Topic 36: Equations of motion*, you should be able to

- recall and use the three equations of motion for solving problems, and

- draw, interpret and use velocity–time and distance–time graphs to solve problems.

After studying *Topic 37: Falling bodies*, you should be able to

- describe the behaviour of falling objects and solve problems on them, and

- determine experimentally the acceleration due to gravity.

After studying *Topic 38: Newton's laws of motion*, you should be able to

- describe an experiment to investigate the relationship between force, mass and acceleration,

- define the unit of force,

- state Newton's three laws of motion and use them to solve problems,

- define the strength of the earth's gravitational field, and

- describe the motion of an object falling in air.

After studying *Topic 39: Momentum*, you should be able to

- define 'momentum',

- describe experiments to demonstrate the principle of conservation of momentum,

- state and use the principle of conservation of momentum to solve problems,

- understand the action of rocket and jet engines, and

- state the relationship between force and rate of change of momentum and use it to solve problems.

After studying *Topic 40: K.E. and P.E.*, you should be able to

- define 'kinetic energy' (k.e.),

- perform calculations using $E_k = \frac{1}{2}mv^2$,

- define 'potential energy' (p.e.),

- calculate changes in p.e. using $\Delta E_p = mgh$, and

- apply the principle of conservation of energy to mechanical systems, e.g. a swinging pendulum.

After studying *Topic 41: Circular motion*, you should be able to

- explain circular motion in terms of an unbalanced centripetal force, and

- discuss the motion of a satellite in orbit.

Heat and energy

After studying *Topic 43: Thermometers*, you should be able to

- define the fixed points on the Celsius scale,

- recall the properties of mercury and alcohol as thermometric liquids,

- describe the clinical and thermocouple thermometers, and

- distinguish between heat and temperature and recall that temperature decides the direction of heat flow.

After studying *Topic 44: Expansion of solids and liquids*, you should be able to

- describe uses of expansion, including the bimetallic strip,

- describe precautions taken against expansion,

- define linear expansivity and describe how it can be measured, and

- recall that water has its maximum density at 4 °C and explain why a pond freezes at the top first.

After studying *Topic 45: The gas laws*, you should be able to

- describe experiments to study the relations between the pressure, volume and temperature of a gas,

- explain the establishment of the Kelvin (absolute) temperature scale from graphs of pressure or volume against temperature and recall the equation connecting the Kelvin and Celsius scales, i.e. $T = 273 + t$,

- state Charles' law, the pressure law and Boyle's law and use them to solve problems,

- recall that $pV/T =$ constant and use it to solve problems, and

- explain the behaviour of gases using the kinetic theory.

After studying *Topic 46: Specific heat capacity*, you should be able to

- define 'specific heat capacity', c,

- solve problems on specific heat capacity using the heat equation $Q = m \times \Delta\theta \times c$,

- describe experiments to measure the specific heat capacity of metals and liquids by electrical heating,

- explain the importance of the high specific heat capacity of water, and

- discuss world energy resources including alternative sources.

After studying *Topic 47: Latent heat*, you should be able to

- describe an experiment to show that during a change of state the temperature stays constant,

- define 'specific latent heat of fusion', l_f,

- describe an experiment to measure l_f for ice,

- define 'specific latent heat of vaporization', l_v,

- describe an experiment to measure l_v for water,

- explain latent heat using the kinetic theory, and

- solve problems on latent heat using $Q = ml$

After studying *Topic 48: Melting and boiling*, you should be able to

- describe the effect of pressure and impurities on the melting point,

- describe the effect of pressure and impurities on the boiling point,

- distinguish between 'evaporation' and 'boiling',

- describe an experiment to show that evaporation causes cooling,

- explain cooling by evaporation using the kinetic theory, and

- explain how a refrigerator works.

After studying *Topic 49: Conduction and convection*, you should be able to

- describe experiments to show the different conducting powers of various substances,
- name good and bad conductors and state uses for each,
- explain conduction using the kinetic theory,
- describe experiments to show convection in fluids (liquids and gases),
- understand how household hot-water and central-heating systems work, and
- relate convection to phenomena such as land and sea breezes.

After studying *Topic 50: Radiation*, you should be able to

- describe experiments to study factors affecting the absorption and emission of radiation,
- recall that good absorbers are also good emitters,
- explain how a knowledge of heat transfer affects the design of a vacuum flask,
- explain how a greenhouse acts as a 'heat-trap', and
- investigate how rate of cooling depends on the ratio of surface area to volume.

After studying *Topic 51: Heat engines*, you should be able to

- explain the principles of operation of the two- and four-stroke petrol and diesel engines, the jet engine, rockets and the steam turbine.

Electricity and magnetism

After studying *Topic 53: Permanent magnets*, you should be able to

- state the properties of magnets,
- recall the test for a magnet,
- describe different methods of magnetizing and demagnetizing,
- explain magnetic induction,
- compare the magnetic properties of iron and steel, and
- explain magnetization and demagnetization in terms of domains.

After studying *Topic 54: Magnetic fields*, you should be able to

- recall that a magnetic field is the region round a magnet where a magnetic force is exerted and is represented by lines of force whose direction at any point is the direction of the force on a N pole,
- map magnetic fields (by the plotting compass and iron filings method) round (*a*) one magnet, (*b*) two magnets,
- recall that at a neutral point the field due to one magnet cancels that due to the other, and
- define 'declination'.

After studying *Topic 55: Static electricity*, you should be able to

- describe how positive and negative charges are produced by rubbing,
- recall that like charges repel, unlike charges attract,
- explain the charging of objects in terms of the motion of negatively-charged electrons,
- describe the gold-leaf electroscope.

- describe and explain simple experiments using an electroscope to study the sign of a charge and to compare electrical conductivities of different materials, and
- explain the differences between insulators and conductors.

After studying *Topic 56: More electrostatics*, you should be able to

- describe how a conductor can be charged by induction,
- explain how a charged object can attract uncharged objects,
- describe how an electroscope can be charged by induction,
- define an electric field as a region in which an electric charge experiences a force,
- draw the electric field pattern around and between point charges and between parallel plates, and
- explain the action of a lightning conductor.

After studying *Topic 57: Capacitors*, you should be able to

- state the unit of electric charge,
- state what a capacitor does and what its capacitance depends on,
- state the unit capacitance, and
- name some types of practical capacitor.

After studying *Topic 58: Electric current*, you should be able to

- describe an experiment which shows that an electric current is a flow of charge,
- recall that an electric current in a metal is a flow of negative electrons from the negative to the positive terminal of the battery round a circuit,
- state the three effects of an electric current,
- state the unit of electric current and recall that current is measured by an ammeter,
- define the unit of charge in terms of the unit of current,
- recall the relation $Q = It$ and use it to solve problems,
- use circuit symbols for wires, cells, switches, ammeters and lamps,
- draw and connect simple series and parallel circuits, observing correct polarities for meters,
- recall that the current in a series circuit is the same everywhere in the circuit,
- recall that the sum of the currents in the branches of a parallel circuit equals the current entering or leaving the parallel section, and
- distinguish between electron flow and conventional current.

After studying *Topic 59: Potential difference*, you should be able to

- describe simple experiments to show the conversion of electrical energy to other forms (e.g. in a lamp),
- recall the definition of the unit of p.d. and that a p.d. is measured by a voltmeter,
- demonstrate that the sum of the p.d.s across any number of components in series equals the p.d. across all of those components,
- demonstrate that the p.d.s across any number of components in parallel are the same, and
- work out the p.d.s of cells connected in series and parallel.

After studying *Topic 60: Resistance*, you should be able to

- define 'resistance' and state the factors on which it depends

- recall the unit of resistance,

- solve simple problems using $R = V/I$,

- describe experiments using the ammeter–voltmeter method to measure resistance and study the relationship between current and p.d. for (*a*) metallic conductors, (*b*) semiconductor diodes, (*c*) filament lamps,

- plot *I*–*V* graphs from the results of such experiments and draw appropriate conclusions from them, and

- use the formulae for resistors in series and in parallel to solve problems.

After studying *Topic 61: Electromotive force*, you should be able to

- define 'electromotive force', e.m.f.,

- distinguish between 'e.m.f.' and 'terminal p.d.', and

- use the 'circuit equation' to solve problems involving internal resistance.

After studying *Topic 63: Electric power*, you should be able to

- recall the relation $W = ItV$ and $P = IV$ and use them to solve simple problems on energy changes,

- describe experiments to measure electric power, and

- describe electric lamps, heating elements and fuses.

After studying *Topic 64: Electricity in the home*, you should be able to

- describe with the aid of diagrams a house wiring system and explain the functions and positions of switches, fuses, circuit breakers and earth,

- wire a mains plug and recall the international insulation colour code, and

- perform calculations of the cost of electrical energy in joules and kilowatt-hours.

After studying *Topic 65: Electric cells*, you should be able to

- draw and describe a simple cell and a carbon–zinc (dry) cell,

- state the differences between primary and secondary cells, and

- describe a lead–acid cell and how it should be maintained.

After studying *Topic 66: Electromagnets*, you should be able to

- describe the magnetic fields round current-carrying straight and circular conductors and solenoids and draw sketches of them,

- recall the right-hand screw and grip rules for relating current direction and magnetic field direction,

- make a simple electromagnet,

- describe uses of electromagnets, and

- explain the action of an electric bell, a relay, a reed switch and a telephone.

After studying *Topic 67: Electric motors*, you should be able to

- describe a demonstration to show that a force acts on a current-carrying conductor in a magnetic field and recall that it increases with the strength of the field and the size of the current,

- draw the resultant field pattern for a current-carrying conductor which is at right angles to a uniform magnetic field,

- explain why a rectangular, current-carrying coil experiences a couple in a uniform magnetic field,

- draw a diagram of a simple d.c. electric motor and explain how it works.

- describe a practical d.c. motor, and

- draw a diagram of a moving coil loudspeaker and explain how it works,

After studying *Topic 68: Electric meters*, you should be able to

- draw a diagram of a simple moving coil galvanometer and explain how it works,

- explain how a moving coil galvanometer can be modified for use as an ammeter and a voltmeter, and

- explain why (*a*) an ammeter should have a very low resistance and (*b*) a voltmeter should have a very high resistance.

After studying *Topic 69: Generators*, you should be able to

- describe experiments to show electromagnetic induction,

- recall Faraday's explanation of electromagnetic induction,

- predict the direction of the induced e.m.f. using Lenz's law or Fleming's right-hand rule,

- draw a diagram of a simple a.c. generator and sketch a graph of its output,

- draw a diagram of a simple d.c. generator and sketch a graph of its output, and

- describe practical a.c. generators.

After studying *Topic 70: Transformers*, you should be able to

- explain the principle of the transformer,

- recall the transformer equation $V_s/V_p = N_s/N_p$ and use it to solve problems,

- recall that for an ideal transformer $V_p \times I_p = V_s \times I_s$, and use the relation to solve problems,

- recall the energy losses in practical transformers,

- explain why high voltage a.c. is used for transmitting electrical power,

- describe the car ignition system, and

- explain how eddy currents arise and how they are used in a car speedometer and in electromagnetic damping.

After studying *Topic 71: Alternating current*, you should be able to

- distinguish between direct and alternating currents,

- represent d.c. and a.c. on current- and p.d.-time graphs and deduce peak values and frequencies from them,

- recall the meaning of 'r.m.s. values',

- describe an a.c. meter,

- describe the behaviour of capacitors in d.c. and a.c. circuits,

- describe, with the aid of graphs, the charge and discharge of a capacitor through a resistor, and

- describe the behaviour of inductors in d.c. and a.c. circuits.

Electrons and atoms

After studying *Topic 73: Electrons*, you should be able to

- explain the terms 'thermionic emission' and 'cathode rays',
- describe experiments to show that cathode rays are deflected by magnetic and electric fields,
- explain the terms 'specific charge', e/m, and 'electronic charge', e,
- describe with the aid of a diagram the jobs done in a C.R.O. by the electron gun, the X- and Y-plates, the fluorescent screen and the time base, and
- describe how the C.R.O. is used to measure p.d.s, to display waveforms and to measure time intervals and frequency.

After studying *Topic 74: Radioactivity*, you should be able to

- recall that the radiation emitted by a radioactive substance can be detected by its ionizing effect,
- explain the principle of operation of the Geiger–Müller tube and the diffusion cloud chamber,
- recall the nature of α, β and γ rays,
- describe experiments to compare the range and penetrating power of α, β and γ rays in different materials,
- recall the ionizing abilities of α, β and γ rays and relate them to their ranges,
- predict how α, β and γ rays will be deflected in magnetic and electric fields,
- define the term 'half-life',
- describe an experiment from which a radioactive decay curve can be obtained,
- show from graphs that random decay processes have a constant half-life,
- solve simple problems on half-life,
- recall that radioactivity is (*a*) random process, (*b*) due to nuclear instability, (*c*) independent of external conditions,
- recall some uses of radioactivity, and
- discuss the dangers of radioactivity and safety precautions necessary.

After studying *Topic 75: Atomic structure*, you should be able to

- describe how Rutherford and Bohr contributed to views about the structure of the atom,
- describe the Geiger–Marsden experiment which established the nuclear model of the atom,
- recall the charge, relative mass and location in the atom of protons, neutrons and electrons,
- define the terms 'proton number' (Z), 'neutron number' (*N*) and 'nucleon number' (*A*) and use the equation $A = Z + N$,
- explain the terms 'isotope' and 'nuclide' and use symbols to represent them, e.g. $^{35}_{17}\text{Cl}$,
- write equations for radioactive decay and interpret them,
- outline the modern view of the atom,
- connect the release of energy in a nuclear reaction with a change of mass according to the equation $E = mc^2$,
- describe the process of fission,
- describe the release of nuclear energy in a reactor, and
- outline the process of fusion.

After studying *Topic 76: Electronics*, you should be able to

- distinguish between p- and n-type semiconductors,
- explain how a p–n junction behaves in forward and reverse bias and draw its characteristic curve,
- explain how diodes can be used as half- and full-wave rectifiers,
- explain how a capacitor 'smoothes' a rectified output,
- describe the action of a transistor,
- describe the action of an L.D.R., a thermistor, a thyristor, an L.E.D. and a photodiode,
- describe the action of the transistor as a switch,
- with the aid of diagrams explain the operation of light-, temperature-, time- and sound-operated transistor alarm circuits,
- recall the three types of multivibrator circuits and what they do,
- describe the action of NOT, OR, NOR, AND, NAND and exclusive OR logic gates and recall their truth tables,
- recall simple transistor amplifier circuits,
- draw circuits for the op amp as a voltage amplifier and voltage comparator, and
- count from 0 to 15 in binary and use L.E.D.s to represent a 4-bit number.

After studying *Topic 77: Radio and television*, you should be able to

- recall the ways in which radio waves travel,
- give an outline of radio,
- recall how electrical oscillations are produced,
- explain the action of the tuning and detector circuits in a radio receiver, and
- outline how black-and-white and colour television receivers work.

Other topics

After studying *Topic 79: Materials and Structures*, you should be able to

- classify common materials as metals, natural non-metals or synthetic materials,
- explain why certain materials are used for certain jobs,
- appreciate that the reserves of materials are limited and recall how some shortages can be overcome, and
- state the results of investigations on (i) stretching a copper wire, a rubber band and a strip of polythene, (ii) bending 'beams', (iii) compressing 'pillars' and (iv) building a model 'bridge'.

After studying *Topic 80: Insulation of buildings: U-values*, you should be able to

- understand and use U-values to calculate rate of heat loss from buildings.

After studying *Topic 81: Resistivity; cells and recharging*, you should be able to

- define resistivity and use the equation $R = \rho l / A$,
- describe fuel and photovoltaic cells and their uses,
- draw a circuit diagram for recharging accumulators, and
- understand the term 'capacity of a cell'.

Answers

1 Light rays
1. Larger, less bright
2. (a) 4 images (b) brighter but blurred
3. Shadow to left is darker.

2 Reflection of light
1. (a) $50°$ (c) $50°, 40°, 40°$ (d) parallel
2. B
3. Top half

3 Plane mirrors
2. D
3. 4 m towards mirror

4. Curved mirrors
1. Concave, parabolic

5. Refraction of light
3. B
4. 250 000 km/s
5. 0.8 m
6. D

7 Lenses
1. A
2. (c) Image 9 cm from lens, 3 cm high

8 The eye
2. E
3. (a) (i) No (ii) Yes (iii) No
   (b) Concave
4. (a) (i) No (ii) Yes
   (b) Convex

9 Colour
2. D
3. (a) (i) White (ii) White
   (b) Red
   (c) Red

10 Simple optical instruments
1. Close object
2. (i) larger, blurred, less bright (ii) moved closer to the slide
3. (a) 4 cm (b) 8 cm behind lens, virtual, $m = 2$

11 Microscopes and telescopes
1. (a) D (b) B
2. (a) Objective, 100 cm (b) 100 cm (c) Eyepiece
   (d) Real image formed by A; 5 cm (e) 105 cm

12 Additional questions
1. B
3. B
4. C
5. A
6. (a) Ray passes into air
   (b) Total internal reflection occurs in water
7. E
8. E
9. C
10. (a) 4 (b) 3.2 cm
12. Lens gets fatter and pupil contracts
13. (a) Long sight (b) Convex lens
16. (b) Distance between lens and film decreases
21. 3.4×10^3 km
23. 14 m
24. $100°, 80°, 100°$
25. 6 cm
26. $30°$

27. A: convex, $f = 20$ cm; B: convex, $f = 10$ cm
28. (d) Object at 8.0 cm: image real, 8.0 cm from lens, 1.5 cm high
    Object at 7.0 cm: image real, 9.3 cm from lens, 2.0 cm high
    Object at 2.0 cm: image virtual, 4.0 cm behind lens, 3.0 cm high
31. 40 cm:40 cm

13 Water waves
1. (a) 1 cm (b) 1 Hz (c) 1 cm/s
3. (a) Speed of ripple depends on depth of water
   (b) AB since ripples travel more slowly towards it, therefore water shallower in this direction.
4. (b) (i) 3 mm (ii) 15 mm/s (iii) 5 Hz

14 Light waves
1. (a) $S_1O = S_2O$ (b) $S_1O_1 - S_2O_1 = \lambda/2$
   (c) $S_1O_2 - S_2O_2 = \lambda$
3. (a) Fringes closer together
   (b) Fringes farther apart and dimmer
   (c) Fringes farther apart
   (d) Central fringe white, all others coloured
5. 0.5 μm

15 Electromagnetic spectrum
2. D
4. (a) 3 m (b) 2×10^{-4} s

16 Sound waves
1. D
2. (a) When he sees flash of light from pistol
   (b) Time taken by light to travel 110 m is negligible
   (c) Error $= 110/330 = 1/3$ s; this should be *added* to his timing
3. (a) $v = f\lambda$ (b) 300 m (c) 25 cm
4. 1650 m (about 1 mile)
5. (a) $2 \times 160 = 320$ m/s
   (b) $240/(\frac{3}{4}) = 320$ m/s
   (c) 320 m

17 Musical notes
1. (b) (i) 1.0 m (ii) 2.0 m

18 Additional questions
1. $v = 33/1.5 = 22$ cm/s, $\lambda = 4$ cm, $\therefore f = v/\lambda = 22/4 = 5.5$ Hz
2. (b) Refraction (c) 0.75 cm; 0.5 cm (d) Equal
   (e) Speed in A greater than B (f) A
4. E
5. (a) B (b) A (c) D
7. E
8. B
10. X down, Y and Z up
13. (a) 3×10^8 m/s (b) 15/4 m (c) 0.2 MHz (200 kHz)
14. (c) 340 m/s

19 Measurements
1. (a) 10 (b) 40 (c) 5 (d) 67 (e) 1000
2. (a) 3.00 (b) 5.50 (c) 8.70 (d) 0.43 (e) 0.1
3. (a) 1×10^5; 3.5×10^3; 4.28×10^8; 5.04×10^2; 2.7056×10^4
   (b) 1000; 2 000 000; 69 000; 134; 1 000 000 000
4. (a) 1×10^{-3}; 7×10^{-5}; 1×10^{-7}; 5×10^{-5}
   (b) 5×10^{-1}; 8.4×10^{-2}; 3.6×10^{-4}; 1.04×10^{-3}
5. 10 mm
6. (a) two (b) three (c) four (d) two
7. 24 cm^3
8. 40 cm^3; 5
9. 80
10. (a) 250 cm^3 (b) 72 cm^3

20 Density
1. (a) (i) 0.5 g (ii) 1 g (iii) 5 g
   (b) (i) 10 g/cm^3 (ii) 3 kg/m^3
   (c) (i) 2.0 cm^3 (ii) 5.0 cm^3

2. (a) 8.0 g/cm^3 (b) $8.0 \times 10^3 \text{ kg/m}^3$
3. $15\,000$ kg
4. 130 kg
5. (a) 200 cm^3 (b) 200 cm^3 (c) 3.0 g/cm^3

21 Weight and springs

1. (a) 1 N (b) 50 N (c) 0.50 N
2. (a) 120 N (b) 20 N
3. (a) 15 cm (b) 10 cm (c) 30 g
4. (a) 19 cm (b) 35 g

22 Molecules

1. E
3. (a) 1.0 cm (b) 1.0×10^{-7} cm
4. (a) 2×10^{-6} kg (b) $2 \times 10^{-6}/800 = 2.5 \times 10^{-9} \text{ m}^3$
   (c) $2.5 \times 10^{-9}/0.20 = 1.3 \times 10^{-8}$ m
5. The forces between the layers are weak

24 Additional questions

1. 0.88 g/cm^3
2. (b) density of block + lead $= (7.5 + 11)/16 = 18.5/16 \text{ g/cm}^3$; it sinks in water since density greater than 1.0 g/cm^3
   (c) milk
3. (b) 8 cm (c) 2 cm
4. 50 g
11. (a) 70.5 cm^3 (b) 20 cm^3 (c) 1.6 g/cm^3
13. (c) Yes (d) 5 N
14. 1.6×10^{-5} cm

25 Moments and levers

1. (a) Turns anticlockwise (b) Remains horizontal
2. B
3. $X = 20$ N; $X = 5$ N
4. (i) (c) (ii) (a) (iii) (b)
5. Fishing rod

26 Centres of gravity

2. (a) B (b) A (c) C
3. Tips to right
4. (a) (i) 50 cm (iii) 1.2 N

27 Adding forces

1. (a) 50 N making $37°$ with 40 N
   (b) 60 N making $25°$ with 40 N
2. D
3. 200 N

28 Work, energy, power

1. 180 J
2. 1.5×10^5 J
3. (a) Electric lamp (b) Microphone (c) Battery
   (d) Water turbine (e) Electric motor
4. A
5. (a) 150 J (b) 10 W
6. B
7. 2.0 kW
8. 8.0 s

29 Machines

1. (a) 100 J (b) 150 J (c) 67%
2. (a) 5000 J (b) 7500 J (c) 67%
3. (a) (i) $10/7$ (ii) 2 (iii) 71%
   (b) (i) 2 (ii) 3 (iii) 67%
   (c) (i) $10/3$ (ii) 6 (iii) 56%
4. D
5. Distance-multiplier
6. $2\frac{1}{2}$

30 Pressure in liquids

2. (a) (i) 25 Pa (iii) 0.50 Pa (iii) 100 Pa (b) 30 N
3. 230 Pa
4. (a) 200 m^3 (b) $400\,000$ kg (c) $4\,000\,000$ N
   (d) $200\,000$ Pa (e) (i) none (ii) less
5. (a) 100 Pa (b) 200 N
7. $1\,150\,000$ Pa (1.15×10^6 Pa) (ignoring air pressure)
8. 20 m

294

31 Atmospheric pressure

1. (a) (i) 860 mm or 100 mm over atmospheric (ii) 860 mm
2. (c) 74 cm
3. 10 m

32 Pumps and pressure

1. (a) nothing (b) one-third previous volume
   (c) three times previous pressure
2. (a) 15 cm^3 (b) 6 cm^3
3. D
4. 10 m

33 Floating, sinking and flying

1. (a) 10 N (b) 8 N (c) 6 N (d) 5 N
2. (a) 480 kg (b) 4800 N (c) $(4800 - 1600) = 3200$ N
3. $25\,000 - 20\,000 = 5000$ N
4. (a) 30 g (b) 30 g (c) 30 cm^3
6. (i) 10 cm^3 (ii) 10 g (iii) 10 g (iv) 5 g (v) $25/4$ cm

34 Additional questions

1. (a) (ii) 500 N
2. (a) Moments about A: of $P = P \times 0 = 0$, of $500 \text{ N} = 500 \times 1 \text{ N m}$ (clockwise), of $200 \text{ N} = 200 \times 2 \text{ N m}$ (clockwise), of $Q = Q \times 3$ (anticlockwise);
   (b) 300 N (c) 700 N (d) 400 N
3. $T_1 = T_2 = 15$ N; $T_3 = 10$ N; $T_4 = 20$ N
4. (b) (ii) Q (iii) (a) 600 N (b) 150 N
5. Resultant $= 46$ N; $P = 23$ N
6. D
7. (a) 1300 J (b) 1300 J (c) 13 W (e) 8 J (f) 8 J
   (g) Changed to heat and sound energy (h) None (i) 100 N
8. 3.5 kW
12. (c) 150 Pa
13. (a) 3 N (b) 0.3 kg (c) 0.0003 m^3
15. (a) 80% (b) (i) 2400 J (ii) 1.0 m
16. (b) Mass of boy $= 60$ kg
17. Tension in AC $= 140$ N; $W = 193$ N
18. 35 N at $30°$ to each force; 17 N
19. 2×10^5 Pa
20. (a) 4×10^3 Pa (b) (i) 4×10^3 Pa (ii) 400 N
    (c) $M = 40$ kg (d) (i) No (ii) M larger
21. (a) $4:1$ (b) $2:1$
22. 1.63 km
23. (a) 77 cm (b) 75 cm (c) 15 cm
24. $4/3 \text{ cm}^3$
25. D

35 Velocity and acceleration

1. (a) 20 m/s (b) $6\frac{1}{4}$ m/s
2. (a) 15 m/s (b) 900 m
3. 2 m/s^2
4. (a) 1 s
   (b) (i) 10 cm/tentick^2 (ii) 50 cm/s per tentick
   (iii) 250 cm/s^2
   (c) 0
5. 50 s
6. (a) 6 m/s
   (b) 14 m/s
7. 4 s
8. (a) Uniform acceleration (b) 75 cm/s^2

36 Equations of motion

1. 2.5 m/s^2
2. (a) 10 m/s (b) 25 m (c) 10 s
3. 96 m
4. (a) 4 s (b) 24 m
6. (a) 100 m (b) 20 m/s (c) Slows down
7. (a) $5/4 \text{ m/s}^2$ (b) (i) 10 m (ii) 45 m (c) 22 s

37 Falling bodies

1. (a) (i) 10 m/s (ii) 20 m/s (iii) 30 m/s (iv) 50 m/s
   (b) (i) 5 m (ii) 20 m (iii) 45 m (iv) 125 m
2. 5 s: 31 m
3. B
4. (a) 3 s; 30 m/s (b) $30.(02)$ m/s

5. (a) 10 s (b) 2000 m
6. B

38 Newton's laws of motion

1. D
2. 20 N
3. (a) 5000 N (b) 15 m/s^2
4. (a) 2 m/s^2 (b) 1000 N
5. (a) 4 m/s^2 (b) 2 N
6. (i) 0.5 m/s^2 (ii) 2.5 m/s (iii) 25 m
7. $(1650 - 400)$ N/1000 kg $= 1.25$ m/s^2
8. (a) 1000 N (b) 160 N
9. (a) 5000 N (b) 20 000 N; 40 m/s^2

39 Momentum

1. (a) 50 kg m/s (b) 2 kg m/s (c) 100 kg m/s
2. 2 m/s
3. 4 m/s
4. 0.5 m/s
5. 2.5 m/s
6. (a) 40 kg m/s (b) 80 kg m/s (c) 20 kg m/s^2 (d) 20 N
7. (a) 10 000 N (b) 10 m/s^2

40 K.E. and P.E.

1. (a) 2 J (b) 160 J (c) 100 000 $= 10^5$ J
2. (a) 20 m/s (b) (i) 150 J (ii) 300 J
3. (a) 1.8 J (b) 1.8 J (c) 6 m/s (d) 1.25 J (e) 5 m/s
4. (a) (i) 1000 J (ii) 20 m
5. B
6. 3.5×10^9 W $= 3500$ MW

41 Circular motion

2. (a) Sideways friction between tyres and road
   (b) (i) Larger (ii) smaller (iii) larger
3. (a) Gravity (b) Reduces it (c) (i) Decrease (ii) Decrease

42 Additional questions

1. (a) 14 m/s (b) 26 m/s (c) 1200 m (d) -1 m/s^2
2. D
3. A
4. C
5. C
6. (b) (i) 0.5 m/s^2 (ii) $1.3 \times 10^4 (12\,775)$ m (iii) 20 m/s
7. 300 (304) m
8. (i) 45 m/s (ii) 100 m (iii) 5 m/s
9. (b) (i) -1.25 m/s^2 (ii) -625 N
10. 30 N northwards; 21 m
11. A
12. (a) 5 m/s (b) 0.4 kg m/s (c) 1 J (d) 0.2 J
13. (a) 5 m/s (b) 3 m/s (c) 37.5 J : 22.5 J
14. (a) Towards centre of circle (b) 6 N

43 Thermometers

1. (a) 1530 °C (b) 19 °C (c) 0 °C (d) -12 °C (e) 37 °C
2. C
3. E

44 Expansion of solids and liquids

2. Aluminium
4. (a) Aluminium (b) 1.009 m
5. (a) 0.1 m (b) 0.0004 m (c) 0.3 m
6. (a) Water (b) 4 °C

45 The gas laws

1. (a) 4 m^3 (b) 1 m^3
2. 546 K (273 °C)
3. (i) 100 cm^3 (ii) 400 cm^3 (iii) 180 (178) cm^3
4. C

46 Specific heat capacity

1. 15 000 J
2. E
3. A $= 2000$ J/(kg °C); B $= 200$ J/(kg °C); C $= 1000$ J/(kg °C)
4. 88 000 J
5. (a) 20 000 J (b) 120 000 J

6. (a) 6000 J (b) 6000 J/(kg °C)
7. 1050 W
9. 45 kg

47 Latent heat

1. (a) 3400 J (b) 6800 J
2. (a) $5 \times 340 + 5 \times 4.2 \times 50 = 2750$ J (b) 1700 J
3. 680 s
4. (a) 0 °C (b) 45 g
5. (a) 9200 J (b) 25 100 J
6. 157 g
7. 2500 J/g

49 Conduction and convection

4. E

50 Radiation

1. E
6. D

52 Additional questions

1. A 50 °C B 200 °C C 100 °C
2. B
3. A
4. B
5. C
6. B
7. 0.50 J/(g °C)
8. (a) 1 005 000 J (b) 504 000 J (c) 40 °C
9. (a) AB and CD (b) BC (c) DE
11. D
15. (a) 1.4 mm (b) 35 mm (c) 65 °C
16. 363 K (90 °C)
17. (i) 180 mm (ii) 264 K (-9 °C)
18. B
19. D
20. 300 °C; 0.10 kg (*Hint.* The work done in pulling the wire through the hole equals the energy that causes the temperature rise.)
21. D
22. 1440 J/min
23. (a) 2300 J/g (b) 105 J/s

53 Permanent magnets

1. C
2. D
3. (a) Steel (b) N
4. B
7. (a) A (b) A

55 Static electricity

1. D
3. *a, b, c*

56 More electrostatics

1. (a) B (b) Electrons repelled to earth leaving sphere positively charged

58 Electric current

1. (a) 5 C (b) 50 C (c) 1500 C
2. (a) 5 A (b) 0.5 A (c) 2 A
4. B
5. (*b*)
6. (*d*)
7. All read 0.25 A
8. C

59 Potential difference

1. (a) 12 J (b) 60 J (c) 240 J
2. (a) 6 V (b) (i) 2 J (ii) 6 J
3. A
4. B
5. (b) Very bright (c) Normal brightness (d) No light
   (e) Brighter than normal (f) Normal brightness
6. (a) 6 V (b) 360 J
7. $x = 18, y = 2, z = 8$

60 Resistance

1. $3\,\Omega$
2. $20\,V$
3. C
4. D
5. $A = 3\,V; B = 3\,V; C = 6\,V$
6. $3/2 = 1.5\,\Omega$
7. (a) $15\,\Omega$ (b) $1.5\,\Omega$
8. D
9. (a) $3\,\Omega$ (b) $\frac{1}{2} = 0.5\,A$ (c) $\frac{1}{4} = 0.25\,A$
10. $4/3\,\Omega$; $2\,A$
11. B, D, F = 1.5 A; C, E = 3 V; G = 6 A; H, J = 3 A; I, K = 6 V; L = 2 A; M = 4 V; N, P = 1 A; O = 2 V
12. (b) $2\,\Omega$ and $6\,\Omega$

61 Electromotive force

1. $3\,A$; $9\,V$
2. C
3. (a) $0.5\,\Omega$ (b) $4.5\,\Omega$
4. (a) 3.0 (b) $2.6\,V$ (c) $2.6\,V$ (d) $0.40\,V$
   (e) $2\,\Omega$ (f) $13\,\Omega$

62 Additional questions

1. E
3. C
4. B
5. B
6. (a) $\frac{1}{4}\,A$ (b) $1\,A$ (c) $\frac{1}{2}\,A$
7. (a) (ii) $2\,\Omega$
8. B
9. B
12. (i) (a) both lamps normal brightness
    (b) both lamps dimmer than normal
    (ii) (a) 0.3 A; (b) 0.15 A
    (iii) (a) 0.6 A; (b) 0.15 A
13. (a) $6\,\Omega$ (b) $2\,A$ (c) $12/7\,A$ (d) $4/3\,A$
14. (a) 1/4 A, 5/4 V (b) 2/3 A, 0 V (c) 0 A, 2 V
15. (a) $5.0\,A$ (c) $0.12\,V$ (d) $0.06\,\Omega$
16. (a) $18\,\Omega$ (b) $2\,\Omega:9\,\Omega$
17. $2.5\,A$
18. $2.0\,V$; $1\,\Omega$
19. (a) $1.5\,V$ (b) $0.60\,A$ (c) $3.5\,\Omega$ (d) $3.0\,A$ (e) 0

63 Electric power

1. (a) $100\,J$ (b) $500\,J$ (c) $6000\,J$
2. (a) $24\,W$ (b) $3\,J/s$
3. D
4. $3.1\,kW$
5. (a) $8\,V$ (v) $16\,W$
6. (a) $3\,A$ (b) $4\,\Omega$
7. (a) $12\,A$ (b) $250/12 = 21\,\Omega$

64 Electricity in the home

1. Fuse is in live wire in (a) but not in (b)
3. (a) 3 A (b) 13 A (c) 13 A
4. (a) Parallel (b) Yes, current = 25 A (c) 15 p
6. (a) 20 units (b) 100 p
7. 20 p
8. (b) (i) 20 (ii) 1200 W (c) 420 p

66 Electromagnets

2. S
5. (a) B (b) Relay switch (c) B, C

67 Electric motors

1. E
3. Anticlockwise
4. E

70 Transformers

2. B
3. (a) 24 (b) 10 A
4. (b) 12 000

71 Alternating current

3. (a) 12 V (b) 17 V

72 Additional questions

1. (a) (i) 3 kW (ii) 60 W (iii) 750 W (b) 4 A
2. (a) 3 A (b) 36 W (c) 10 800 J (d) 26 °C
3. 27 kWh; 135 p
7. (a) To complete the circuits to the battery negative
   (b) One contains the starter switch and relay coil; the other contains the relay contacts and starter motor
   (c) Carries much larger current to starter motor
   (d) Allows wires to starter switch to be thin since they only carry the small current needed to energize the relay
11. B
14. (b) 53%
15. (a) E (b) B (c) C (d) E
16. (a) 5 A (b) 400 W (c) 80 V
18. (b) (i) 60 (ii) 0.1 A
19. 3.3 A
20. (a) $25\,\mu F$ (b) $20\,\mu F$

73 Electrons

1. A
2. (a) A − ve, B + ve (b) down

74 Radioactivity

4. 25 minutes
5. D

75 Atomic structure

2. (a) 27 protons and 32 neutrons in nucleus; 27 electrons outside

76 Electronics

3. (a) $V_1 = V_2 = 3\,V$ (b) $V_1 = 1\,V, V_2 = 5\,V$
   (c) $V_1 = 4\,V, V_2 = 2\,V$
5. (c) Current gain $= 60 = I_c/I_b = 30/I_b$
   $\therefore I_b = 30/60 = 0.5\,mA$
7. A: AND, B: OR, C: ex-OR, D: NAND, E: NOR

9.

A	B	D	C	F
0	0	0	0	0
0	1	0	0	0
1	0	0	0	0
1	1	1	0	1
0	0	0	1	1
0	1	0	1	1
1	0	0	1	1
1	1	1	1	1

13. (a) 10 (b) (i) ±10 V (ii) ±15 V
15. (b) See Fig. A1

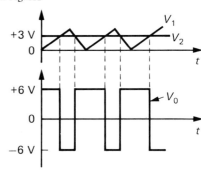

Fig. A1

77 Radio and television

1. (a) (ii) 3 m

78 Additional questions

2. 6 minutes
3. E
4. (i) β (ii) β (iii) α: A, D
7. (b) M_1 (c) collector-base
11. (c) 1/50 s
12. (b) 1.8×10^{11} C/kg
13. (a) (i) 10 V   (ii) 7 V   (b) 50 Hz
14. 1/16

80 Insulation of buildings: *U*-values

1. (a) (i) 2.16 MJ (ii) 1.08 MJ
   (b) 300 W

81 Resistivity; cells and recharging

1. 30 Ω
2. 86 Ω
3. 2.0 m (approx.)

Core level revision questions

1. E
2. C
3. (a) 60°   (b) 30°
4. B
5. (a) refraction   (b) POQ
   (c) towards   (d) 40°   (e) 90 − 65 = 25°
6. C
7. A
8. D
9. B
10. C
11. A
12. (a) dispersion   (b) (i) red   (ii) violet
13. A
14. A
15. B
16. C
17. (a) 30/6 = 5 cm   (b) 4 Hz   (c) $v = f\lambda$ (d) 20 cm/s
18. E
19. (a) Circular, (b) and (c) no change
20. (a) (i) troughs from one arrive at same time as crests from other
   (ii) crests from one arrive at same time as crests from other
   (b) (i) larger wave   (ii) cancel out   (iii) larger wave
21. D
22. E
23. E
24. (a) longitudinal   (b) (i) compression   (ii) rarefaction
25. D
26. B
27. A
28. D
29. C
30. B
31. D
32. E
33. A
34. C
35. C
36. A
37. B
38. B
39. D
40. E
41. E
42. D
43. A
44. B
45. A
46. C
47. (a) 20 J (no work is done horizontally against gravity)
   (b) 100 N × 5 m = 500 J
48. D
49. E
50. B
51. C
52. A
53. A
54. C
55. (a) 80 N   (b) 80 N   (c) 8 kg · (d) 0.01 m³   (e) 0.01 m³
56. A
57. E
58. D
59. (a) 480 m   (b) 6 m/s   (c) 8/20 = 0.4 m/s²
60. E
61. C
62. B
63. D ($F = 20 − 12 = 8$ N: $m = 2$ kg: $a = F/m = 8/2 = 4$ m/s²)
64. D
65. E
66. C
67. D
68. (a) B   (b) A
69. A
70. E
71. C
72. D
73. B
74. (a) (i) 280 K   (ii) 298 K   (iii) 250 K
   (b) (i) 27 °C   (ii) 300 °C   (iii) −73 °C
75. B
76. (a) E   (b) C   (c) D   (d) E
77. C
78. A
79. D
80. A
81. B
82. C
83. C
84. E
85. D
86. D
87. (a) 3 Ω (b) 2 A   (c) 4 V across 2 Ω and 2 V across 1 Ω
88. (a) 1 Ω (b) 3 A   (c) 6 V
89. D
90. (a) B   (b) D   (c) A   (d) C
91. D
92. B
93. C
94. (a) D   (b) A   (c) E
95. B
96. C
97. D
98. E
99. B (answer to 3 is 4000)
100. (a) E   (b) A   (c) C   (d) B   (e) D
101. C
102. (a) A   (b) C
103. E
104. B
105. B
106. B
107. C (symbol is $^7_3$Li)
108. (a) C   (b) A   (c) B   (d) A
109. (a) D   (b) E
110. C (a rectifier changes a.c. to d.c.)
111. E

Further level revision questions

1. D
3. B
6. D
7. A
8. (a) transverse, stationary   (b) longitudinal, progressive
   (c) transverse progressive   (d) longitudinal, stationary

9. **(c)** **(i)** BP − AP = even number of half-wavelengths
   (ii) BQ − AQ = odd number of half-wavelengths
10. B
11. C
13. 60 m/s
15. D
16. C
17. **(a)** Yes, 1 mm = 0.001 m   **(b)** E
18. **(i)** 90 kg   **(ii)** 4.5×10^4 Pa (N/m^2)
19. C
20. **(i)** 3 cm   **(ii)** 2 s   **(iii)** 2.3 cm
21. B
22. E
23. 4500 N at 9° to AB
24. B
25. **(b)** 120 J   **(c)** 30 W   **(d)** 75%
26. B
27. B
28. 80 N
29. **(b)** **(i)** 8.16×10^4 N/m^2 **(ii)** 2.04×10^4 N/m^2
    (iii) 1.7×10^3 m
30. **(c)** 1.33 m^3
31. **(a)** **(i)** 2.16×10^4 kg   **(ii)** 2.16×10^5 N   **(iii)** 2.7×10^5 N
    (iv) 5.4×10^4 N
    (b) **(i)** 2.16×10^5 N   **(ii)** 21.6 m^3
32. D
33. 1.5×10^6 m
34. **(b)** 1.6 m/s^2 **(c)** 10.4 m   **(d)** 0.32 N
35. B
36. A
37. 1.5×10^3 N
38. **(a)** **(i)** 15 m/s   **(ii)** 11 m   **(b)** **(i)** 45 000 N   **(ii)** 506 000 J
40. velocity, momentum
41. 20 °C
42. 2.4×10^{-5} K^{-1}
43. A
44. E
45. 0.47 m^3
46. B
47. 20 minutes
48. 2.1×10^{-3} kg
49. E
51. **(a)** D   **(b)** A
52. 3000 C
53. D
54. **(a)** **(i)** 1.5 V   **(ii)** 4 A   **(iii)** 0.4 Ω   **(iv)** 6 W
    (b) 120 J
55. D
56. **(a)** 4 V   **(b)** 20 Ω   **(c)** 0.13 A
57. 20 W, 15 J/s
58. C
59. A
60. 0.8 A; from B to D
61. B
62. B
63. **(e)** **(i)** 1 V   **(ii)** 6 A
64. D
65. A
66. C
67. B
68. A
69. **(b)** 1830
70. C
72. C
74. 15 counts/second
75. **(d)** 1/16

76. **(a)** 3   **(b)** 3; 4
78. C
79. E
80. D
81. A
82. **(b)** 1.4
83. 60 mm
85. 4 cm
86. **(iii)** 5 mm   **(iv)** halve separation   **(v)** Q, S
88. **(b)** 280 N; 820 N
89. **(i)** 5500 N   **(ii)** 5500 N   **(iii)** 2600 N in each wire
90. **(c)** **(i)** 2 s   **(ii)** 8 s   **(iii)** 80 m/s   **(iv)** 320 m
    (v) 300 m   **(vi)** 500 m
91. **(b)** 1.2 m/s
93. **(i)** 13 A   **(ii)** 13 A   **(iii)** 3 A
94. **(i)** 200 turns   **(ii)** 0.15 A   **(iii)** 10 800 J
95. 4×10^3 N; 12 J/s
96. **(b)** D

Mathematics for physics

1. **(a)** 3   **(b)** 5   **(c)** 8/3   **(d)** 20   **(e)** 12   **(f)** 6
   (g) 2   **(h)** 3   **(i)** 8

2. **(a)** $f = \dfrac{v}{\lambda}$ **(b)** $\lambda = \dfrac{v}{f}$ **(c)** $I = \dfrac{V}{R}$ **(d)** $R = \dfrac{V}{I}$ **(e)** $m = d \times V$
   (f) $V = \dfrac{m}{d}$ **(g)** $s = vt$ **(h)** $t = \dfrac{s}{v}$ **(i)** $a = \dfrac{F}{m}$
   (j) $I = \dfrac{Q}{t}$ **(k)** $Q = VC$ **(l)** $A = \dfrac{F}{p}$ **(m)** $p_1 = p_2 \times \dfrac{V_2}{V_1}$
   (n) $a = \dfrac{x}{bc}$ **(o)** $t = \dfrac{W}{IV}$ **(p)** $\Delta\theta = \dfrac{Q}{mc}$

3. **(a)** $I^2 = \dfrac{P}{R}$ **(b)** $I = \sqrt{\dfrac{P}{R}}$ **(c)** $a = \dfrac{2s}{t^2}$
   (d) $t^2 = \dfrac{2s}{a}$ **(e)** $t = \sqrt{\dfrac{2s}{a}}$ **(f)** $v = \sqrt{2gh}$
   (g) $y = \dfrac{D\lambda}{a}$ **(h)** $\rho = \dfrac{AR}{l}$ **(i)** $V_1 = V_2 \times \dfrac{T_1}{T_2}$
   (j) $T_1 = \dfrac{V_1}{V_2} \times T_2$ **(k)** $V_2 = \dfrac{p_1}{p_2} \times \dfrac{T_2}{T_1} \times V_1$
   (l) $T_2 = \dfrac{p_2}{p_1} \times \dfrac{V_2}{V_1} \times T_1$

4. **(a)** 10   **(b)** 34   **(c)** 2/3   **(d)** 1/10   **(e)** 10   **(f)** 3×10^8
5. **(a)** 2.0×10^5 **(b)** 10   **(c)** 8   **(d)** 2.0×10^8 **(e)** 20
   (f) 300   **(g)** 10^{-7} **(h)** 15   **(i)** 5
   (j) 0.000 02 = 2×10^{-5} **(k)** 10   **(l)** 5/3
6. **(a)** 4   **(b)** 2   **(c)** 5   **(d)** 8   **(e)** 2/3   **(f)** $-\frac{3}{4}$
   (g) 13/6   **(h)** −16   **(i)** 1

7. $a = \dfrac{v-u}{t}$: **(a)** 5   **(b)** 60   **(c)** 75

8. $a = \dfrac{v^2 - u^2}{2s}$

9. **(b)** Extension $\propto$ mass since the graph is a straight line through the origin
10. **(b)** No: graph is straight line but does not pass through the origin
    (c) 32
11. **(d)** 9.0 g/cm^3
12. **(a)** is a curve;   **(b)** is a straight line through the origin, therefore $s \propto t^2$ or s/t^2 = a constant = 2

Index